The Biology of Water

The Biology of Water

Edited by:

Shivsanjeevi Sripathi

Delve
PUBLISHING

www.delvepublishing.com

The Biology of Water

Shivsanjeevi Sripathi

Delve Publishing
224 Shoreacres Road
Burlington, ON L7L 2H2
Canada
www.delvepublishing.com

Email: orders@arclereducation.com

ISBN: 978-1-77469-045-1 (Hardcover)

DECLARATION

Some content or chapters in this book are open access copyright free published research work, which is published under Creative Commons License and are indicated with the citation. We are thankful to the publishers and authors of the content and chapters as without them this book wouldn't have been possible.

ABOUT THE EDITOR

ShivSanjeevi Sripathi completed his Masters in Biotechnology from Mumbai University in 2008. He was awarded for academic excellence in both his Bachelors and Masters for securing second rank in Mumbai University in 2006 and first rank in his college: Kishinchand chellaram College. For his Masters he secured first rank in his college KET's V.G.Vaze College. He qualified CSIR and NET and and TOEFL in September 2008. He then worked on a stem cell project at the Specialized Centre for Cell Based Therapy (SCCT),KEM Hospital at Mumbai on a project entitled, : Isolation & detection of stem cells from Human Umbilical cord/ amniotic membrane" following which he worked at Junior Research Fellow at Microbiology & Cell Biology Department, Indian Institute of Science, Bangalore on cloning of cell wall genes and transcription factors in E.coli & M.smegmatis. As a writer, he has authored and co-authored 35 books on various aspects of biology such as bionics, molecular wires, cloning, hypertension, the epidemics of the 21st century, handling depression, camouflage, hygiene, immunology and many more with international publishers. He loves to read and share on interesting aspects of life sciences in books. In his free time he loves to travel and explore and give talks on spirituality, ancient customs and traditions.

TABLE OF CONTENTS

LIST OF CONTRIBUTORS

Shivsanjeevi Sripthi

Vladyslav V. Goncharuk
A.V. Dumansky Institute of Colloid Chemistry and Chemistry of Water, National Academy of Sciences, 42 Academician Vernadskoho Boulevard, Kyiv 03680, Ukraine

Lawrence E. Armstrong
Human Performance Laboratory and Department of Nutritional Sciences, University of Connecticut, Storrs, CT 06269-1110, USA

Stavros A. Kavouras
Arizona State University, College of Health Solutions, Hydration Science Lab, Phoenix, AZ 85004, USA

Stephen C. Bondy
Center for Occupational and Environmental Health, Department of Medicine, University of California, Irvine, CA 92617-1830, USA

Arezoo Campbell
Department of Pharmaceutical Sciences, Western University of Health Sciences, Pomona, CA 91766-1854, USA

João P. S. Cabral
Center for Interdisciplinary Marine and Environmental Research (C. I. I. M. A. R.), Faculty of Sciences, Oporto University, Rua do Campo Alegre, 4169-007 Oporto, Portugal

Stephen B Hladky
Department of Pharmacology, University of Cambridge, Cambridge CB2 1PD, UK

Margery A Barrand
Department of Pharmacology, University of Cambridge, Cambridge CB2 1PD, UK

Agnes M. Schitter
Department of TCM/Acupuncture, Institute of Complementary Medicine IKOM, University of Bern, Imhoof-Pavillon, Inselspital, 3010 Bern, Switzerland

Marko Nedeljkovic
Department of TCM/Acupuncture, Institute of Complementary Medicine IKOM, University of Bern, Imhoof-Pavillon, Inselspital, 3010 Bern, Switzerland

Heiner Baur
Department of Health, Bern University of Applied Sciences, Murtenstrasse 10, 3008 Bern, Switzerland

Johannes Fleckenstein
Department of TCM/Acupuncture, Institute of Complementary Medicine IKOM, University of Bern, Imhoof-Pavillon, Inselspital, 3010 Bern, Switzerland

Luigi Raio
Department of Obstetrics and Gynecology Inselspital, University Hospital of Bern, Effingerstrasse 102, 3010 Bern, Switzerland

Aan Fleur Terrens
Movement Disorder Program, Peninsula Health, VIC, Australia
Department of Physiotherapy, Monash University, VIC, Australia

Sze-Ee Soh
Department of Physiotherapy, Monash University, VIC, Australia
Department of Epidemiology and Preventative Medicine, Monash University, VIC, Australia

Prue Morgan
Department of Physiotherapy, Monash University, VIC, Australia

LIST OF ABBREVIATIONS

AD	Alzheimer's disease
APP	Amyloid precursor protein
AVP	Arginine vasopressin
ACIA	Ashong colonic irrigation apparatus
BBS	Berg Balance Scale
BM	Bowel movements
BSS	Bristol Stool Scale
CDC	Center for Disease Control and Prevention
CNS	Central nervous system
CSF	Cerebrospinal fluid
CGU	Colonies of generatrices units
CBPs	Chlorination by-products
CUMS	Chronic unpredictable mild stress
CFU	Colony-forming units
CHDs	Coronary heart diseases
DNA	Deoxyribonucleic acid
DRI	Dietary Reference Intakes
ECG	Electrocardiogram
ECV	External cephalic version
HR	Hazard ratio
HREC	Human Research Ethics Committee
ISF	Interstitial fluid
ICP	Intracranial pressure
IBS	Irritable bowel syndrome
LPS	Lipopolysaccharide
LOELs	Lowest Observed Effect Levels
MDD	Major depressive disorder
MCVs	Maximum water concentration levels

MMSE	Mini Mental State Exam
MDMQ	Multidimensional Mood Questionnaire
MS	Multiple sclerosis
Mac	Mycobacterium avium complex
NHANES	National Health and Nutrition Examination Survey
PBN	Parabrachial nucleus
PD	Parkinson's disease
PSS	Perceived Stress Scale
PAS	Periarterial spaces
PET	Positron emission tomography
PTD	Preterm delivery
PI	Pulsatility indexes
ROS	Reactive oxygen species
RGVs	Regulatory guidance levels
SARS-CoV-2	Severe acute respiratory syndrome coronavirus-2
SGA	Small-for-gestational-age
SON	Supraoptic nucleus
SIADH	Syndrome of inappropriate ADH excess
TBW	Total body water
TCP	Toxin co-regulated pili
TCM	Traditional Chinese Medicine
TEWL	Transepidermal water loss
TCE	Trichloroethylene
TNF	Tumor necrosis factor
UVOL	Urine volume
VAS	Visual Analogue Scales
WASH	Water, sanitation, and hygiene
WTR	Water turnover rates
WHO	World Health Organization

PREFACE

"How much water should we have?"-this is a question that can get one a slew of answers!

"Water is life's matter and matrix, mother, and medium. There is no life without water."

– Albert Szent-Gyorgyi, M.D, Discoverer of Vitamin C.

We have been told to "drink lots of water" whether in health or illness. Nothing can be more relaxing for more people like a bath or a nice soak in water. This precious and essential nutrient of life that forms the basis of the majority of metabolic reactions within this body is being shown by research to emerge as a "healer" of diseases.

This book is one of the first to show the biological aspects of water in a concise manner backed by scientific research. "How much water should we have?"-read this book can offer insights as to the "dynamic complexity" of the homeostasis of water in the human body to guide what is the recommended intake for regular and pregnant individuals. Research has also been presented as to how water can influence our cognition and mood positively.

According to the World Health Organization (WHO), "264 million people of all ages suffer from depression" while the year 2017 saw 10.7 % (792 million people) of the world's population suffer from a mental disorder (*Hannah Ritchie and Max Roser (2018) - "Mental Health". Published online at OurWorldInData.org*). This becomes more pertinent today as we are grappling with the COVID-19 epidemic and its aftermath in today's already competitive world. Based on research of 1441 respondents from during the COVID-19 pandemic and 5065 respondents from before the pandemic, the extent of depression was found to have multiplied by a factor of 3" (*Ettman CK, Abdalla SM, Cohen GH, Sampson L, Vivier PM, Galea S. Prevalence of Depression Symptoms in US Adults Before and During the COVID-19 Pandemic. JAMA Netw Open. 2020;3(9):e2019686. doi:10.1001/jamanetworkopen.2020.19686*).

This book presents research where hydrogen-rich water has shown promise in addressing the inflammatory changes seen in the brain due to depressive conditions in mice opening up the avenue of using the humble glass of water to address this grave condition.

Research also shows the promise of "hydrotherapy" in several conditions such as insomnia and multiple sclerosis and the ability of water to improve the skin and accelerate wound healing.

Feeling blue? Taking a shower can help-Yes indeed! Research also shows the physical and mental benefits of bathing.

Thus, here we present "The biology of water" where interesting and vital aspects of water have been presented.

Let us "dive" into this ocean of knowledge and benefit from it.

Note: This book is not intended to be used for diagnosis or treatment-please take medical advice for any approach that shows promise in alleviating disease.

"In time and with water, everything changes"-Leonardo da Vinci

Editor

Chapter 1

WATER: From Birth Weight to Moods

Shiv Sanjeevi Sripathi[1]

[1]Shiv Sanjeevi Sripathi secured second rank in Mumbai University and first rank in his college at his undergrad level and in his college during his Masters and has worked on a stem cell project at the Specialized Centre for Cell Based Therapy (SCCT), KEM Hospital at Mumbai, following which he was a Junior Research Fellow at Microbiology & Cell Biology Department, Indian Institute of Science, Bangalore from August 2009 to November 2011. The field of work involved cloning of cell wall genes and transcription factors in E. coli & M. smegmatis. This was followed by a stint as a Research and Development Executive at Transasia Biomedicals, Mumbai where he worked on patent writing and Immunoassays from 2011 till 2013. In 2013, he entered the teaching field where he has been tutoring biology and chemistry to the students of 11th and 12th grades along with English grammar and communication skills to people of all age groups. Shiv also loves helping people write scientific manuscripts and has co-edited several books on life sciences and aspects of biotechnology with Apple Academic Press.

1.1 WATER, THE EARTH, AND THE BODY

Accounting for 55%–65% of the body weight in a person, as the body cannot produce sufficient water, the consumption of this essential factor should be throughout the day. Based on the amount of water lost from the body in the form of measurable or immeasurable losses (that are roughly at par with each other and refer to sensible and insensible losses, respectively), the consumption of water should be balanced. Excretion (stool/urine) accounts for sensible water loss while the outgoing breath and sweat account for insensible loss.

Several functions are attributed to this molecule ranging from serving as a solvent for amino acids, glucose, vitamins, and minerals, maintenance of body temperature, cell shape control, a medium to dissolve waste products, comprises of vital fluids of the body to also assists in digestion and subsequent absorption of food.

The major sources of this precious element are surface water and groundwater. The former is inclusive of reservoirs, lakes, and rivers while the latter includes aquifers from rocks that are accessed via pumps (Dudek, 2001; Montgomery, 2002).

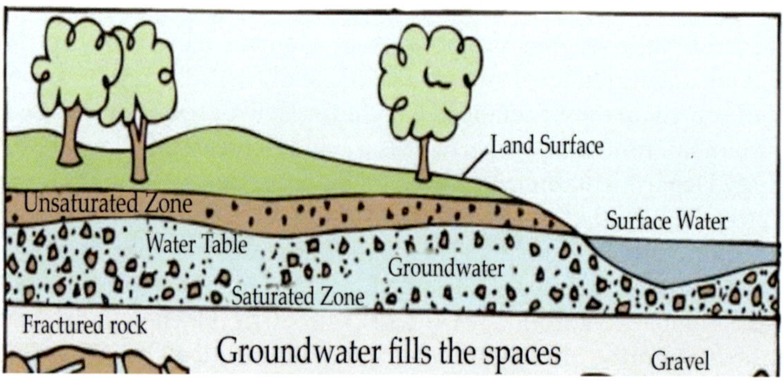

Figure 1.1: below shows the scheme of both these sources of water (Afrose et al, 2018; *This work is licensed under Creative Commons Attribution 4.0 License*).

If we look at the various types of bottled water available, there are a slew of brands available in the market. These are subjected to analyses of the quality and labeling. Figure 1.2 below summarizes the various types of bottled water:

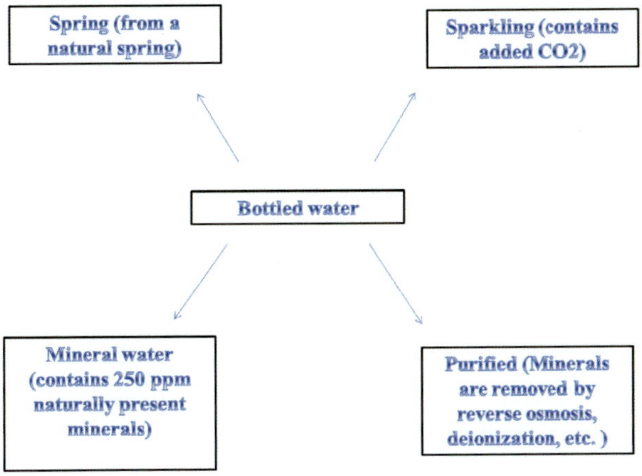

Figure 1.2.

While the effect of beverages on health has received much limelight, the effect of water is yet to be addressed in detail. Analyses have been suggestive of a lowered energy intake with the replacement of beverages containing sugar with water.

In the case of elderly people, the extent of thirst and fluid consumption is lower than what is seen in younger people. The amount of water consumed by the elderly is lesser for replenishing the lost water following fluid deprivation.

The production of sweat to account for the loss of water from the body facilitates the cooling of the body. The range of water loss from the skin can be 2.0 L/h in heat or activity to 0.3 L/h in sedentary conditions. While sweating allows for the body to be cooled, paucity in water intake to account for this loss can result in hypohydration.

The process of thermoregulation displays variation across age groups with a higher susceptibility to illness arising due to heat than adults. In the case of children, a higher core temperature increase is suggested to sweat less while research also points out the loss of dry heat from children (Popkin et al, 2010).

1.2 HOW MUCH WATER SHOULD WE HAVE?

While a universal concept of euhydration is yet to see the light of the day, the term hydration is utilized when the P_{OSM} is in the range of 285–295 mOsm/kg.

Table 1.1: Reports from the U.S. National Academy of Medicine (NAM; 2004) and the European Food Safety Authority (EFSA: 2010) as recommended Adequate Intakes. This refers to "represent an amount that should meet the needs of almost everyone in a specific life-stage group who is healthy, consumes an average diet, and performs moderate levels of physical activity" (Armstrong and Johnson, 2018; Licensee MDPI, Basel, Switzerland. *This article is an open-access article distributed under the terms and conditions of the Creative Commons Attribution (CC BY) license (http://creativecommons.org/licenses/by/4.0/). You are free to: Share — copy and redistribute the material in any medium or format; Adapt — remix, transform, and build upon the material for any purpose, even commercially*): b: values refer to total water intake (TWI = plain water + beverages + food moisture)

Life Stage & Sex	Age	European Food Safety Authority, Parma, Italy [b] 2010 (ml/day)	National Academy of Medicine, USA 2004 [b] (ml/day)
Infants	0–6 months	680 via milk	700
	6–12 months	800–1,000	800
Children	1–2 years	1100-1200	1300
	2–3 years	1300	
	4–8 years	1600	1700
	9–13 years, boys	2100	2400
	9–13 years, girls	1900	2100
	14–18 years, boys	2500	3300
	14–18 years, girls	2000	2300
Adults			
Men		2500	3700
Women		2000	2700
Pregnant Women	≥ 19 year	2300	3000
Lactating Women	≥ 19 year	2600–2700	3800
Elderly		same as adults	same as adults

In the case of water balance among other physiological functions in the body, steady-state maintenance involves the CNS and neuroendocrine hormones. There is a level of both complexity and dynamism in the regulation of water in the human body. There is a change in the intake (food/water), the amount of urine produced, and water loss from the body. The connections among the organs, fluids present in the body and body spaces, and various processes such as active transport or diffusion or osmosis all create regulatory pathways that are subject to alterations.

Table 1.2: The water balance in humans over 24 hours involves a slew of factors ranging from the diet, physiology, behavior, and metabolism. (Armstrong and Johnson, 2018; Licensee MDPI, Basel, Switzerland. *This article is an open-access article distributed under the terms and conditions of the Creative Commons Attribution (CC BY) license (http://creativecommons.org/licenses/by/4.0/). You are free to: Share — copy and redistribute the material in any medium or format; Adapt — remix, transform, and build upon the material for any purpose, even commercially*):

Key: "a, TWI, total water intake = (plain water + water in beverages + food moisture); b, water generated during substrate oxidation; c, greatly influenced by diet composition; d, in a 24-h sample; e, NRWL includes eccrine sweat, transdermal, respiratory and stool water losses; f, FWR = (24-h urine volume, L/day) − (obligatory urine volume, L/day). The latter term is the water volume necessary to excrete the 24-h solute load, hypothetically calculated as (830 mOsm/kg) − (3–4 mOsm/kg per year > 20 years of age). Hydration status is inadequate if FWR is negative. Abbreviations: TBW, total body water; NES, neuroendocrine system (central nervous system + hormones); ECV, extracellular volume; ICV, intracellular volume; MRCA, maximal renal concentrating ability; CNS, central nervous system (brain + spinal cord); GI, gastrointestinal organs; GE, Germany; FR, France; USA, United States of America"

	Total water intake [a] (L/24h)	**Intracellular metabolic water production [b]**	**Total solute load [c] (mOsm/24h)**	**Urine osmolality [d] (mOsm/kg)**	**Maximal renal concentrating ability (mOsm/kg)**	**Urine volume (L/24h)**	**Non-renal water loss (L/24h) [e]**	**Free water reserve [f] (L/24h)**
Functions and characteristics	Contributes to TBW	Product of human metabolism	Metabolized and digested products excreted in urine	Regulates TBW and ECV-ICV osmolality	Inherent quality of the kidneys	Regulates TBW and ECV-ICV osmolality	Excretory and secretory processes	Calculated index of euhydration, based on population statistics
Influential factors	Meal timing and contents, idiosyncratic thirst, physical activity, body size, cultural and learned preferences	Metabolic rate and substrates, physical activity, diet macronutrient and energy content, NES responses	Metabolic products, dietary contents, body size, idiosyncratic hunger, learned food preferences	TWI, MRCA, solute load, NRWL, physical activity, NES responses	Life-stage group, male or female sex	TWI, total solute load, NRWL, physical activity, NES responses	Diet, ventilatory rate, physical activity, body size	TWI, total solute load, NRWL, physical activity
Organs involved	GI, CNS, NES, mouth and throat	CNS, NES	GI, CNS, NES	Kidneys, CNS, NES	Kidneys, CNS, NES	GI, kidneys, CNS, NES	Skin, GI, respiratory organs	GI, kidneys, CNS, NES

Conscious or behavioral influence?	Yes, habitual 24-h water intake	No	Yes, solid food consumption	Yes, secondary to TWI and food contents	No	Yes, secondary to TWI and food contents	Yes, eccrine sweat loss during labor or exercise	Yes, secondary to water and food intake
Representative mean, median, or range of values for sedentary adults	♀, 1.8–2.0 and ♂, 1.9–2.4 (FR, UK); ♀&♂, 1.5–2.5 (13countries); ♀, 2.3 (range: 0.8–4.5) (USA); ♂, 3.0 (range: 1.4–7.7) and ♀, 2.5 (range: 1.2–4.6) (USA); ♀, 1.9 and ♂, 2.3 (GE); ♀&♂, 0.2–3.9 (FR) L/24h	♀, 0.2–0.3; ♂, 0.3–0.4; ♂, 0.4 L/24h	♀, 669–781 and ♂, 915–992 (GE); ♂, 951 (USA); ♀&♂, 362–1365 (4 countries); ♂, 750 (USA); ♀, 752 and ♂, 941 (GE) mOsm/24h	♀&♂, 120–1250 (FR); ♀&♂, 555 (UK) mOsm/kg	♀&♂, 1430 (UK) mOsm/kg; ♀&♂ range, 1100–1300 (GE); ♀&♂, 1010–1330 (USA)	♀&♂, 0.2–3.9 (FR); ♀&♂, 1.9 (*n* = 8, UK) L/24h	♂, 0.3–0.4 (UK); ♀, 0.5–0.7 and ♂, 0.7–1.3 (GE) L/24h	♀, 0.4–0.5 and ♂, 0.2–0.3 (GE); ♀, 0.5 and ♂, 0.3 (GE) L/24h

Based on the scenario, the water homeostasis also demonstrates changes depending on whether the lifestyle is sedentary or during exercises. Among subjects, there are also variations within subjects. For example, research by Leiper and team (2001) reported that there was a variation in the sweat produced among athletes. The analyses of the total body water (TBW) and water turnover rates (WTR) of 6 competitive male cyclists vs. that of sedentary men over 7 days revealed that the former group displayed a TBW of 70.1 [65.5-73.9] % vs. 63.5 [52.7-71.0]% for the sedentary group. It was concluded that water loss in cool environments involves loss due to sweat and respiration throughout exercise.

The TWI for small individuals is lesser than that of large individuals (Institute of Medicine, 2004). There are also other factors that enter the scenario ranging from the amount of sodium in the diet (given that dietary preferences display a high variation) to beverage consumption and the amount of water present in foods.

The hormone called arginine vasopressin (AVP: from the hypothalamus) facilitates the maintenance of the P_{OSM} or plasma osmolality in prescribed ranges thus, serve as the major hormone to influence water balance. The levels of this hormone are boosted when P_{OSM} is high due to dehydration. If one looks at the threshold to allow the production of AVP in the plasma, complexity can be observed in terms of this water content regulation with the value between 276–291 mOsm/kg for an individual.

Table 1.3: Representation of the aspects of the hormone AVP and its role in maintaining the water balance (Armstrong and Johnson, 2018; Licensee MDPI, Basel, Switzerland. *This article is an open-access article distributed under the terms and conditions of the Creative Commons Attribution (CC BY) license (http://creativecommons.org/licenses/by/4.0/). You are free to: Share — copy and redistribute the material in any medium or format; Adapt — remix, transform, and build upon the material for any purpose, even commercially*)

Biological functions
Regulates body water and sodium homeostasis by acting on renal nephrons to decrease urine volume and increase the concentration of urine Maintains plasma osmolality within narrow limits Encourages vasodilation in vascular smooth muscle cells by inducing NO production Affects liver metabolism (e.g., gluconeogenesis, glycogenolysis) Stimulates the pancreas to produce either insulin or glucagon, depending on blood glucose concentration
Factors which influence neurohypohysial AVP release and plasma AVP concentration
Plasma osmolality Angiotensin II Oropharyngeal environment Water restriction and consumption Osmolar content of diet, especially sodium Hypoglycemia Blood volume and pressure Upright posture Emotional stress Exercise Circadian rhythmicity Hypoxia Nausea Pain
Diseases and disorders that involve AVP dysfunction
Diabetes insipidus and diabetes mellitus Syndrome of inappropriate ADH excess (SIADH) Sepsis Severe bleeding, hemorrhage Chronic hypernatremia Primary polydipsia syndrome, compulsive water drinking Kallmann's syndrome Autosomal dominant polycystic kidney disease

Another factor that should be considered is age: with water intake and thirst displaying a dip in people who are more than 65 years of age. The ability of such adults in sensing the lowering of blood volume and altered water satiation due to changes in the brain accounts for this lowered water intake.

(Armstrong and Johnson, 2018)

Research by Johnson and team (2016) assessed the levels of AVP and the osmolality of blood in women whose water consumption displays variation. The approach described as "a single-blind controlled experiment" entailed the measurement of the TWI or total water intake (water + beverages + food moisture) over 24 hours in 120 women. Following the identification of the lowest (LOW, 1.6 ± 0.5 L·day^{-1}) and the highest (HIGH, 3.2 ± 0.6 L·day^{-1}) TWI, variables were assessed.

Figure 1.3 below displays ***TOP:*** the AVP levels in the plasma of the HIGH vs. LOW groups throughout the ad libitum baseline (three days), controlled water intake (four days), and ad libitum recovery (one day). Vertical dotted lines are indicative of the different phases ***BOTTOM:*** the association of thirst and the AVP levels in both the groups HIGH (squares) and LOW (circles). The differences in the y-intercepts are indicative of the thirst status influencing the hormone for conserving water (Johnson et al, 2016: *2016 by the authors; licensee MDPI, Basel, Switzerland. This article is an open-access article distributed under the terms and conditions of the Creative Commons Attribution (CC-BY) license (http://creativecommons.org/licenses/by/4.0/): You are free to: Share — copy and redistribute the material in any medium or format; Adapt — remix, transform, and build upon the material for any purpose, even commercially*):

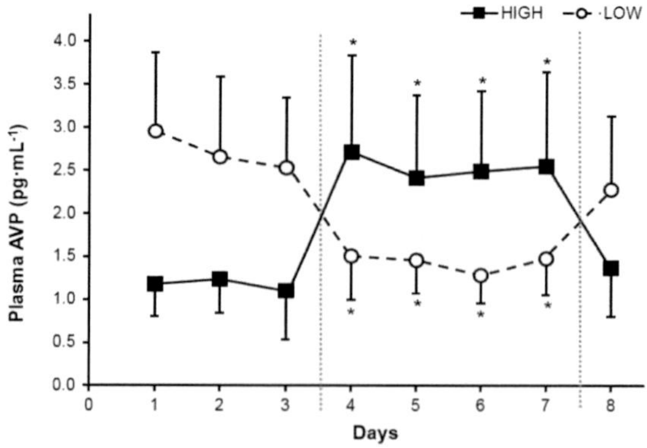

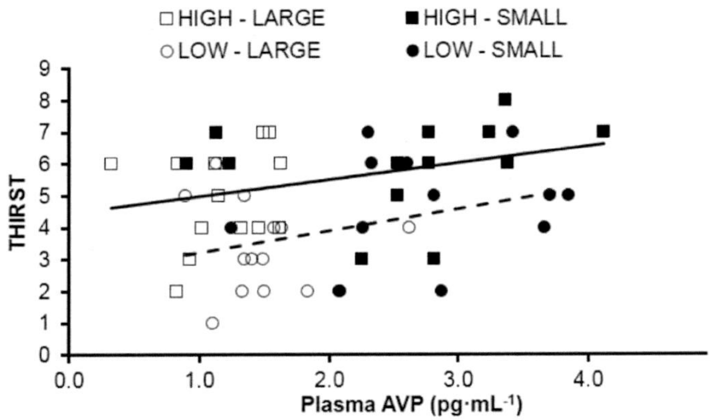

Figure 1.3.

Overall, it was concluded that the fluid balance is maintained in both those who consume large amounts of water vs. those who consume low amounts in free-living conditions. The involvement of the perception of thirst to drive the intake thus emerges as a possibility (Johnson et al, 2016).

Based on research findings, the changes in the volume and concentration of urine were assessed using the plasma AVP levels and P_{OSM}. A moderately strong relationship was described between urine osmolality and TWI while a strong relation was noticed between urine volume and the TWI. As per the NAM report, a minimum intake is not recommended as it can manifest as health consequences.

Figure 1.4 below displays the daily water intake vs. Urine volume (UVOL), plasma osmolality (POSM), urine osmolality (UOSM, 24h), and plasma AVP. The vertical column displays the EFSA and NAM water Adequate Intakes. The horizontal shaded rows are indicative of the AVP levels during water restriction (Armstrong and Johnson, 2018; *licensee MDPI, Basel, Switzerland. This article is an open-access article distributed under the terms and conditions of the Creative Commons Attribution (CC-BY) license (http://creativecommons.org/licenses/by/4.0/): You are free to: Share — copy and redistribute the material in any medium or format; Adapt — remix, transform, and build upon the material for any purpose, even commercially*):

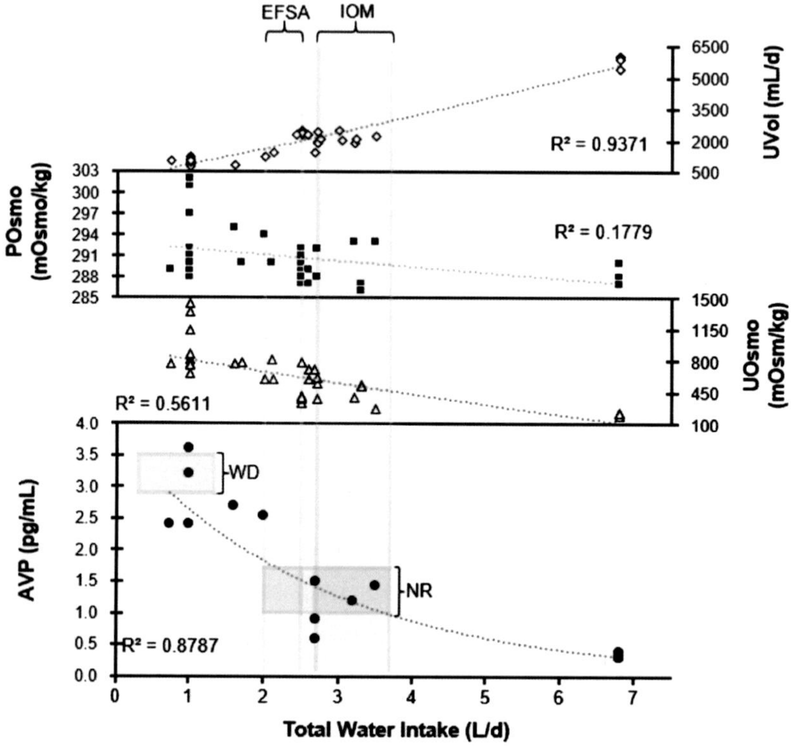

Figure 1.4.

Given the vital involvement of water in circulation, homeostasis, transport, and metabolism, a precise consensus for human water requirements across various demographic groups is yet to be ascertained. Given the variations among individuals and the involvement of the nervous system, the water system in the human body is conserved by dynamic mechanisms. Figure 1.5 below represents the various factors that are involved in the water balance in our bodies (Armstrong and Johnson, 2018; Licensee MDPI, Basel, Switzerland. *This article is an open-access article distributed under the terms and conditions of the Creative Commons Attribution (CC BY) license (http://creativecommons.org/licenses/by/4.0/). You are free to: Share — copy and redistribute the material in any medium or format; Adapt — remix, transform, and build upon the material for any purpose, even commercially*):

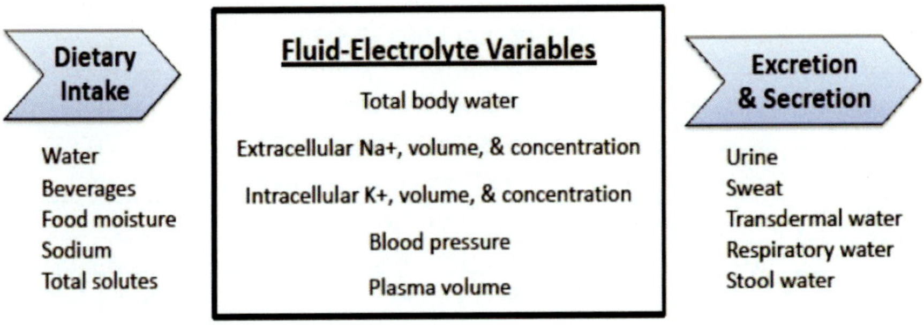

Figure 1.5.

Overall, studies and analyses point out to an initiation of a neuroendocrine defense in the body when the TWI is < 1.8 L over 24 hours. Care should be taken to take in not less than this amount to circumvent disease (Armstrong and Johnson, 2018).

1.3 WATER LOSS FROM THE BODY

Early studies dating back to the 1900s described how the deprivation of water by 2 researchers resulted in "dry mouths, husky voices, and difficulty swallowing," to "temperamental, hollow, and pale" highlighting the importance of the essential molecule (Black et al, 1944).

If we look at the consumption of water in terms of trends, researchers Drewnowski and team (2013) assessed the consumption of water (tap and bottled) across US adult subjects on the basis of sex, economic status, and race. The intake of water from the National Health and Nutrition Examination Survey (NHANES) data was subjected to comparison to that of Adequate Intake (AI) values from the US Dietary Reference Intakes (DRI).

Interestingly, the total dietary water was sourced from "plain water, soda, coffee, tea, milk, alcohol, fruit drinks, and fruit juices".

Table 1.4: below shows the amount of water obtained from various sources (Drewnowski et al, 2013: *This is an open-access article distributed under the terms of the Creative Commons Attribution License (http://creativecommons.org/licenses/by/2.0), which permits unrestricted use, distribution, and reproduction in any medium, provided the original work is properly cited*)

	20-50y		**51-70y**		**≥71y**	
	Mean (SE)	**% of total water**	**Mean (SE)**	**% of total water**	**Mean (SE)**	**% of total water**
Water	1294 (22)	37.1	1020 (20)	32.0	669 (12)	30.1
Soda	466 (9)	13.4	312 (8)	9.8	140 (5)	6.3
Diet soda	164 (4)	4.7	178 (6)	5.6	62 (3)	2.8
Regular soda	310 (8)	8.9	135 (3)	4.2	79 (3)	3.6
Coffee	297 (6)	8.5	515 (11)	16.2	406 (8)	18.3
Alcohol	280 (9)	8.0	156 (6)	4.9	50 (3)	2.3
Milk	142 (4)	4.1	137 (4)	4.3	153 (4)	6.9
Milk (no cereal)	111 (4)	3.2	108 (4)	3.4	106 (3)	4.8
Tea	197 (5)	5.6	292 (7)	9.2	158 (6)	7.1
Fruit drinks	90 (3)	2.6	74 (4)	2.3	32 (2)	1.4
Diet fruit drinks	16 (2)	0.5	36 (3)	1.1	13 (2)	0.6
Regular fruit drinks	74 (3)	2.1	38 (3)	1.2	19 (2)	0.9
Fruit juice	74 (3)	2.1	59 (2)	1.8	59 (2)	2.7
Sports/energy	54 (3)	1.5	24 (2)	0.7	6 (1)	0.3
Water from food	590 (5)	16.7	598 (11)	18.3	547 (11)	24.4

Water from beverages[1]	2940 (20)	83.3	2665 (23)	81.7	1693 (15)	75.6
Total daily water	3563 (24)	-	3229 (27)	-	2251 (17)	-

Further, the AI value was not met by older men while the value was exceeded by women below 70 years of age. The total water intake was not met by 42.7% of men and 40.6% of women (20 to 50 years old) while the value was not met by 59.1% of men and 44.9% of women (51-70 years) and 94.7% of men and 82.6% of women above 71 years of age.

Figure 1.6 below shows this insufficient water consumption trend across the populations studied (Drewnowski et al, 2013: *This is an open-access article distributed under the terms of the Creative Commons Attribution License (http://creativecommons.org/licenses/by/2.0), which permits unrestricted use, distribution, and reproduction in any medium, provided the original work is properly cited*):

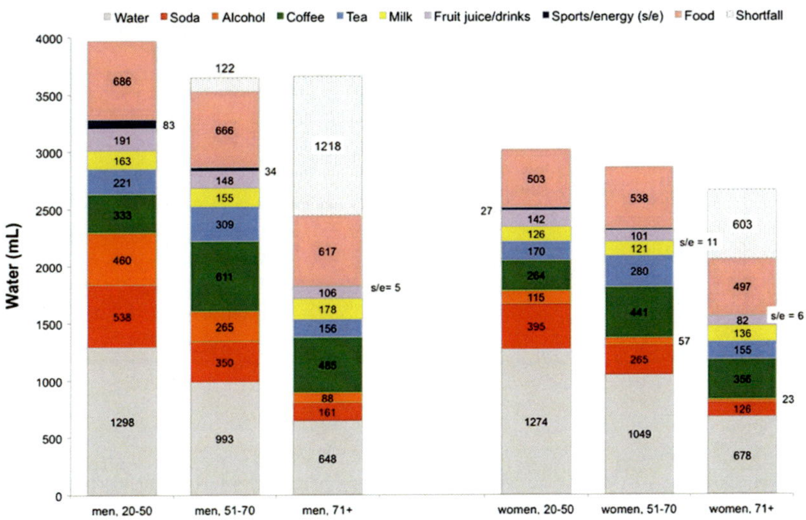

Figure 1.6.

Overall, the lack of optimum water intake was highlighted that becomes a cause of concern in the elderly population (Drewnowski et al, 2013).

The risk of several diseases ranging from cardiovascular disease to augmented levels of serum- and glucocorticoid-inducible kinase 1 activity (SGK1), in turn, involved in stroke, cardiac fibrosis, and hypertension. The condition of hypohydration defined as a deficit in the water content of the body due to acute or chronic dehydration is linked to several conditions such as affected cognition, lowered exercise, foul mood, aberrant glycemic regulation, and disrupted thermoregulatory functions (Watso and Farquhar, 2019).

Figure 1.7 below displays the effect of water dehydration on our physiology (healthy humans) (Watso and Farquhar, 2019): *Licensee MDPI, Basel, Switzerland. This article is an open-access article distributed under the terms and conditions of the Creative Commons Attribution (CC BY) license (http://creativecommons.org/licenses/by/4.0/). You are free to: Share — copy and redistribute the material in any medium or format; Adapt — remix, transform, and build upon the material for any purpose, even commercially*):

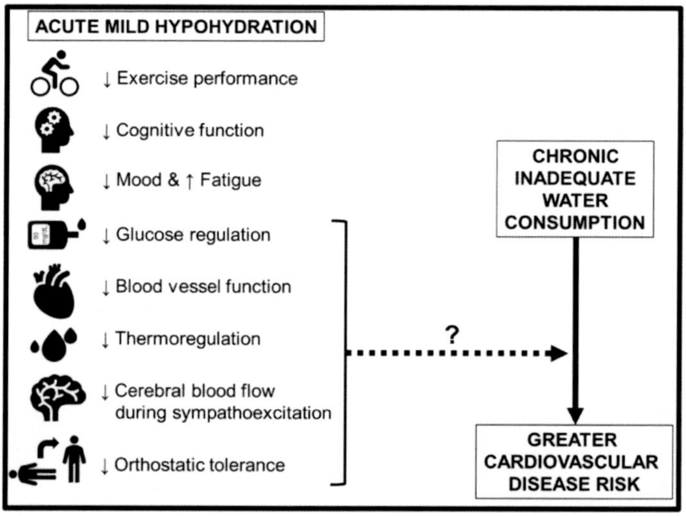

Figure 1.7.

Given that the UNICEF Water, Sanitation and Hygiene Annual Report 2012 highlighted the access of 53% of schools in low-income countries to safe and all-round water supply, the grim scenario enters given that dehydration due to the paucity of water can impact the cognition of school children. The lowering of short-term and long-term memory, attention, and efficiency due to dehydration in adults from heat or exercise has been documented in experiments. In children, short-term memory was impacted adversely due to dehydration (Chard et al, 2019).

Early work in 2009 by Edmonds and Burford involved analyzing the cognition of 7-9-year-old children (n=58). The performance of letter cancellation, spot the difference memory tasks was better in children who received water than those who did not. The conclusion was, "even children in a state of mild dehydration, not induced by intentional water deprivation or by heat stress and living in a cold climate, can benefit from drinking more water and improve their cognitive performance" (Edmonds and Burford, 2009).

In another study, by Benton (2011) reported that consumption of water had a positive effect on 22 children (9 years) who performed tasks. A particular task was evaluated upon an electronic bleep over time that is shown below in Table 1.5 with the children taking lesser time on a task post-drinking water when the percentage of time spent was assessed (Benton, 2011: *This article is an open-access article distributed under the terms and conditions of the Creative Commons Attribution license: http://creativecommons.org/licenses/by/3.0/: You are free to: Share — copy and redistribute the material in any medium or format; Adapt — remix, transform, and build upon the material for any purpose, even commercially.*

Table 1.5

Minutes of observation	Water	No water
1 to 5	85.4 ± 1.8	57.2 ± 3.8
6 to 10	82.1 ± 1.9	55.8 ± 3.6
11 to 15	79.1 ± 2.8	55.2 ± 3.0
16 to 20	79.4 ± 2.8	54.1 ± 3.5
21 to 25	73.9 ± 3.1	48.3 ± 3.7
26 to 30	72.7 ± 2.3	47.6 ± 3.2

Researchers Chard and team (2019) described the analyses of the effect of water shortage on the cognition in schoolchildren in Mali, West Africa. Students (mean age= 11.6 (1.0) years in School 1 and 12.1 in School 2) were analyzed in a cross-over study where the urine specific gravity (U_{sg}), urine color (U_{col}), and self-reported thirst (hydration status), as well as, visual attention, visual memory, short-term memory, and visuomotor skills (cognition). The latter involved canceling non-target letters, differences among images, recalling number sequences in both forward and reverse formats, and line tracing to assess the cognition status.

Statistical analyses revealed that the reverse number recall test showed a better result with hydration while the line tracing showed the opposite trend. Dehydration had a more pronounced effect on boys rather than girls in terms of performance. Overall, statistical significance was observed only for few cognitive test performances affected by dehydration.

The results of the association of cognitive performance with dehydration are presented in Figure 1.8 below (Chard et al, 2019: *This is an open-access article distributed under the terms of the Creative Commons Attribution License, which permits unrestricted use, distribution, and reproduction in any medium, provided the original author and source are credited*):

Cognitive Test	Treatment			Dehydrated*			Urine Specific Gravity (U_{SG})		
	Beta	95% CI	p	Beta	95% CI	p	Beta	95% CI	p
Letter cancellation	0.36	-0.81, 1.53	0.545	-1.95	(-4.23, 0.32)	0.092	-64.98	(-177.47, 47.51)	0.258
Visit (ref: Visit 1)	***7.43***	***6.26, 8.59***	***<0.001***	***7.40***	***(6.26, 8.54)***	***<0.001***	***7.28***	***(6.12, 8.44)***	***<0.001***
Image difference, direct	0.25	-0.10, 0.60	0.163	-0.47	(-1.05, 0.10)	0.103	-18.22	(-46.28, 9.84)	0.203
Visit (ref: Visit 1)	***0.90***	***0.55, 1.25***	***<0.001***	***0.87***	***(0.52, 1.22)***	***<0.001***	***0.84***	***(0.49, 1.19)***	***<0.001***
Pupil grade (ref: 5th grade)	***0.77***	***0.36, 1.19***	***<0.001***	—	—	—	—	—	—
Image difference, indirect	0.26	-0.09, 0.61	0.151	0.30	(-0.29, 0.89)	0.323	6.81	(-22.33, 35.95)	0.647
Visit (ref: Visit 1)	***0.74***	***0.39, 1.09***	***<0.001***	***0.71***	***(0.36, 1.06)***	***<0.001***	***0.72***	***(0.37, 1.08)***	***<0.001***
Pupil grade (ref: 5th grade)	***0.70***	***0.25, 1.15***	***0.002***	—	—	—	—	—	—
Number recall total, forward	-0.01	-0.32, 0.30	0.943	0.22	(-0.32, 0.77)	0.420	17.35	(-9.31, 44.02)	0.202
Visit (ref: Visit 1)	***0.52***	***0.21, 0.83***	***0.001***	***0.52***	***(0.21, 0.83)***	***0.001***	***0.55***	***(0.24, 0.86)***	***0.001***
Number recall maximum digit span, forward	0.01	-0.20, 0.22	0.936	0.17	(-0.20, 0.55)	0.359	4.61	(-13.70, 22.93)	0.621
Visit (ref: Visit 1)	***0.23***	***0.01, 0.44***	***0.038***	0.22	(0.01, 0.44)	0.040	0.23	(0.02, 0.45)	0.034
Number recall total, reverse	***0.61***	***0.19, 1.03***	***0.005***	0.19	(-0.48, 0.87)	0.577	-8.29	(-41.39, 24.80)	0.623
Visit (ref: Visit 1)	0.18	-0.25, 0.60	0.412	0.10	(-0.33, 0.53)	0.648	0.09	(0.35, 0.53)	0.687
Number recall maximum digit span, reverse	0.17	-0.15, 0.48	0.296	0.08	(-0.39, 0.54)	0.746	3.12	(-19.64, 25.89)	0.788
Visit (ref: Visit 1)	0.15	-0.16, 0.47	0.346	0.13	(-0.18, 0.44)	0.415	0.14	(0.18, 0.45)	0.399
Line trace	***-4.48***	***-5.87, -3.08***	***<0.001***	1.29	(-1.38, 3.96)	0.344	79.16	(-53.10, 211.41)	0.241
Visit (ref: Visit 1)	***-4.48***	***1.93, 4.73***	***<0.001***	***3.81***	***(2.20, 5.43)***	***<0.001***	***3.96***	***(2.34, 5.58)***	***<0.001***

*Pupils with a U_{SG} >1.019 classified as dehydrated

Bold values indicate a significant association at $\alpha = 0.006$, the level of 95% significance after correcting for multiple comparisons

Models include a random intercept at the pupil level to account for clustering

https://doi.org/10.1371/journal.pone.0210568.t003

Figure 1.8.

While the effects of dehydration on cognitive performance may have involved practice given that the performance of many students showed an improvement with practice, hydration was improved by augmenting the water supply showing the relevance of water, sanitation, and hygiene (WASH) in schools. (Chard et al, 2019)

1.4 WATER AND PREGNANCY

For facilitating circulation in the fetus and the amniotic fluid, the net fluid intake during pregnancy should be augmented. 8–10 glasses of water are recommended. The optimum amount of water consumption facilitates the circumventing of constipation seen during pregnancy and blood loss during delivery. For a breastfeeding mother, the water consumption should be slightly more than that required to slake one's thirst. The recommendation is a glass of water with meals and also while feeding the baby.

For infants, research points out at the direct association of the percentage of body weight that is water and the age with younger infants made of higher water content. The amount of water in breast milk or commercial milk formulae contain the requisite extent of water that an infant loses.

(Whitney and Rolfes, 2002; Montgomery, 2002)

Let us now look at a few studies that explored the biology of water in pregnant women:

Early work in 1990 by Shaw and team using telephonic interviews explored the association between congenital cardiac disorders and the consumption of tap water (exposed to a contamination incident). Analyses of the data from 145 mothers who bore infants with cardiac disorders vs. that of 176 mothers who bore infants with no such disorder revealed that the use of bottled water lowered the risk of such anomalies as against the use of tap water.

In 1995, Savitz and team explored the effect of specific toxic molecules present in water as contaminants on reproductive health. The risk of the presence of trihalomethane (THM) in the water source was examined utilizing data from an analysis in North Carolina of miscarriage, preterm delivery, and low birth weight. The authors reported that any of these aforementioned pregnancy outcomes were not evidently associated with the water source or THM. Interestingly, the risk of these adverse effects was lowered conspicuously by augmenting the consumption of water: from 1.5 for 0 glasses to 0.8 for more than 4 glasses of water per day.

Swan et al (1998) reported that the risk of spontaneous abortion was lowered in women who consumed bottled water than those who used tap water.

Aggazzotti and team (2004) reported a case-control study where an association between adverse pregnancy outcomes, small for gestational age at term (term-SGA) and preterm delivery and chlorination by-products (CBPs) in water was assessed (n= 1,194 subjects). A dose-response relationship was observed for CBPs and term-SGA while this was not reported for preterm delivery. The authors put forth the necessity of handling chlorite levels in water with care to circumvent such adverse pregnancy outcomes.

Lewis and team (2007) analyzed the relationship between preterm births and products used for disinfecting drinking water. Using data from records of 27 Massachusetts communities (a wise racial spectrum) that used a single public utility as the source of drinking water, the impact of total trihalomethanes (TTHM) was subject to evaluation on pregnancy.

The team reported that exposure to ≥ 60 μg/L TTHM during the 4 weeks before birth had a "small" impact on the risk for delivering a preterm baby birth [hazard ratio (HR) = 1.13; 95% confidence interval (CI), 0.95–1.35]. This risk showed a spike (HR = 1.39; 95% CI, 1.06–1.81) when the women depended on a governmental source of payment for prenatal care.

The results of the study are displayed below in Tables 1.6 and 1.7 below (reproduced from Lewis et al, 2007: *Publication of EHP lies in the public domain and is therefore without copyright. All text from EHP may be reprinted freely. Use of materials published in EHP should be acknowledged (for example, ? Reproduced with permission from Environmental Health Perspectives?); pertinent reference information should be provided for the article from which the material was reproduced*):

Table 1.6: Adjusted HRs (95% CIs) for TTHM exposure and preterm birth on the basis of race and ancestry (Lewis et al, *Environmental Health Perspectives*, 2007):

TTHM (μg/L)	1st Trimester	2nd Trimester	4-Week risk sets	Pregnancy average
All races (n = 2,813)				
< 40	1.00	1.00	1.00	1.00
40 to < 60	1.02 (0.92–1.13)	0.87 (0.77–0.99)	1.00 (0.87–1.15)	0.92 (0.82–1.02)
≥ 60	1.00 (0.88–1.14)	0.82 (0.71–0.94)	1.13 (0.95–1.35)	0.85 (0.74–0.97)
Per 10 μg/L	0.98 (0.95–1.02)	0.95 (0.92–0.99)	1.01 (0.97–1.06)	0.95 (0.91–0.99)
Caucasians (n = 1,429)				
< 40	1.00	1.00	1.00	1.00
40 to < 60	0.97 (0.84–1.12)	0.93 (0.77–1.11)	0.97 (0.79–1.18)	0.92 (0.79–1.06)
≥ 60	1.04 (0.87–1.24)	0.86 (0.71–1.05)	1.14 (0.90–1.45)	0.81 (0.67–0.99)
Per 10 μg/L	0.99 (0.94–1.05)	0.95 (0.90–1.00)	1.04 (0.97–1.10)	0.96 (0.91–1.03)
African American (n = 607)				
< 40	1.00	1.00	1.00	1.00
40 to < 60	1.27 (1.00–1.60)	0.73 (0.56–0.96)	0.78 (0.57–1.07)	0.95 (0.74–1.20)
≥ 60	1.13 (0.84–1.52)	0.62 (0.46–0.84)	1.00 (0.69–1.45)	0.77 (0.57–1.05)
Per 10 μg/L	1.01 (0.93–1.10)	0.89 (0.82–0.96)	0.98 (0.89–1.08)	0.90 (0.81–0.99)
Hispanic (n = 541)				
< 40	1.00	1.00	1.00	1.00
40 to < 60	0.97 (0.75–1.24)	1.07 (0.80–1.43)	1.59 (1.15–2.21)	0.87 (0.68–1.12)
≥ 60	0.86 (0.63–1.17)	1.07 (0.78–1.46)	1.63 (1.09–2.43)	1.00 (0.73–1.36)
Per 10 μg/L	0.95 (0.87–1.04)	1.04 (0.96–1.14)	1.05 (0.95–1.17)	1.01 (0.91–1.11)

Asian (*n* = 233)				
< 40	1.00	1.00	1.00	1.00
40 to < 60	0.94 (0.65–1.34)	0.65 (0.42–1.02)	0.80 (0.49–1.29)	0.89 (0.62–1.26)
≥ 60	0.77 (0.49–1.22)	0.72 (0.44–1.20)	0.68 (0.35–1.29)	0.86 (0.54–1.37)
Per 10 μg/L	0.90 (0.79–1.03)	0.95 (0.82–1.09)	0.88 (0.74–1.04)	0.86 (0.74–1.01)

Table 1.7: shows the crude and adjusted HRs (95% CIs) for TTHM exposure and preterm birth on the basis of the payment source for prenatal care (Lewis et al, *Environmental Health Perspectives*, 2007):

	2nd Trimester		**4 Weeks before birth**	
TTHM (μg/L)	**Government[b] (*n* = 1,129)**	**Private[c] (*n* = 1,684)**	**Government[b] (*n* = 1,129)**	**Private[c] (*n* = 1,684)**
Crude				
< 40	1.00	1.00	1.00	1.00
40 to < 60	0.88 (0.75–1.02)	0.92 (0.81–1.04)	1.01 (0.84–1.23)	0.92 (0.79–1.08)
≥ 60	0.84 (0.71–0.99)	0.88 (0.77–1.01)	1.19 (0.98–1.44)	0.94 (0.80–1.10)
Adjusted				
< 40	1.00	1.00	1.00	1.00
40 to < 60	0.88 (0.72–1.07)	0.87 (0.74–1.03)	1.07 (0.85–1.34)	0.96 (0.80–1.15)
≥ 60	0.83 (0.67–1.04)	0.82 (0.69–0.99)	1.39 (1.06–1.81)	1.00 (0.79–1.25)
Per 10 μg/L	0.96 (0.91–1.02)	0.95 (0.90–1.00)	1.03 (0.96–1.11)	1.00 (0.95–1.06)

The authors overall concluded that mothers of lower socioeconomic status were associated with a plausible association of preterm delivery with trihalomethane exposure in the third-trimester while a "negative association during the second trimester" was presented as the outcome.

2010-published research by Wright and team assessed the association of small-for-gestational-age (SGA) and the risk of preterm delivery (PTD) with that of the intake of water. The prospective cohort study entailed the analyses of infant details, the health of the mother, and water consumption (the quantity/ cup sizes/total water consumed inclusive of drinks made using water).

This work reported as " this is the first study to examine the relationship between water intake and birth weight in a prospective epidemiological study" showed that the adjusted mean birth weight was 27 when > 51-78 ounces of water was consumed that increased to 39 for > 78-114 ounces of water and 50 for > 114 ounces/day.

Table 1.8: below shows the birth weights for the various types of water consumed (Wright et al, 2010: *This article is published under license to BioMed Central Ltd. This is an Open Access article distributed under the terms of the Creative Commons Attribution License (http://creativecommons.org/licenses/by/2.0), which permits unrestricted use, distribution, and reproduction in any medium, provided the original work is properly cited*):

1. *[a]Adjusted for race, education level, annual household income, smoking, pre-pregnancy BMI, vitamin use, parity, employed during last 4 months, infant gender, and study site*
2. *[b]Categorical exposure cutpoints for average data results were slightly higher than the follow-up data*
3. *BMI= body mass index.*

	Follow-up data					**Average of baseline and follow-up data**	
		Unadjusted		**Adjusted[a]**		**Adjusted[a]**	
Exposure	**n (%)**	**ß**	**95% CI**	**ß**	**95% CI**	**ß**	**95% CI**
Bottled water							
None	448 (25%)	Ref		Ref		Ref	
Any	1329 (75%)	12	(-38, 63)	31	(-20, 82)	43	(-27, 113)
Cold tap water (*ounces*)[b]							
0-27	424 (24%)	Ref		Ref		Ref	
> 27-53	455 (25%)	12	(-50, 74)	9	(-53, 72)	25	(-38, 88)
> 53-91	444 (25%)	74	(11, 136)	52	(-11, 116)	44	(-19, 107)
> 91	453 (26%)	77	(14, 139)	49	(-14, 111)	65	(2, 128)
Per 20 ounce		12.3	(3.9, 20.7)	8.5	(0.1, 16.9)	8.5	(-1.5, 18.5)
Total tap water (*ounces*)[b]							
0-30	443 (25%)	Ref		Ref		Ref	
> 30-61	445 (25%)	41	(-21, 103)	44	(-18, 106)	10	(-52, 73)
> 61-96	496 (28%)	100	(40, 161)	78	(17, 139)	34	(-30, 97)
> 96	392 (22%)	85	(21, 149)	43	(-21, 107)	46	(-17, 109)
Per 20 ounce		12.0	(4.0, 20.1)	6.8	(-1.3, 15.0)	5.3	(-4.3, 15.1)
Total water (*ounces*)[b]							
0-51	452 (25%)	Ref		Ref		Ref	
> 51-78	439 (25%)	60	(-1, 122)	27	(-34, 87)	10	(-50, 71)
> 78-114	442 (25%)	67	(5, 128)	39	(-22, 99)	55	(-6, 116)
> 114	441 (25%)	83	(21, 145)	50	(-11, 111)	37	(-25, 98)
Per 20 ounce		9.8	(1.6, 18.1)	7.3	(-0.8, 15.5)	5.2	(-4.7, 15.2)

The risk of SGA was found to be lowered by increasing the amount of water consumed while the RR= risk ratio was higher for women drinking the least amount of water.

Table 1.9: shows the water consumed and Small-for-gestational-age results (Wright et al, 2010: *This article is published under license to BioMed Central Ltd. This is an Open Access article distributed under the terms of the Creative Commons Attribution License (http://creativecommons.org/licenses/by/2.0), which permits unrestricted use, distribution, and reproduction in any medium, provided the original work is properly cited*):

1. *[a]Term SGA models restricted to infants born to non-Hispanic White, non-Hispanic Black, or Hispanic women*
2. *[b]Adjusted for race, education level, annual household income, smoking, pre-pregnancy BMI, vitamin use, parity, employed during last 4 months, infant gender, and study site*
3. *[c]Categorical exposure cutpoints for average data results were slightly higher than the follow-up data*
4. *BMI= body mass index; RR= risk ratio*

	Follow-up data			**Average of baseline and follow-up data**
Exposure	**n (%)**	**Unadjusted RR (95% CI)**	**Adjusted RR (95% CI)[b]**	**Adjusted RR (95% CI)[b]**
Bottled water				
None	448 (26%)	1	1	1
Any	1258 (74%)	1.1 (0.7, 1.7)	0.9 (0.5, 1.4)	1.4 (0.6, 3.0)
Cold tap water (*ounces*)[c]				
0-27	403 (24%)	1	1	1
> 27-53	439 (26%)	0.9 (0.5, 1.5)	1.2 (0.6, 2.3)	1.1 (0.6, 2.2)
> 53-91	429 (25%)	0.9 (0.5, 1.6)	1.3 (0.7, 2.4)	1.4 (0.7, 2.6)
> 91	439 (26%)	0.7 (0.4, 1.2)	0.9 (0.5, 1.9)	1.1 (0.5, 2.1)
Per 20 ounce		1.0 (0.9, 1.0)	1.0 (0.9, 1.1)	1.0 (0.9, 1.1)
Total tap water (*ounces*)[c]				
0-30	423 (25%)	1	1	1
> 30-61	426 (25%)	0.7 (0.4, 1.3)	0.9 (0.5, 1.7)	1.2 (0.6, 2.2)
> 61-96	476 (28%)	0.7 (0.4, 1.2)	0.8 (0.5, 1.6)	1.3 (0.7, 2.6)
> 96	380 (22%)	0.7 (0.4, 1.2)	0.9 (0.5, 1.9)	1.1 (0.6, 2.2)
Per 20 ounce		1.0 (0.9, 1.1)	1.0 (0.9, 1.1)	1.0 (0.9, 1.1)

Total water (*ounces*) [c]				
0-51	434 (25%)	1	1	1
> 51-78	418 (25%)	0.7 (0.4, 1.2)	0.8 (0.4, 1.4)	0.9 (0.5, 1.6)
> 78-114	424 (25%)	0.6 (0.3, 1.0)	0.6 (0.3, 1.0)	0.5 (0.2, 1.0)
> 114	427 (25%)	0.8 (0.5, 1.4)	0.9 (0.5, 1.6)	1.0 (0.6, 1.8)
Per 20 ounce		1.0 (0.9, 1.1)	1.0 (0.9, 1.1)	1.0 (0.9, 1.1)

Table 1.10: below shows the water consumed and Preterm delivery results (Wright et al, 2010: *This article is published under license to BioMed Central Ltd. This is an Open Access article distributed under the terms of the Creative Commons Attribution License (http://creativecommons.org/licenses/by/2.0), which permits unrestricted use, distribution, and reproduction in any medium, provided the original work is properly cited*):

1. *[a]Adjusted for race, education level, annual household income, smoking, pre-pregnancy BMI, vitamin use, parity, employed during last 4 months, infant gender, and study site*
2. *[b]Categorical exposure cutpoints for average data results were slightly higher than the follow-up data*
3. *BMI= body mass index; RR= risk ratio*

	Follow-up data			**Average of baseline and follow-up data**
Exposure	**n (%)**	**Unadjusted RR (95% CI)**	**Adjusted RR (95% CI)[a]**	**Adjusted RR (95% CI)[a]**
Bottled water				
None	483 (25%)	1	1	1
Any	1458 (75%)	1.2 (0.9, 1.8)	1.2 (0.8, 1.8)	0.8 (0.5, 1.4)
Cold tap water (*ounces*)[b]				
0-27	472 (24%)	1	1	1
> 27-53	494 (25%)	0.8 (0.5, 1.2)	0.8 (0.5, 1.3)	0.9 (0.6, 1.4)
> 53-91	485 (25%)	0.9 (0.6, 1.3)	0.9 (0.6, 1.4)	0.9 (0.6, 1.4)
> 91	490 (25%)	0.8 (0.5, 1.1)	0.8 (0.5, 1.3)	1.0 (0.6, 1.5)
Per 20 ounce		1.0 (0.9, 1.0)	0.9 (0.7, 1.3)	0.9 (0.6, 1.3)
Total tap water (*ounces*)[b]				
0-30	493 (25%)	1	1	1
> 30-61	483 (25%)	0.8 (0.5, 1.2)	0.8 (0.5, 1.2)	1.0 (0.6, 1.5)
> 61-96	535 (28%)	0.7 (0.5, 1.1)	0.8 (0.5, 1.2)	1.2 (0.8, 1.9)
> 96	429 (22%)	0.9 (0.6, 1.3)	1.0 (0.6, 1.5)	0.9 (0.6, 1.4)
Per 20 ounce		1.0 (0.9, 1.0)	0.9 (0.7, 1.2)	0.9 (0.6, 1.3)

Total water (*ounces*)[b]				
0-51	491 (25%)	1.0	1.0	1.0
> 51-78	478 (25%)	1.0 (0.7, 1.6)	1.2 (0.7, 1.9)	1.1 (0.7, 1.7)
> 78-114	481 (25%)	1.0 (0.7, 1.6)	1.1 (0.7, 1.8)	1.0 (0.6, 1.5)
> 114	478 (25%)	1.2 (0.8, 1.8)	1.4 (0.9, 2.2)	1.2 (0.8, 1.9)
Per 20 ounce		1.0 (1.0, 1.1)	1.1 (0.9, 1.4)	1.1 (0.8, 1.5)

Overall, it was concluded that augmenting the intake of water during pregnancy was associated with small increases in the birth weight of infants to open up new avenues of research in terms of the impact of water on pregnancy (Wright et al, 2010).

1.5 COVID-19 AND WATER

The year 2020 shall remain etched in the memories of all: with the severe acute respiratory syndrome coronavirus-2 (SARS-CoV-2) causing an epidemic and shut down of routine life and the economy. Among the various sites where the virus has been isolated from, the viral particles were suggested to be present in the wastewater of humans. Reports highlighted the detection of the virus in the Amsterdam Schiphol Airport wastewater and also from near the site where the first case was reported. These findings bring in the picture of making the risk of the environmental transmission potential of SARS-CoV-2. This becomes pertinent in developing countries.

The organism *Naegleria fowleri* that causes primary amoebic meningoencephalitis is associated with water containing the pathogen entering the nose due to swimming or washing. This infection is often a challenge to detect in rural settings that warrant water to be "safe". The passage of this pathogen from the nose where it enters to enter the brain to cause disease is known. The entry of SARS-CoV-2 via the nose also brings in the use of filters prior to using tap water. Fungi, protozoans, and bacteria have been prevented by filters that however are a challenge for viruses given the size of the filters. This can be addressed by nanofilters or charged surfaces to trap the negatively charged viral particles.

Given the challenges associated with public health in areas where the awareness is limited, the monitoring for such infectious agents including that of SARS-CoV-2 in addressing health outbreaks.

(Siddiqui et al, 2020)

1.6 WATER AND THE MOOD

Given that the various effects of water on the human system have been presented, this chapter is concluded with a study that examined the effect of water on mood. Researchers Pross and team (2014) analyzed the effects of altering water consumption in 22 habitual high-volume (HIGH; 2-4 L/d) and 30 low-volume (LOW; <1.2 L/d) drinkers. The water intake in the former was lowered while it was increased in the latter.

Initially, 2.5 L of water was consumed by the HIGH group while 1 L was the intake in the LOW. This was the baseline. The values were interchanged for 2 controlled days and the mood was assessed employing the Karolinska Sleepiness Scale, Bond & Lader Visual Analog Scale (VAS), Thirst & Emotional VAS, and Profile of Mood States.

At the baseline, the HIGH group displayed elevated depression scores than that of the LOW group. The lowering of water intake in the HIGH resulted in an evident lowering of contentedness, calmness, and positive emotions while vigor and fatigue were lowered. In the LOW group, augmenting the water intake resulted in lowered confusion-bewilderment, vigor activity, and fatigue-inertia.

Tables 1.11 and 1.12 show the effects of decreasing the water intake and increasing the water intake on the drinkers, respectively:

Table 1.11: The mean and standard deviation for the mood in the HIGH group with lowered water intake (Pross et al, 2014: *This is an open-access article distributed under the terms of the Creative Commons Attribution License, which permits unrestricted use, distribution, and reproduction in any medium, provided the original author and source are properly credited*):

Outcome parameters	***Time point***					***p- value (for main intervention effect)***
	07:00	14:00	16:00	18:00	20:00	
POMS						
Tension						
Day 1	−0.7±3.1	−2.3±1.5	−2.1±1.2	−1.7±1.6	−1.9±1.6	Ns
Day 3-4-5	−0.9±1.5	−1.8±1.2	−1.3±1.5	−1.0±2.0	−1.7±1.6	
Depression						
Day 1	1.2±2.6	1.3±1.9	0.9±2.2	0.5±1.1	1.1±2.5	Ns
Day 3-4-5	0.7±1.5	0.8±1.4	1.0±2.3	1.0±2.1	0.9±1.8	

Anger						
Day 1	1.0±2.2	0.6±0.9	0.5±1.0	0.2±0.6	0.2±0.8	Ns
Day 3-4-5	1.0±1.2	0.4±0.9	0.6±1.4	0.5±1.3	0.6±1.1	
Fatigue						
Day 1	5.1±2.3	3.6±2.9	2.5±2.1	2.4±2.1	3.7±3.8	p<0.05
Day 3-4-5	3.9±2.5	2.5±2.0	2.7±2.2	2.1±1.7	2.2±1.8	
Confusion						
Day 1	−0.6±2.1	−1.2±1.9	−1.4±1.5	−1.3±1.6	−1.0±3.0	Ns
Day 3-4-5	−0.5±1.6	−1.1±1.3	−1.1±1.6	−0.9±1.3	−1.1±1.1	
Vigor						
Day 1	10.3±4.8	12.8±4.9	11.2±4.4	11.3±5.4	8.6±4.1	p<0.001
Day 3-4-5	5.7 ±3.4	8.4±4.3	8.4±4.1	8.1±3.8	7.5±4.2	
KSS						
Day 1	3.6±1.1	2.8±1.3	2.7±1.1	2.8±1.6	3.2±1.2	Ns
Day 3-4-5	4.2±1.0	2.6±1.2	2.6±1.0	2.5±0.9	2.7±0.9	
eVAS						
Day 1	67.4±16.8	76.3±14.9	74.3±12.2	71.8±13.4	69.5±8.3	p<0.05
Day 3-4-5	59.8±12.0	69.1±13.2	72.0±10.7	69.0±12.1	70.9±11.3	
B&L VAS						
Alertness						
Day 1	36.4±10.0	30.9±9.8	30.5±9.7	31.6±8.9	37.1±13.4	Ns
Day 3-4-5	41.6±12.8	32.9±12.3	33.0±13.6	32.4±11.8	33.0±11.9	
Calmness						
Day 1	25.7±12.7	24.3±10.7	21.8±9.6	26.4±10.9	22.8±10.2	p<0.01
Day 3-4-5	25.9±11.0	29.5±11.6	28.6±11.3	30.4±13.8	28.2±11.8	
Contented-ness						
Day 1	27.2±11.0	25.5±11.0	24.4±11.4	24.9±11.8	26.6±10.6	p<0.05
Day 3-4-5	32.3±11.5	27.2±10.6	28.9±11.0	28.9±12.5	27.9±11.3	
tVAS						
Day 1	59.7±20.9	55.6±16.6	59.7±19.5	60.0±18.5	68.3±17.7	p<0.001
Day 3-4-5	80.5±13.3	75.8±16.2	79.8±13.5	76.2±13.0	80.4±13.4	

Table 1.12: Mean and standard deviation of the effect of increasing water intake in the LOW group (Pross et al, 2014: *This is an open-access article distributed under the terms of the Creative Commons Attribution License, which permits unrestricted use, distribution, and reproduction in any medium, provided the original author and source are properly credited*):

Outcome parameters	***Time point***					***p value (for main intervention effect)***
	07:00	14:00	16:00	18:00	20:00	
POMS						
Tension						
Day 1	−0.3±4.1	−1.9±1.9	−2.3±1.6	−1.6±1.8	−1.8±1.5	Ns
Day 3-4-5	−1.3±1.4	−1.8±1.6	−1.7±1.6	−1.7±1.7	−1.8±1.7	
Depression						
Day 1	1.2±1.7	0.8±1.4	0.5±1.0	0.2±0.6	0.5±1.2	Ns
Day 3-4-5	0.5±0.9	0.6±1.0	0.5±1.1	0.5±1.1	0.7±1.7	
Anger						
Day 1	1.8±2.8	1.1±1.5	0.5±0.8	0.4±0.6	0.5±1.0	Ns
Day 3-4-5	0.9±1.6	0.6±1.1	0.4±0.9	0.7±1.4	0.6±1.1	
Fatigue						
Day 1	5.0±3.7	3.0±2.4	2.1±2.9	1.8±2.3	2.7±3.2	$p<0.001$
Day 3-4-5	2.3±2.3	1.8±1.8	1.9±1.9	1.7±1.7	1.8±1.8	
Confusion						
Day 1	0.8±4.0	−0.8±2.1	−1.0±2.2	−1.0±2.1	−0.5±2.0	$p=0.05$
Day 3-4-5	−0.4±1.6	−1.2±1.8	−1.2±1.9	−1.1±1.7	−1.0±2.1	
Vigor						
Day 1	11.6±6.2	11.7±5.7	11.7±6.0	11.1±6.3	10.1±6.9	$p<0.05$
Day 3-4-5	6.7±5.2	10.2±6.2	10.8±6.3	10.4±6.2	9.7±6.4	
KSS						
Day 1	4.3±1.8	3.2±1.3	3.0±1.8	3.2±1.5	3.2±1.7	$p=0.0729$
Day 3-4-5	4.6±1.4	2.8±1.4	2.7±1.2	2.5±1.1	2.7±1.1	
eVAS						
Day 1	68.7±14.9	71.0±13.2	68.1±16.1	72.6±13.5	70.2±16.4	Ns
Day 3-4-5	61.1±12.0	71.5±12.9	74.3±14.4	75.6±14.2	75.5±13.6	
B&L VAS						
Alertness						
Day 1	37.0±17.5	30.4±14.8	30.5±16.8	28.4±15.0	32.9±16.6	Ns
Day 3-4-5	41.6±16.7	28.4±15.0	27.2±15.6	26.7±13.9	28.3±15.3	

Calmness						
Day 1	26.8±20.1	23.4±15.1	25.0±15.2	26.1±13.4	24.7±13.7	Ns
Day 3-4-5	24.1±23.7	25.3±25.2	24.8±22.0	26.5±22.3	24.4±21.7	
Contented-ness						
Day 1	26.9±17.0	24.5±12.1	22.2±10.6	23.2±11.7	24.4±13.1	Ns
Day 3-4-5	28.8±12.9	23.7±14.4	22.2±13.7	22.1±12.4	22.5±12.6	
tVAS						
Day 1	43.7±23.7	43.1±20.4	53.8±22.5	38.8±21.7	49.4±21.3	p<0.001
Day 3-4-5	55.8±20.4	31.8±18.7	34.4±19.6	30.0±20.7	33.5±20.5	

Overall, the authors concluded that when habitual drinkers (LOW) increase their water content, the thirst, fatigue, sleepiness, and confusion were lowered. The lowering of the water intake in high volume habitual drinkers lowered the positive emotions to thus, show the ability of water to impact the mood (Pross et al, 2014).

Thus, while in high school the effects of water were limited to being a solvent, research is opening up new vistas where water plays a role in overall health, thinking, moods, performance, and even birth weight. The biology of water warrants more studies!

1.7 REFERENCES

1. Dudek S. G. 2001. *Nutritional essentials for nursing practice* (4th ed). Philadelphia: Lippincott.

2. Montgomery K. S. (2002). Nutrition Column An Update on Water Needs during Pregnancy and Beyond. *The Journal of perinatal education*, *11*(3), 40–42. https://doi.org/10.1624/105812402X88830.

3. Tahmina Afrose, Anthony Leela, Kevin Fernandez and Swe Swe Latt. Global Access to Clean Water: The Role of Rain Water Harvesting. Ann Rev Resear. 2018; 2(2): 555585.

4. Popkin, B. M., D'Anci, K. E., & Rosenberg, I. H. (2010). Water, hydration, and health. *Nutrition reviews*, *68*(8), 439–458. https://doi.org/10.1111/j.1753-4887.2010.00304.x

5. Armstrong, L. E., & Johnson, E. C. (2018). Water Intake, Water Balance, and the Elusive Daily Water Requirement. *Nutrients*, *10*(12), 1928. https://doi.org/10.3390/nu10121928

6. Leiper J.B., Pitsiladis Y., Maughan R.J. Comparison of water turnover rates in men undertaking prolonged cycling exercise and sedentary men. Int. J. Sports Med. 2001;22:181–185. doi: 10.1055/s-2001-15912.

7. Institute of Medicine . Dietary reference intakes for water, potassium, sodium, chloride, and sulfate. Institute of Medicine; Washington, DC, USA: 2004.

8. Johnson, E. C., Muñoz, C. X., Jimenez, L., Le Bellego, L., Kupchak, B. R., Kraemer, W. J., Casa, D. J., Maresh, C. M., & Armstrong, L. E. (2016). Hormonal and Thirst Modulated Maintenance of Fluid Balance in Young Women with Different Levels of Habitual Fluid Consumption. *Nutrients*, *8*(5), 302. https://doi.org/10.3390/nu8050302.

9. Black D.A., McCance R.A., Young W.F. A study of dehydration by means of balance experiments. J. Physiol. 1944;102:406–414. doi: 10.1113/jphysiol.1944.sp004046.

10. Drewnowski, A., Rehm, C. D., & Constant, F. (2013). Water and beverage consumption among adults in the United States: cross-sectional study using data from NHANES 2005-2010. *BMC public health*, *13*, 1068. https://doi.org/10.1186/1471-2458-13-1068.

11. Watso, J. C., & Farquhar, W. B. (2019). Hydration Status and Cardiovascular Function. *Nutrients*, *11*(8), 1866. https://doi.org/10.3390/nu11081866.

12. Edmonds C.J., Burford D. Should children drink more water?: The effects of drinking water on cognition in children. Appetite. 2009;52:776–779.
13. Benton D. (2011). Dehydration influences mood and cognition: a plausible hypothesis?. *Nutrients*, *3*(5), 555–573. https://doi.org/10.3390/nu3050555.
14. Chard AN, Trinies V, Edmonds CJ, Sogore A, Freeman MC (2019) The impact of water consumption on hydration and cognition among schoolchildren: Methods and results from a crossover trial in rural Mali. PLoS ONE 14(1): e0210568. https://doi.org/10.1371/journal.pone.0210568.
15. Whitney E. N, Rolfes S. R. 2002. *Understanding nutrition* (9th ed.). Belmont, CA: Wadsworth/Thomson Learning.
16. Shaw GM, Swan SH, Harris JA, Malcoe LH. Maternal water consumption during pregnancy and congenital cardiac anomalies. Epidemiology. 1990 May;1(3):206-11. doi: 10.1097/00001648-199005000-00005. PMID: 2081254.
17. Savitz, D. A., Andrews, K. W., & Pastore, L. M. (1995). Drinking water and pregnancy outcome in central North Carolina: source, amount, and trihalomethane levels. *Environmental health perspectives*, *103*(6), 592–596. https://doi.org/10.1289/ehp.95103592.
18. Swan SH, Waller K, Hopkins B, Windham G, Fenster L, Schaefer C, Neutra RR: A prospective study of spontaneous abortion: relation to amount and source of drinking water consumed in early pregnancy. Epidemiology. 1998, 9: 126-133. 10.1097/00001648-199803000-00005.
19. Aggazzotti G, Righi E, Fantuzzi G, Biasotti B, Ravera G, Kanitz S, Barbone F, Sansebastiano G, Battaglia MA, Leoni V, et al: Chlorination by-products (CBPs) in drinking water and adverse pregnancy outcomes in Italy. J Water Health. 2004, 2: 233-247.
20. Lewis, C., Suffet, I. H., Hoggatt, K., & Ritz, B. (2007). Estimated effects of disinfection by-products on preterm birth in a population served by a single water utility. *Environmental health perspectives*, *115*(2), 290–295. https://doi.org/10.1289/ehp.9394.
21. Wright, J.M., Hoffman, C.S. & Savitz, D.A. The relationship between water intake and foetal growth and preterm delivery in a prospective cohort study. *BMC Pregnancy Childbirth* **10,** 48 (2010). https://doi.org/10.1186/1471-2393-10-48.

22. Siddiqui, R., Khamis, M., Ibrahim, T., & Khan, N. A. (2020). SARS-CoV-2: The Increasing Importance of Water Filtration against Highly Pathogenic Microbes. *ACS chemical neuroscience*, *11*(17), 2482–2484. https://doi.org/10.1021/acschemneuro.0c00468.
23. Pross, N., Demazières, A., Girard, N., Barnouin, R., Metzger, D., Klein, A., Perrier, E., & Guelinckx, I. (2014). Effects of changes in water intake on mood of high and low drinkers. *PloS one*, *9*(4), e94754. https://doi.org/10.1371/journal.pone.0094754.

Chapter 2

Water for Health

Shiv Sanjeevi Sripathi[1]

[1]Shiv Sanjeevi Sripathi secured second rank in Mumbai University and first rank in his college at his undergrad level and in his college during his Masters and has worked on a stem cell project at the Specialized Centre for Cell Based Therapy (SCCT), KEM Hospital at Mumbai, following which he was a Junior Research Fellow at Microbiology & Cell Biology Department, Indian Institute of Science, Bangalore from August 2009 to November 2011. The field of work involved cloning of cell wall genes and transcription factors in E. coli & M. smegmatis. This was followed by a stint as a Research and Development Executive at Transasia Biomedicals, Mumbai where he worked on patent writing and Immunoassays from 2011 till 2013. In 2013, he entered the teaching field where he has been tutoring biology and chemistry to the students of 11th and 12th grades along with English grammar and communication skills to people of all age groups. Shiv also loves helping people write scientific manuscripts and has co-edited several books on life sciences and aspects of biotechnology with Apple Academic Press.

This chapter looks at how water can be used for healing or treatment based on scientific literature.

2.1 WOUND HEALING

Early research published in 2006 analyzed available data as to the effect of employing saline or water for cleaning wounds. Many studies reported that the rate of infection while using saline or water to rinse wounds. It was suggested in one study that showering wounds with water made patients feel better over utilizing a bottle containing spirit (Beam, 2006).

The process of wound healing involves synchronous coordination entailing the processes of inflammation, the proliferation of cells, deposition of the matrix deposition, and remodeling of tissue. Growth factors, chemokines, and cytokines are vitally involved in the repair with the net result of healing showing up as structural and functional tissue regeneration.

Research has pointed out the activation of the nuclear factor-E2 related factor 2 (Nrf2) pathway by hydrogen that further activates genes functioning as anti-oxidants. This becomes significant for conditions such as recurrent aphthous stomatitis where inflammation caused by reactive oxygen species (ROS) and cytokines play a key role.

Scientists Tamaki et al (2016) assessed the effects of hydrogen-rich water on Wistar rats (considered clinically reproducible according to the researchers) where oral palatal region wounds were created. In the animals that were administered hydrogen-rich water, there was an acceleration in wound healing as against controls (that received distilled water). The extent of oxidative stress (as evidenced by markers ROM and OXY) was lowered while the levels of proinflammatory cytokine (IL-6, TNF-β) also displayed an evident dip due to the hydrogen-rich water. Further, the levels of genes involved in wound healing: TGF-β1, FGF7, VEGF, α-SMA, and type 1 collagen were also augmented by the water treatment.

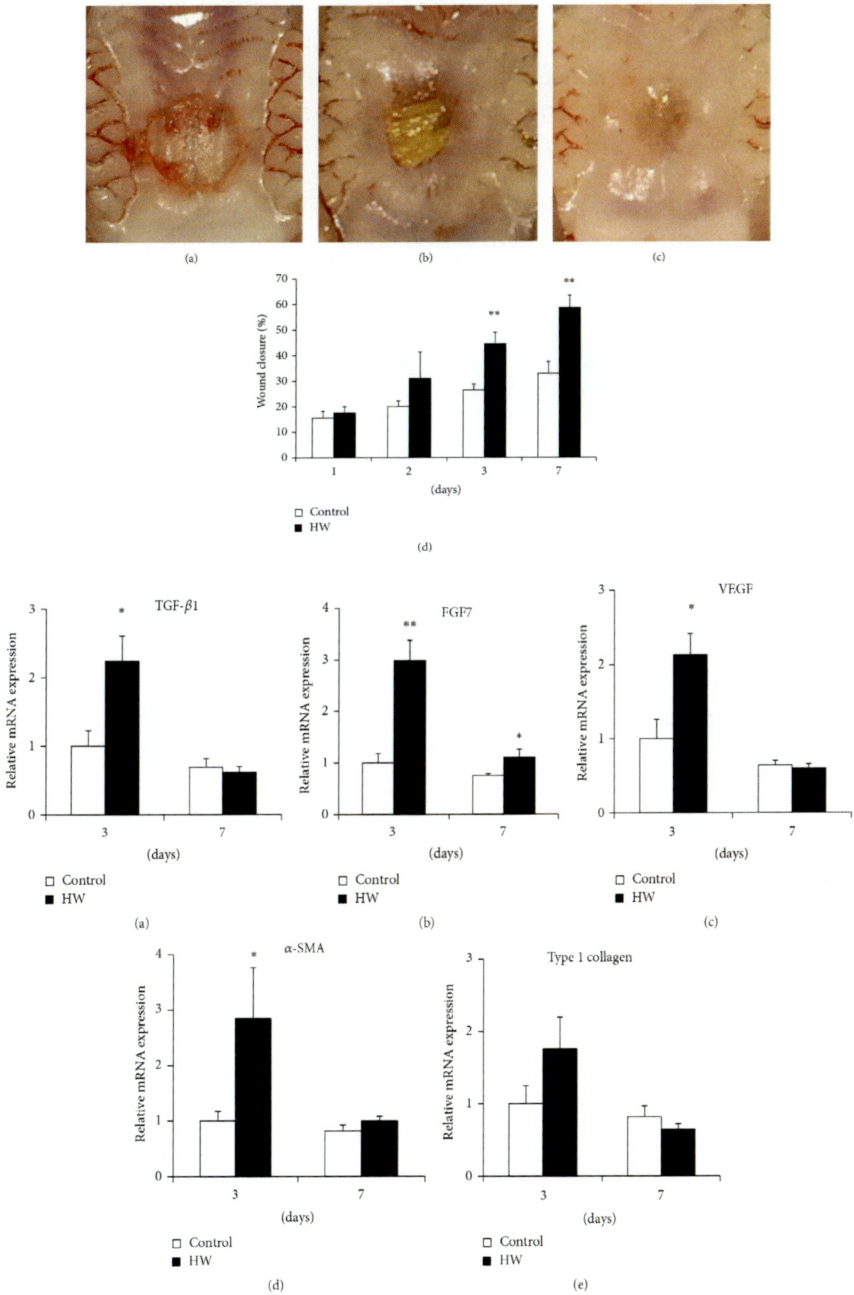

Figure 2.1: displays the positive effect of hydrogen-rich water on rat oral wounds: ***TOP:*** the wound healing process (a) the wound Photos taken 7 days-post treatment in the (b) control and (c) hydrogen-rich water group (d) Graphi-

cal display of the wound healing. BOTTOM: Increased levels of genes involved in healing with hydrogen-water treatment (Tamaki et al, 2016; *This is an open-access article distributed under the Creative Commons Attribution License, which permits unrestricted use, distribution, and reproduction in any medium, provided the original work is properly cited*).

Given the targeting of inflammatory conditions associated with pain or disease, the initial research highlighted the benefits of drinking hydrogen-rich water for healing wounds (Tamaki et al, 2016).

2.2 WATER SHOWS PROMISE FOR ADDRESSING DEPRESSION

According to a systematic analysis for the Global Burden of Disease Study 2017, depression strikes 264 million people globally. The worst effect of this condition is with 800000 lives lost due to this act (WHO fact sheet). Studies on the major depressive disorder (MDD) have revealed that elevated levels of tumor necrosis factor (TNF), interleukin-1 (IL-1), and interleukin-6 (IL-6) are involved in neurotransmitter and neuroendocrine alterations to cause depression. Stressful conditions further add to these depressive states by augmenting IL-1β levels. While the use of anti-depressants is linked to adverse effects and limited efficacy: 50% success (Lee et al, 2010), the quest is on to target this condition. Given the involvement of ROS in MDD, researchers Zhang and team assessed the utility of hydrogen-rich water for depression employing the chronic unpredictable mild stress (CUMS) mouse model.

The spike in IL-1β due to stress was found to be lowered by hydrogen-rich water that also decreased the caspase-1 activity in the hippocampus, associated with inflammation and stress. Overproduced ROS in the hippocampus of the mice was evidently targeted by the hydrogen-rich water.

The results of the utility of hydrogen-rich water to target depression are shown in Figure 2.2. ***TOP***: Increased IL-1β levels due to stress were lowered by hydrogen-rich water in the (a) hippocampus (b) cortex. ***BOTTOM***: The hydrogen-rich water also lowered the overproduced ROS associated with depression in the (a) hippocampus and (b) prefrontal cortex (Zhang et al, 2016; *this work is licensed under a Creative Commons Attribution 4.0 International License. The images or other third party material in this article are included in the article's Creative Commons license unless indicated otherwise in the credit line; if the material is not included under* *the Creative Commons license, users will need to obtain permission from the license holder to reproduce the material. To view a copy of this license, visit http://creativecommons.org/licenses/by/4.0/; You are free to: Share — copy and redistribute the material in any medium or format; Adapt — remix, transform, and build upon the material for any purpose, even commercially*):

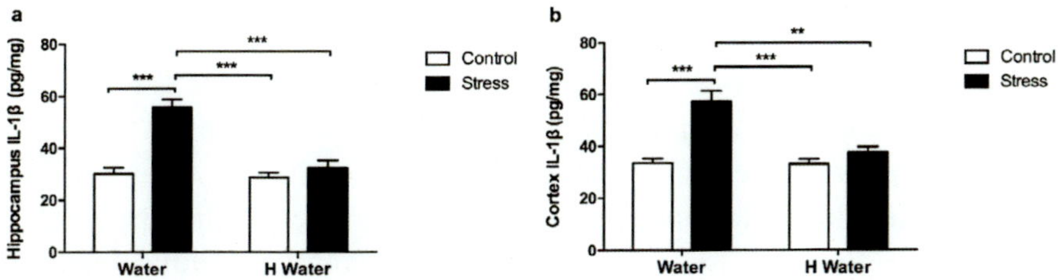

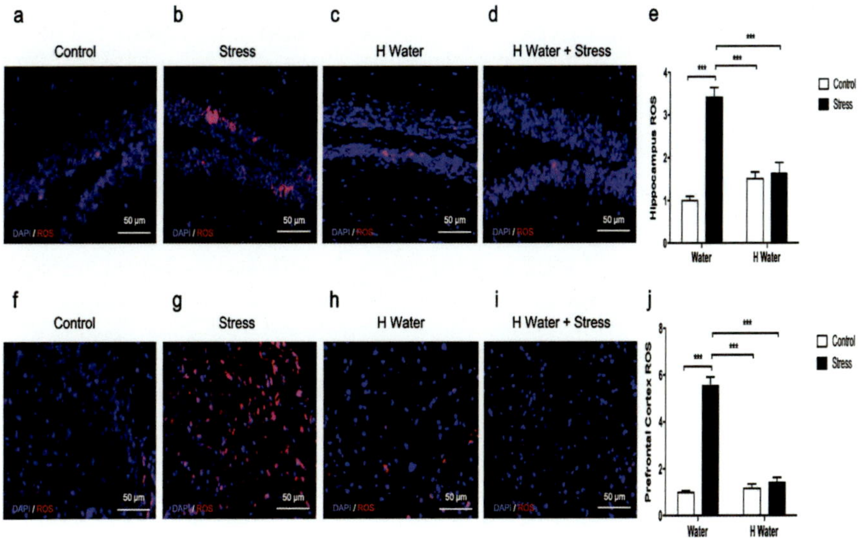

Overall, the authors concluded that MDD can be targeted by looking at the inflammasome pathway that is activated in stress to boost the levels of IL-1β and ROS associated with neuroinflammation. Further research can ascertain the utility of hydrogen-rich water to treat depression (Zhang et al, 2016).

A cross-sectional study published in 2018 analyzed the effect of consuming water on anxiety and depression in 3327 Iranian general adults. Analyses using the validated Iranian version of the Hospital Anxiety and Depression Scale showed that against the baseline of 5 or more glasses of water a day, the risk of depression and anxiety escalated by a factor of 2 when the consumption of water was lower (2 glasses a day). An inverse association was revealed between such low water intake and depression. The risk was augmented by 54% in ladies and 73% in men due to low water intake opening up the potential of utilizing water for maintaining mental health (Haghighatdoost et al, 2018).

2.3 HYDROTHERAPY

Based on the Greek word "hydrotherapia" or using water for treatment, hydrotherapy looks at employing water to address a health issue. The introduction of a reflex effect stimulated by nerve endings by water accounts for changes in blood flow and metabolism. The use of this approach can be traced back to pages in history where Asian counties: India, Japan, and China reported the use of water to heal the soul and body. The concept of lavish baths existed in Egypt for ceremonies and enjoyment. The term hydrotherapy can be attributed to Sebastian Kneipp with the year 1844 seeing the first clinic in the United States. There are several forms attributed to this approach now:

- Watsu is based on warm water-based massage when an individual is floating for relaxing the body and emotions.
- Sitz bath that looks at addressing health issues by immersion of the hips in water.
- Finnish steam saunas where steaming is performed in wooden rooms for addressing respiratory and cardiovascular concerns.
- Colon hydrotherapy involves the use of water to flush out the colon.
- Wet wrap therapy involves employing cloth or gauze for wrapping.
- Steam inhalation to address respiratory concerns and nasal congestion.
- Contrast hydrotherapy looks at hot and cold water treatment in an alternate format.

(Almassmoum et al, 2018)

Multiple sclerosis (MS) accounts for cognitive challenges and an inability to achieve coordination with a perfect cure still being sought after. Other approaches sought after by MS patients are massage, reflexology, meditation, nutrient supplements, and Tai-Chi. Given the blocking of nociceptors by warm water to alleviate pain, relax the muscles by augmenting the blood flow, and lowering swelling, research has been reported for using hydrotherapy for MS.

A randomized controlled trial involving 73 MS patients reported the benefits of hydrotherapy with the controls undergoing exercises and breathing while the test received Ai-Chi exercises in a swimming pool. The latter approach resulted in a lowering of pain intensity for 10 weeks: the median visual analog scale scores dipped by 50%. The level of autonomy and fatigue also displayed an improvement. The authors concluded that "an Ai-Chi aquatic exercise program improves pain, spasms, disability, fatigue, depression, and autonomy in MS patients".

(Castro-Sánchez et al, 2012)

Figure 2.3 below shows the beneficial effects of hydrotherapy for improving the life quality of MS patients. An improvement was observed for Mean Pain Visual Analogue Scale scores with the pain lowered by 50% in MS patients undergoing hydrotherapy-based treatment as against the controls (Castro-Sánchez et al, 2012: *This is an open-access article distributed under the Creative Commons Attribution License, which permits unrestricted use, distribution, and reproduction in any medium, provided the original work is properly cited*):

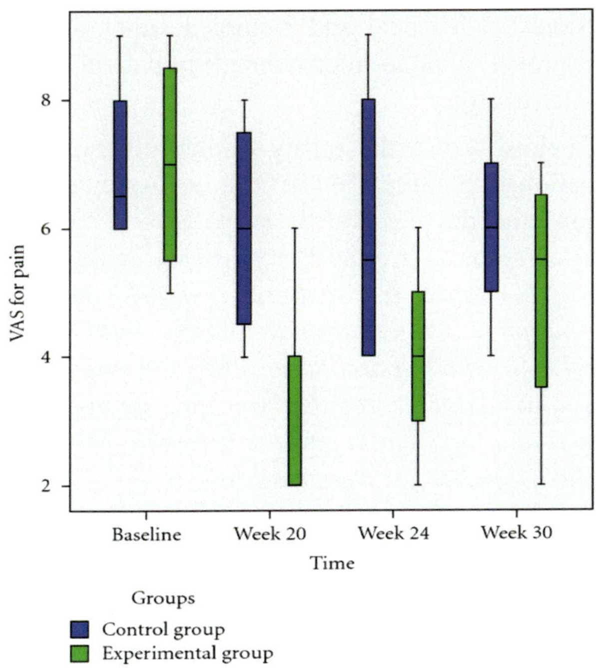

Figure 2.3.

Defined as "a chronic functional disorder of the gastrointestinal system", 11% of the world is affected by irritable bowel syndrome (IBS) (Canavan et al, 2014). The disease involves several conditions ranging from changes in the gut immune system, intestinal permeability, and the microbiome. Treatment involves medical treatment such as antispasmodics, anti-diarrheal compounds, and antidepressants along with dietary changes such as laxatives, fiber, and probiotics.

The promise of colonic irrigation to treat constipation and fecal incontinence promoted Hsu and team (2016) to use the Ashong colonic irrigation apparatus (ACIA) for treating 12 patients with constipation-dominant IBS (IBS-C) and 6 with diarrhea-dominant IBS (IBS-D).

Approved by the Taiwan Food and Drug Administration, the ACIA can be operated by a patient employing a remote control by sitting on a toilet. The patients used the system for 6 days/week twice a day followed by an analysis of the symptoms, bowel movement, fecal consistency, abdominal pain, and how much the patient experienced satisfaction in terms of bowel movements. In the IBS-C group, the stool frequency showed a spike from 1.68 to 3.78/week while the IBS-C group showed a dip from 45 to 13 % in the Bristol Stool Scale type 1 and 2 stool. Further, abdominal pain and distress were improved with an improvement in patient satisfaction paving the way for future research.

Figure 2.4 below shows the utility of hydrotherapy to address IBS. TOP: the irrigation apparatus; BOTTOM: a) Assessing the abdominal pain, satisfaction, and degree of distress b) Bowel movements (BM) c) Consistency of the stool as per the Bristol Stool Scale (BSS) (types 1–7) (Hsu et al, 2016: *This article is distributed under the terms of the Creative Commons Attribution 4.0 International License (http://creativecommons.org/licenses/by/4.0/), which permits unrestricted use, distribution, and reproduction in any medium, provided you give appropriate credit to the original author(s) and the source, provide a link to the Creative Commons license, and indicate if changes were made*):

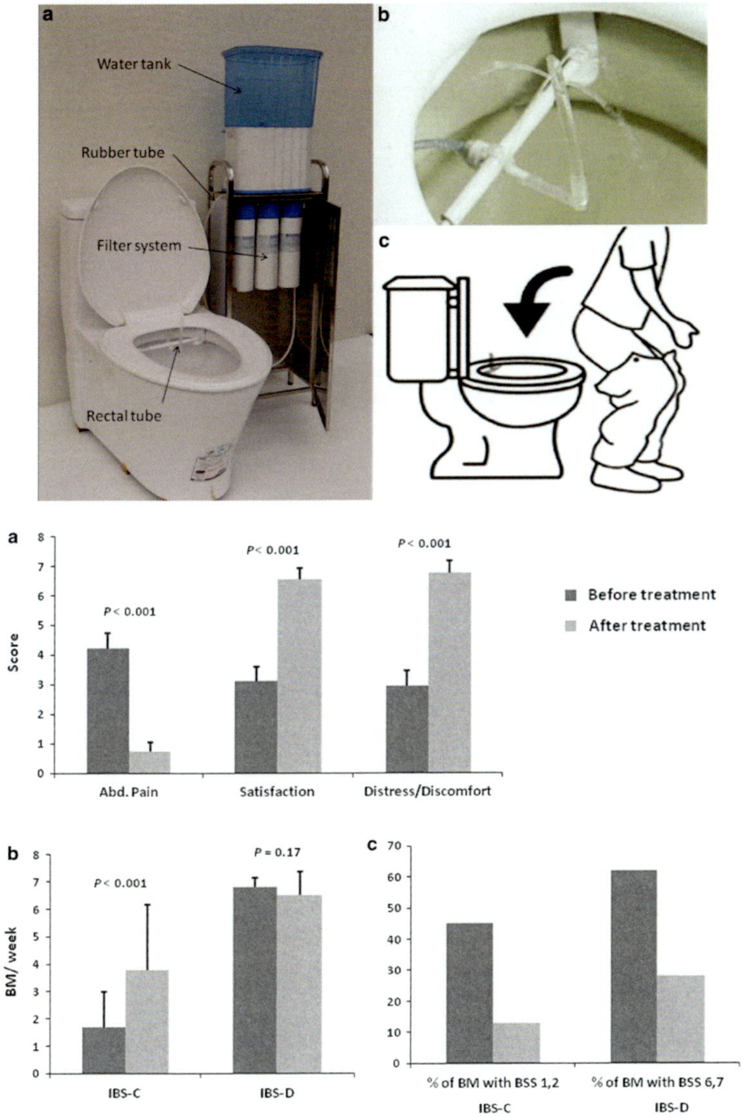

Figure 2.4.

(Hsu et al, 2016)

Other conditions that have been reported in clinical trials (2012-2017) using hydrotherapy have been summarized below in Figure 2.5 (Almassmoum et al, 2018: *This work is licensed under a CC-BY Creative Commons Attribution: Open access: www.balimedicaljournal.org and ojs.unud.ac.id/index.php/bmj*):

Years	Number of volunteers	Disease /condition	Observation	*P*-value	References
2012	73	Multiple sclerosis	Improvement	<0.02	Castro-Sánchez *et al.*, 2012
	42	Static and dynamic balance among elderly men		<0.001	Alikhajeh *et al.*, 2012
	103	Musculoskeletal condition		-	Verhagen *et al.*, 2012.
2013	33	Multiple sclerosis		<0.05	Segura-Jiménez *et al.*, 2013
	30	Fibromyalgia		<0.05	Fathollah *et al.*, 2013
2014	40	Hemophilia results in pain and restricted range of motion		<0.001	Mazloum *et al.*, 2014[30]
	31	Burn conditions		-	Moufarrij *et al.*, 2014[5]
	9	Hereditary spastic paraparesis (HSP)		-	Zhang *et al.*, 2014[31]
	26	Physical functional Fitness and Health Perception.		-	Pereira *et al.*, 2014[32]
	30	shoulder abduction, shoulder pain and disability in post mastectomy female patients		<0.05	Mohammed *et al.*, 2014[33]
	34	Balance dysfunction in Parkinsonian patients		-	Volpe *et al.*, 2014[34]
2015	2447	Irrigation of open Fracture Wounds	Decrease reoperation rates	0.01	Bhandari *et al.*, 2015[35]
	30	Fat embolism in total knee arthroplasty (TKA)	Improvement	0.016	Zhao *et al.*, 2015[36]
2016	40	Migraine		0.01	Sujan *et al.*, 2016[37]
	39	Wound healing		0.03	Enblom *et al.*, 2016[39]
	1	Musculoskeletal disorders		-	Atkin and Ousey, 2016[38]
2017	45	Postpartum pain		<0.001	Batten *et al.*, 2017[40]
	21	Parkinson disease		<0.01	Carroll *et al.*, 2017[41]
	40	Heart failure		0.005	Hägglund *et al.*, 2017[42]

Figure 2.5.

2016-published data reported that a cross-sectional study of an outpatient department (OPD) in a hospital showed the prevalence of **chronic insomnia** to be prevalent in 33% of the adult population sampled. Conventional treatment entails the use of hypnotic drugs that can display unwanted side effects.

Among the various nonpharmacological treatments for relaxation, researchers Ichiba and team (2019) looked at electrocardiogram (ECG)-based analyses of warm steam inhalation before bedtime to address mild insomnia. The study described as a "single-blind, placebo-controlled, randomized cross-over design" involved the use of a warm steam-generating mask or sham mask 15 minutes prior to bed in their homes. Electroencephalogram (EEG) and visual analog scale-based analyses revealed that subjective apprehension of the next day was lowered in subjects who used steam. Further, the subjective sleep quality and delta power density during the first third of sleep episodes was augmented. Overall, the authors concluded that the usage of hydrotherapy in the form of steam can be looked at to boost relaxed sleep.

Figure 2.6 below shows the outcome of the effect of steam inhalation on sleep quality Ichiba et al, 2019: (*This is an open-access article distributed under the Creative Commons Attribution License, which permits unrestricted use, distribution, and reproduction in any medium, provided the original work is properly cited*):

	Sham	Steam inhalation	
Bedtime (h : min)	0 : 46±0 : 51	0 : 45±0 : 54	0.807
Wake time (h : min)	7 : 02±0 : 38	6 : 59±0 : 26	0.221
SPT (min)	380.1±60.7	370.5±63.5	0.221
TST (min)	348.0±71.3	342.8±66.9	0.311
Sleep latency (min)	20.2±25.9	14.1±22.5	0.600
WASO (min)	11.9±10.7	13.6±23.8	0.861
Sleep efficiency (%)	91.2±8.5	92.4±7.8	0.807
Deep NREM (%)	20.2±6.4	22.2±10.1	0.753
Deep NREM latency (min)	23.8±16.9	17.7±12.1	0.045
REM latency (min)	48.6±22.8	67.2±34.2	0.196

Values are expressed as means±standard deviations. Comparisons are relative to the sham condition (Wilcoxon signed-rank test). SPT: sleep period time; TST: total sleep time; WASO: wake after sleep onset; REM: rapid eye movement; NREM: non-REM.

Delta, theta, alpha, and beta bands: EEG data in the sham and steam-inhalation subjects:

Normal-ized EEG	First NREM-REM cycle			Second NREM-REM cycle			Third NREM-REM cycle		
	Sham	Steam		Sham	Steam		Sham	Steam	
Delta (%) (1–4 Hz)	89.6±36.0	121.0±45.1	0.039	105.7±34.5	103.5±31.3	0.972	101.3±45.9	93.2±22.6	0.552
Theta (%) (4–8 Hz)	90.9±40.1	125.1±58.0	0.028	145.0±44.2	135.9±87.2	0.421	150.2±71.6	187.8±95.7	0.422
Alpha (%) (8–12 Hz)	73.8±32.6	98.2±44.4	0.064	104.8±33.9	104.5±36.0	0.917	95.2±42.0	94.2±23.1	0.807
Beta (%) (16–35 Hz)	70.7±36.1	94.7±61.2	0.075	95.7±34.9	98.6±39.5	0.861	106.3±46.8	101.2±32.3	0.753

Figure 2.6.

2.4 FINAL POINTS

"If there is magic on this planet, it is contained in water"-Loren Eiseley

On the same lines, we have heard the importance of "drinking lots of water" for several addressing ailments and maintaining health. For instance, analyses by Akdeniz and team (2018) revealed that when individuals consuming less water increased their intake, roughness, and dryness of the skin were lowered with "deep" hydration. Scientists Nishimura and team (2019) explained how dry skin is the outcome of humidity changes indoors. Given the distress associated with dry skin in an environment with low humidity, the researchers evaluated the effect of spraying fine water particles on the face of 17 healthy adult women. The skin conductance and skin elasticity of the stratum corneum were high while the transepidermal water loss (TEWL) was lowered when the size of the water particles was 0.5 μm. The lowered loss of water due to the water spraying is shown in Figure 2.7 below (Nishimura et al, 2019; *This is an open-access article under the terms of the Creative Commons Attribution* License, which permits use, distribution and reproduction in any medium, provided the original work is properly cited):

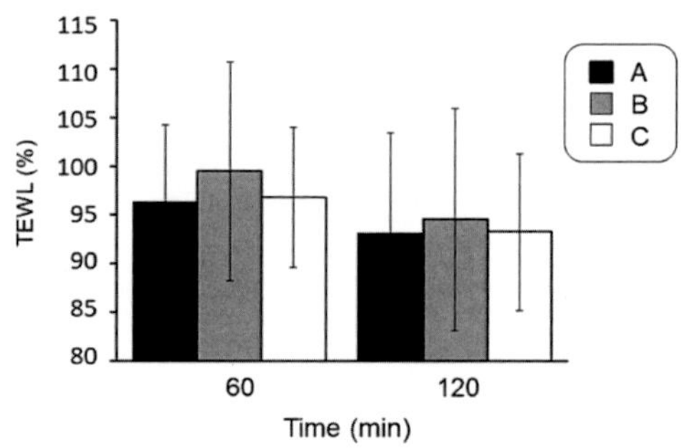

Figure 2.7.

The authors concluded that the moisture content of the skin can be altered favorably using non charged fine water particles (Nishimura et al, 2019).

Aquatic cycling involves pedaling against water resistance to ease pain and augment the fitness of muscles and the cardiovascular system. Rewald et al (2017) examined how this approach has been employed in addressing medical conditions such as obesity, multiple sclerosis, and weight loss. Figure 2.8 below shows the intervention programs based on water cycling (Rewald et al, 2017: *This is an open-access article distributed under the terms of the Creative Commons Attribution License, which permits unrestricted use, distribution, and reproduction in any medium, provided the original author and source are credited*):

UPRIGHT BODY POSITION									
Author	**Year**	**Study design**	**Study aim**	**Sample**[#]	**Exercise parameters**	**Key findings**	**T_{Water}**	**Aquatic bike used**	**Immersion depths**
Boidin [71]	2015	Retro-spective cohort: AC vs. LC	To compare the effects of a lifestyle intervention in addition to AC or LC on cardiometabolic and exercise parameters in obese patients	n = 95 obese people • AC: n = 21 (F/M: 19/2) age: 58±9 • LC: n = 74 (F/M: 55/19) age: 55±7	• Exercise programme: HIIT AC + water-based resistance training, 5x Mediterranean diet counselling • Programme duration: 36 wk • Frequency: 2-3/wk • Duration session: 34min AC, 20min resistance exercise • Intensity: 15 RPE/ 80% MAP • Rpm: NR	• Reduction** in body mass, WC, total and trunk fat mass; no group difference • Improvement* in resting BP, maximal aerobic capacity, resting HR; no group difference • Improvement* in fasting glycaemia, triglyceride levels; no group difference • Improvement* in abdominal and thigh muscle endurance; no group difference	NR	Hydrorider®	NR
Bansi [72, 73]	2013	RCT: AC vs. LC	To investigate the influence of exercise in cytokine response, health-related QoL, fatigue, neurothophin concentrations and cardiorespiratory values	n = 60 MS patients • AC: n = 25 (F/M: 17/8), age: 50, range: 44.6–55.1 • LC: n = 28 (F/M: 18/10), age: 52, range: 46.7–56.3	• Exercise programme: steady AC + usual care rehabilitation • Programme duration: 3 wk • Frequency: 5/wk • Duration session: 30min • Intensity: lactate threshold = 60% VO_{2peak} • Rpm: 50–60	• Short term immune adaptions and increased VO_2 lactate values were associated with improved health-related QoL and reduced fatigue [73] • Improved health-related QoL; no group difference [73] • Improved self-reported physical fatigue; no group difference [73] • Cytokines and neurotrophins showed no change over time and between groups [72] • Fatigue scores associated with baseline and post-intervention exercise tests remained unchanged in both groups [72] • Cardiorespiratory values improved** over time; no group difference [72]	28°C	Aquarider Professional®	1,30m
Young [69, 70]	1995, 1993	Quasi experi-ment: hot vs. cold water	Comparison of metabolic and thermal adaption to endurance training in hot and cold water and its effect on aerobic capacity	n = 18 M • 20°C water: n = 9, age: 20 ±1 • 35°C water: n = 9, age: 20 ±1	• Exercise programme: steady AC in hot or cold water • Programme duration: 8wk • Frequency: 5/wk • Duration session: 60min • Intensity: 60%VO_{2max} • Rpm: 40	• Reduced* post-exercise muscle glycogen use; no group difference [70] • Lactate accumulation was equal for hot and cold water [70] • 13% post-intervention increase** of VO_{2max}; no group difference [69, 70] • 4% increase** of erythrocyte volume; no group difference [69] • Unchanged plasma volume; no group difference [69] • 38% increase** of vastus lateralis citrate synthase activity; no group difference [69]	35°C vs. 20°C	Modified Monark EM (Shapiro)	neck
Avellini [67]	1982	Quasi experi-ment: warm vs. cold water vs. land	To determine how physical training on land compared to warm and cold water training affects heat tolerance	n = 15 M • land: n = 5, age: 23.2± 4.7 • 32°C water: n = 5, age: 20.8±1.8 • 20°C water: n = 5, age: 23.0±4.1	• Exercise programme: steady AC in warm and cold water. Groups were divided based on the maximal exercise capacity, body surface area, and % body fat. • Programme duration: 4wk Frequency: 5/wk • Duration session: 60min • Intensity: 75%VO_{2max} • Rpm: NR	• Similar increase in VO_{2max}; no group difference • Higher* post-training values of T_{rectal} for LC compared to AC • 0.9°C reduction* of post-training mean T_{skin} for LC and warm water AC. • Post- training total body sweat: increased* only warm water AC (25%). • Post-heat acclimation: only cold water AC demonstrated an increase* in sweat rate (25%) • Decrease* in post-training HR in all 3 groups, greatest decline (29 beats·min) in land group. HR in warm and cold-water AC HR decreased 14 and 18 beats·min. • Post-heat acclimation HR: reduction* from post-training values in all groups	20°C vs. 32°C	Modified Monark EM	neck
Sheldahl [68]	1986	RCT: AC vs. LC vs. control	Comparison of land and water training to determine whether the cephalad shift in blood volume due to water immersion affects normal adaptations to aerobic endurance training	n = 22 M • • age: 49±8 • water: n = 9 • land: n = 9 • control: n = 4	• Exercise programme: steady AC Programme duration: 12wk • Frequency: 3/wk • Duration session: 30min • Intensity: 60–80% VO_{2max} • Rpm: NR	• Increase* in SV at submaximal exercise intensities; no difference between AC and LC • Increase** in VO_{2max} in both exercise groups • Decrease** in HR at submaximal exercise intensities; no difference between AC and LC • Decrease* in BP at submaximal exercise intensities in LC and AC group	31°C	Modified Monark EM (Morlock & Dressen-dorfer)	Shoulder
Sheldahl [66]	1982	Single group test- retest	To investigate if AC in cold water leads to weight loss	n = 7 obese F age: 31.4 ±11.1	• Exercise programme: steady AC in cold water. T_{water} was determined in preliminary tests in 31°C, 30°C, 28°C, 24°C and 20°C water. • Programme duration: 8wk • Frequency: 5/wk • Duration session: 90min • Intensity: 30–40%VO_{2max} • Rpm: NR	• No change in body weight, body fat, fat-free body weight • Constant caloric intake throughout the intervention • VO_{2max} did not change	17–22°C	Modified Monark EM (Morlock & Dressen-dorfer)	neck

AC, aquatic cycling; BL, blood lactate; BP, blood pressure; C, Celsius; EM, ergometer; F, female; HIIT, high-intensity interval training; HR, heart rate; LC, land-based cycling; M, male; m, meter; min, minute(s); MS, multiple sclerosis; NR, not reported; QoL, quality of life; RCT, randomized controlled trial; Rpm, revolution per minute; SV, stroke volume; T, temperature; VO_2, oxygen uptake; W, Watts; WC, waist circumference; wk, week(s)

*, significant at p-value <0.05

**, significant at p-value <0.01

[#]If not stated otherwise participants are healthy and age is presented in years as mean±standard deviation.

https://doi.org/10.1371/journal.pone.0177704.t005

Figure 2.8.

Thus, what has been heard from childhood that "drink more water" is gathering proof of principle through research. The "biology of water" is extremely interesting with research showing the therapeutic benefits of water-based approaches to address a vast spectrum of diseases ranging from depression to irritable bowel syndrome.

2.5 REFERENCES

1. Beam J. W. (2006). Wound Cleansing: Water or Saline?. *Journal of Athletic Training*, *41*(2), 196–197.
2. Tamaki, N., Orihuela-Campos, R. C., Fukui, M., & Ito, H. O. (2016). Hydrogen-Rich Water Intake Accelerates Oral Palatal Wound Healing via Activation of the Nrf2/Antioxidant Defense Pathways in a Rat Model. *Oxidative medicine and cellular longevity*, *2016*, 5679040. https://doi.org/10.1155/2016/5679040.
3. GBD 2017 Disease and Injury Incidence and Prevalence Collaborators. (2018). Global, regional, and national incidence, prevalence, and years lived with disability for 354 diseases and injuries for 195 countries and territories, 1990–2017: a systematic analysis for the Global Burden of Disease Study 2017. The Lancet. DOI.
4. https://www.who.int/news-room/fact-sheets/detail/depression
5. Lee, S., Jeong, J., Kwak, Y. & Park, S. K. Depression research: where are we now? Mol Brain 3, 8 (2010).
6. Zhang, Y., Su, W., Chen, Y. *et al.* Effects of hydrogen-rich water on depressive-like behavior in mice. *Sci Rep* 6, 23742 (2016). https://doi.org/10.1038/srep23742.
7. Haghighatdoost, F., Feizi, A., Esmaillzadeh, A., Rashidi-Pourfard, N., Keshteli, A. H., Roohafza, H., & Adibi, P. (2018). Drinking plain water is associated with decreased risk of depression and anxiety in adults: Results from a large cross-sectional study. *World journal of psychiatry*, *8*(3), 88–96. https://doi.org/10.5498/wjp.v8.i3.88.
8. Almassmoum, S.M., Balahmar, E.A., Almutairi, S.T., Albuainain, G,S., Ahmad, R., Naqvi, A.A. 2018. Current clinical status of hydrotherapy; an evidence based retrospective six-years (2012-2017) systemic review. Bali Medical Journal 7(3): 578-586. DOI:10.15562/bmj.v7i3.1159.
9. Hsu, H. H., Leung, W. H., & Hu, G. C. (2016). Treatment of irritable bowel syndrome with a novel colonic irrigation system: a pilot study. *Techniques in coloproctology*, *20*(8), 551–557. https://doi.org/10.1007/s10151-016-1491-x
10. Castro-Sánchez, A. M., Matarán-Peñarrocha, G. A., Lara-Palomo, I., Saavedra-Hernández, M., Arroyo-Morales, M., & Moreno-Lorenzo, C. (2012). Hydrotherapy for the treatment of pain in people with multiple sclerosis: a randomized controlled trial. *Evidence-based*

complementary and alternative medicine : eCAM, 2012, 473963. https://doi.org/10.1155/2012/473963.

11. Canavan, C., West, J., & Card, T. (2014). The epidemiology of irritable bowel syndrome. *Clinical epidemiology*, *6*, 71–80. https://doi.org/10.2147/CLEP.S40245
12. Bhakar, S., Hemavathy, D., & Prasad, S. (2016). Prevalence of chronic insomnia in adult patients and its correlation with medical comorbidities. *Journal of family medicine and primary care*, *5*(4), 780–784. https://doi.org/10.4103/2249-4863.201153.
13. Tomohisa Ichiba, Kenta Kakiuchi, Masahiro Suzuki, Makoto Uchiyama, "Warm Steam Inhalation before Bedtime Improved Sleep Quality in Adult Men", Evidence-Based Complementary and Alternative Medicine, vol. 2019, Article ID 2453483, 7 pages, 2019. https://doi.org/10.1155/2019/2453483.
14. Akdeniz, M, Tomova-Simitchieva, T, Dobos, G, Blume-Peytavi, U, Kottner, J. Does dietary fluid intake affect skin hydration in healthy humans? A systematic literature review. *Skin Res Technol*. 2018; 24: 459– 465. https://doi.org/10.1111/srt.12454
15. Nishimura, N, Inoue, S, Yokoyama, K, Iwase, S. Effect of spraying of fine water particles on facial skin moisture and viscoelasticity in adult women. *Skin Res Technol*. 2019; 25: 294– 298. https://doi.org/10.1111/srt.12648
16. Rewald S, Mesters I, Lenssen AF, Bansi J, Lambeck J, de Bie RA, et al. (2017) Aquatic cycling—What do we know? A scoping review on head-out aquatic cycling. PLoS ONE 12(5): e0177704. https://doi.org/10.1371/journal.pone.0177704

Chapter 3

Water and Bathing: Research Shows The Way

Shiv Sanjeevi Sripathi[1]

[1]Shiv Sanjeevi Sripathi secured second rank in Mumbai University and first rank in his college at his undergrad level and in his college during his Masters and has worked on a stem cell project at the Specialized Centre for Cell Based Therapy (SCCT), KEM Hospital at Mumbai, following which he was a Junior Research Fellow at Microbiology & Cell Biology Department, Indian Institute of Science, Bangalore from August 2009 to November 2011. The field of work involved cloning of cell wall genes and transcription factors in E. coli & M. smegmatis. This was followed by a stint as a Research and Development Executive at Transasia Biomedicals, Mumbai where he worked on patent writing and Immunoassays from 2011 till 2013. In 2013, he entered the teaching field where he has been tutoring biology and chemistry to the students of 11th and 12th grades along with English grammar and communication skills to people of all age groups. Shiv also loves helping people write scientific manuscripts and has co-edited several books on life sciences and aspects of biotechnology with Apple Academic Press.

3.1 CASE STUDY 1

3.1.1 Summary

Across the world, taking a shower is the most common form of bathing, except when it comes to Japan. Here the most popular form of bathing is drawing a bath, which is very common in warm water, about forty degrees centigrade. Such full-body immersion actually exerts sufficient hyperthermic action to initiate vasodilation and accelerate the flow of blood thereby delivering richer blood flow and therefore more oxygen and nutrients to the nodes and the peripheries of the body. A diverse array of studies indicate that there is a better health status as a result of immersion bathing. There was this randomized control test that sought to compare the effects (and advantages) of full immersion bathing with showering.

A sample of 38 participants was enrolled and they were provided with a two-week intervention of ten minutes full immersion bathing regimen in the warm water of around forty degrees, followed by another two-week stint of just showering without any form of full-body immersion with n=19 in each group. The visual scores and the analog scale were set for the parameters of pain, smile, stress, and fatigue. The self-reported health conditions and levels of skin health were much better in the case of full-body immersion when compared with showering.

RAND Corporation's short-form health-related quality of life instrument, the SF-8 Health Survey displayed much better results with respect to general health, displayed better social and emotional functioning, and better mental health scores as well. As far as the scores for anger-hostility, tension-anxiety, depression-dejection, and stress were concerned the Mood State Scores also profiled much lower in the case of full immersion bathing. Since there is an exertion of hyperthermic action in the case of full-body immersion bathing when compared to showering, there is a greater incidence of blood flow to the body's extremities, affording much more efficient metabolic waste elimination, and this, in turn, augurs the feeling of much better physical refreshment. There is little doubt that full-body immersion bathing will improve the emotional tangent of the quality of life index as well as physiological improvements as well.

Lifestyles cannot be painted with a broad brush, in that there will be an enormous gamut in variety with several aspects of lifestyle such as diets, consumption of alcohol, and habits of smoking that will govern overall health and survival. There is a good deal of variety when it comes to the forms of bathing and showering as well, with variations like the steam shower, the bathtub, the sauna, and so on, with showering being the most common form of bathing. However, in Japan, full-body immersion is the most popular form of bathing and it is popularly done in water at about forty degrees centigrade. The benefits of such a full-body immersion in warm water are that the hyperthermic reaction that occurs enhances the blood flow to the peripheries thereby enabling an enhanced circulation of oxygen as well as a better cardiovascular supply of nutrients and more effective removal of waste materials from the body.

In a survey that was conducted of people in Japan, the findings were that eighty percent of the people really enjoyed some form of showering or bathing. And among them, the sensations recorded as a result of full-body immersion bathing were enhanced refreshment, relief from all forms of stress and fatigue, and a general feeling of warmth. Seasons play a vital role in deciding the frequency of bathing: full-body immersion bathing in winter displays a result of 5.0 times per week (at 41.2 degrees centigrade ± 1.2°C for 12.9 ± 9.7 min) and shower bathing at 1.6 times while in summer, in a week full immersion bathing displays a frequency of 3.4 times while showering displays an average 4.4 times (39.4 degrees centigrade ± 1.4°C for 9.4 ± 8.8 min).

In the earlier cross-sectional studies, there was a palpable display of various benefits physically and emotionally, with respect to better sleep, enhanced rest, and higher happiness levels on a subjective parameter, along with noticeable levels of stress reduction. That said, there have been certain facets of full-body immersion bathing, the benefits of which have gone unreported.

The goal of this randomized controlled trial was to assess the impact of the two forms of bathing with respect to the physical and mental states of the participants, and to this effect, a total of 4 weeks of the intervention was performed with 2 weeks of full-body immersion hereinafter referred to as the Bathing Intervention and 2 weeks of showering hereinafter referred to as the Showering Intervention.

3.1.2 Approach

The participants were twenty-six women and twelve men all healthy (making a total of thirty-eight healthy adults with a mean age of 45.7 years with a Standard Deviation of 8.4). The Japanese portal KARADAKARA, which is an internet health-oriented portal with an active membership of about two hundred and fifty thousand people, served as the point of source of the participants. The study and the protocols contained within were approved by the Ethics Committee of Japan Health & Research Institute, and The Helsinki Declaration of 1964 was front and center for the conduction of all work. The study was started only after obtaining the written consent of all the participants. The UMIN Clinical Trials Registry is where this trial is registered and bears the registration number UMIN000006618.

In order to investigate the differences and the advantages of one over the other between two types of bathing, vis-à-vis full-body immersion in warm water around forty degrees centigrade for ten minutes and showering this Study was conducted between the 30th of October and the 26th of November 2011. The cross-over method was employed and the participant group was randomly split into two groups, with them being made to do full-body immersion (taking a bath) for two weeks followed by showering for two weeks without permitting anyone to perform washouts.

A 100 millimeter Visual Analogue Scale with markers of extremely bad at zero on the left and extremely good at 100 on the right was used to record the responses from the participants every day during the intervention period. We should note that the self-reported parameters of skin health, fatigue, stress, pain relief, and smile at reflection are perceived health indices that are actually associated with mortality. As a gross cumulative assessment of perceived and actual states of health and of mood, the standard Japanese SF-8 form which is the eight-item Short Form Health Survey document is filled in after the two week intervention period along with the standard Mood State profile document, The POMS. The Short Form Health Survey SF-8 document comprises of eight items:

(i) Physical functioning, (ii) social functioning, (iii) limitations due to bodily pain, (iv) limitations due to physical health problems, (v) limitations in vitality and energy, (vi) general health, (vii) mental health indices, and (viii) limitations owing to emotional problems. The component summary scores for the Profile of Mood Status is also calculated with regard to both the physical and mental component. As far as indices regarding constructs such as anger hostility, fatigue, depression dejection, tension anxiety, vigor, and confusion, The Profile of Mood States documentation method is the accepted standard worldwide.

In order to compare the states of health between. The analysis involved the deployment of the statistical package SPSS 19.0J (IBM, Tokyo, Japan).

3.1.3 Outcome

Of the participants in the study, 33 of the 38 participants maintained good health and completed all the set parameters and data points. Owing to private reasons that are not a result of health issues, five participants were eliminated from the study.

With regard to the status of self-reported health, the Visual Analogue Scale scores every day for the two week period recorded after the interventions of either bathing or showering are represented in Tables 3.1 and 3.2. With every intervention with regard to the assessed items, a lot of improvements were noticed:

The Visual Analogue Scale scores by way of Self-reported health status before and after bathing intervention (Goto et al, 2018: *This is an open-access article distributed under the Creative Commons Attribution License, which permits unrestricted use, distribution, and reproduction in any medium, provided the original work is properly cited*):

Tables 3.1

	Before			**After**			**Difference**	**p- value**	
	Mean		**SD**	**Mean**		**SD**	**Mean**		
Self-reported health	67.1	±	17.2	77.0	±	13.8	9.9	<0.001	**
Skin condition	64.1	±	16.8	76.0	±	13.8	11.9	<0.001	**
Fatigue	56.3	±	23.0	40.8	±	24.1	−15.5	<0.001	**
Stress	43.9	±	26.3	28.6	±	22.1	−15.3	<0.001	**
Pain	20.6	±	26.8	13.9	±	20.3	−6.8	<0.001	**
Smile	63.2	±	19.3	75.9	±	16.7	12.6	<0.001	**

**p<0.01.

Self-reported health status (VAS score) before and after showering intervention.

Tables 3.2

	Before			**After**			**Difference**	**p value**	
	Mean		**SD**	**Mean**		**SD**	**Mean**		
Self-reported health	67.5	±	17.0	73.0	±	16.1	5.5	<0.001	**

Skin condition	62.8	±	14.2	69.3	±	13.4	6.5	<0.001	**
Fatigue	57.2	±	25.5	48.8	±	25.9	−8.4	<0.001	**
Stress	49.0	±	28.2	38.7	±	24.7	−10.2	<0.001	**
Pain	24.9	±	28.7	21.9	±	26.8	−3.0	<0.001	**
Smile	62.0	±	19.8	70.3	±	20.2	8.3	<0.001	**

**p<0.01.

The data with regard to the self-reported health status for each successfully completed intervention is displayed in the following table 3.3. The Visual Analogue Scale scores were greater for self-reported health and skin condition ($p < 0.10$) and noticeably higher for smile ($p < 0.05$) during bathing intervention than when compared with showering intervention. On the contrary, fatigue, stress, and pain scores were markedly lower during bathing intervention than when compared with showering intervention ($p < 0.05$) (Goto et al, 2018: *This is an open-access article distributed under the Creative Commons Attribution License, which permits unrestricted use, distribution, and reproduction in any medium, provided the original work is properly cited*):

Tables 3.3

	Bathing			**Showering**			**Difference**	**p- value**	
	Mean		**SD**	**Mean**		**SD**	**Mean**		
Self-reported health	76.5	±	9.8	71.8	±	14.5	4.6	0.072	
Skin condition	72.5	±	12.0	67.4	±	12.9	5.1	0.053	
Fatigue	42.0	±	20.9	52.2	±	24.6	−10.2	0.028	*
Stress	39.2	±	23.2	50.7	±	25.5	−11.5	0.008	**
Pain	16.2	±	20.0	23.5	±	26.2	−7.3	0.040	*
Smile	74.3	±	13.2	68.7	±	13.5	5.6	0.016	*

*p<0.05 and **p<0.01.

Post the two-week intervention for bathing and showering, the following table 3.4 shows the assessment measured using the 6-item Profile of Mood States. There is a marked reduction in the scores for tension-anxiety, depression-dejection, and anger-hostility in the full-body immersion intervention as compared with the showering intervention ($p < 0.05$). With regard to confusion, vigor, and fatigue, there were no major variances (Goto et al, 2018: *This is an open-access article distributed under the Creative Commons Attribution License, which permits unrestricted use, distribution, and reproduction in any medium, provided the original work is properly cited*):

Tables 3.4

	Bathing			Showering			Difference	p- value	
	Mean		SD	Mean		SD	Mean		
Tension-Anxiety	43.7	±	6.2	46.5	±	8.1	−2.9	0.010	*
Depression-Dejection	47.5	±	8.7	49.6	±	8.5	−2.1	0.042	*
Anger-Hostility	46.9	±	6.8	50.0	±	9.7	−3.1	0.024	*
Fatigue	46.2	±	7.1	48.1	±	8.6	−1.8	0.124	
Confusion	48.2	±	8.4	49.6	±	8.6	−1.4	0.329	
Vigor	51.6	±	10.8	50.4	±	8.2	1.2	0.495	

* $p<0.05$.

The result of this exhaustive study over a span of two weeks comparing the benefits of full-body immersion with respect to showering with regards to the impact on the states of the physique as well as the mental states has significantly proven that the full-body immersion intervention is much more beneficial in the overall sense than showering intervention. The result of this study has limited power because the size of the sample is very small. That being said since this study is a cross-over study the result could be displayed as one of casual relationship.

With regards to the assessment of health status both before the intervention and after the intervention, the results displayed an improvement in the health status for both interventions; however, the level of improvement with regard to full-body immersion intervention was much larger than that of the showering intervention. The fact is that both forms of intervention are better at improving the physical and mental conditions, as suggested by the results. There is an indisputable sense of cleanliness with regard to the body and therefore a sense of refreshment as a result of showering and bathing.

With regard to the Visual Analogue Scale scores, however, full-body immersion intervention did display better subjective health status when compared to the showering intervention (Goto et al, 2018).

3.2 CASE STUDY 2: PAIN RELIEF AND A BATH

The difficulty in sustaining a voluntary activity and that which is a common symptom of several illnesses, and can also be found in healthy individuals, for the crux of the definition of fatigue, as gleaned from early studies. The need to develop excellent and amazingly effective strategies in order to combat chronic fatigue as well as accumulated acute fatigue is the need of the hour. As far as bathing as a form of combating fatigue is concerned, in the case of comparison between bathing in a mild stream as compared with full-body immersion bathing, the average heart rate was higher, the stiffness of waist muscles and plasma cortisol levels were palpably lower (Lee et al, 2012).

Johnston *et al* (1981) highlighted that there were noticeable electrocardiograph changes and higher levels of oxygen consumption in a showering intervention when compared to a full-body immersion bath.

Japan has witnessed a plethora of studies in the recent past with regard to identifying the benefits of full-body immersion bathing when compared with the showering intervention. Chief among these results, the mist sauna technique has emerged as the home favorite. When we compare with the popular Finnish sauna which is associated with low humidity, the mist sauna is displayed by a low-temperature sauna where there are fine drops of water.

That being said, the psychological effects of mist sauna bathing, when compared with traditional bathing forms, have not yet been studied and compiled. Therefore, the psychological benefits of mist sauna bathing with regard to muscle fatigue and recovery from said fatigue has been investigated in this work.

3.2. 1 Approach

This study consisted of ten male students in good health (22±1.2 years, 172±6.20 cm, 62±8.6 kg). On the day preceding the experiment, the participants were asked to not engage in any form of intense muscular activity. They were also asked to abstain from smoking and drinking coffee or caffeinated beverages in the two hours immediately preceding the experiment.

The Approach is displayed by the following figure 3.1 (Lee et al, 2012: *This is an Open Access article distributed under the terms of the Creative Commons Attribution License (*http://creativecommons.org/licenses/by/2.0*), which permits unrestricted use, distribution, and reproduction in any medium, provided the original work is properly cited*):

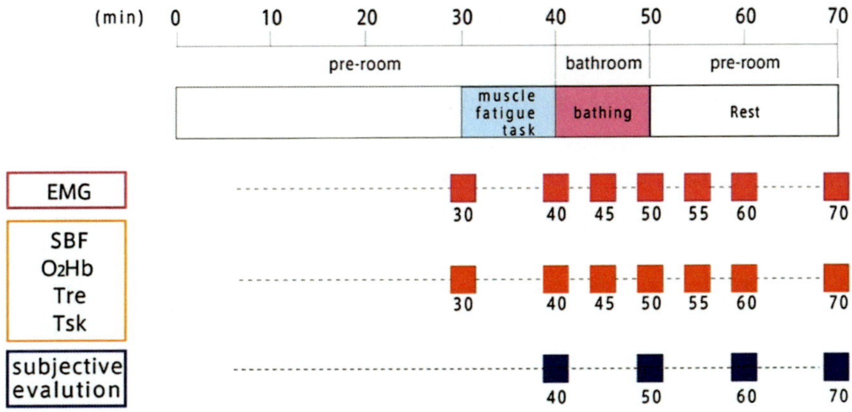

Figure 3.1.

The control was no bathing and the different methods like full-body immersion bathing, mist sauna bathing, and showering were investigated in this study. The following shows the manner in which the participants were controlled in each procedure. In the case of full-body immersion intervention, the participant was immersed in forty degrees centigrade temperature water up to the neck region. The participant was doused with fine droplets of forty degrees centigrade warm water in the case of mist sauna. As for the case of showering, the participants were doused with water as they remained in a seated position outside of the bathtub. The four experiments were conducted on four separate days at the same time.

A right arm voluntary isometric flexion at maximal capacity was done three times in a span of 5s. And for that particular individual, the MVC or the Maximal Voluntary Contraction index value was taken from the maximum value of the rest trial. The Maximal Voluntary Contraction reading was taken two days before the trial as well in order to eliminate deviation as a result of the influence of cumulative fatigue.

After making every participant sit in a chair, the fixed elbow position was arrived at with reference to the different heights of different participants with the aid of scaffolding. The targets were set at 30% on the Visual Feedback display in the case of the task that induces muscle fatigue. The participant was asked to maintain a ninety-degree contraction level and unaltered posture with a constant thirty percentile Maximal Value Contraction. The participants were asked to perform practice contractions at ten percent Maximal Voluntary Contraction for fifteen seconds while positioned in the bathtub, in order to eliminate fatigue and promulgate fatigue recovery in the EMG (the Electromyogram).

Thermistors were used to obtain the thigh, chest, and forearm skin temperatures (Tsk) as well as the rectal temperature (Tre). With regard to rectal temperature, the anal probe was inserted to beyond ten centimeters of the anal sphincter. A data logger recorded the temperature readout data every 2 seconds (LT-8; Gram Corporation). The Ramanathan formula was used to calculate the Tsk or the Mean skin temperature. By placing surface electrodes (TSD150, BIOPAC Systems) on the right biceps brachii abraded with skin cleaned using ethanol, the surface electromyographic signals were recorded.

Using a NIRO-300 (Hamamatsu Photonics, Japan) set at a wavelength of approximately 700 to 1000 nm NIRS, or Near-infrared Spectroscopy was performed on the right biceps brachii and the measurement was monitored continuously. The direct analysis of oxygen saturation in the blood along with the changes reflected in oxyhemoglobin (O2Hb), deoxyhemoglobin (HHB) along with monitoring of hemoglobin concentration (Hb) in the muscle and brain tissues, was performed using the non-invasive diagnostic tool, the Near-infrared Spectroscopy Tool or the NIRS. The Near-infrared Spectroscopy measures the concentration of hemoglobin and therefore helps us in ascertaining the changes in the levels of O2Hb, HHb as well as tHb. By isolating the absorption from scattering, the Near-Infrared Spectroscopy gives a perfect quantification devoid of any ambiguity of tissue oxygenation and there is also a complete report on the tissue as well. O2Hb of course was used by us as the index of blood flow.

A Doppler flow meter was used to measure the ventral blood flow from the forearm (ALF21, Advance Co., Ltd.). In order to avoid artifacts as a result of the movement of the laser Doppler probe during the measurement, the participants were advised to keep their arm stabilized. A 1KHz sampling rate was used to convert all the signals from analog to digital (A/D converter: BIOPAC MP150), and a Toshiba Dynabook EX/56L computer was used to store the data. The Visual Analogue Scale Method was used for evaluating the fatigue levels subjectively.

The physiological parameter Changes (Δ) were computed by deducting from the muscle-fatigue. By dividing the value of the muscle-fatigue inducing task immediately after the activity, the change rate (◇) was calculated.

3.2. 2 Outcome

The bathing method (F (3,27)=12.864, P=0.000) and time displayed noticeable effects in ΔO2Hb concentration. There was a differential factor of anywhere between 55 minutes to 70 minutes in favor of full-body immersion bathing and mist sauna when compared with no bathing intervention and showering. And in addition, the results showed much higher values for full-body immersion when compared with mist sauna as well. In addition to thatΔSBF showed marked benefits for the full-body immersion system (F (3,27)=35.909, P=0.000) and time. The differences in the effects between full-body immersion and mist sauna were quite palpable at anywhere from forty five minutes to seventy minutes.

The following Table 3.4 revels the findings (Lee et al, 2012: *This is an Open Access article distributed under the terms of the Creative Commons Attribution License (*http://creativecommons.org/licenses/by/2.0*), which permits unrestricted use, distribution, and reproduction in any medium, provided the original work is properly cited*):

Table 3.4

Time (min)	Bathing methods			
	Control	**Showering**	**Mist sauna**	**Full immersion bath**
0	0±0	0±0	0±0	0±0
5	12.596±4.638	13.511±2.418	11.683±5.624	13.513±2.301
10	13.994±2.475	14.048±1.662	14.281±2.747	14.056±2.590
15	14.777±2.144[a]	15.076±1.735[b]	21.763±6.825[cd]	23.145±4.413[e]
20	14.159±2.275[a]	15.358±2.961[b]	23.038±3.786[cd]	23.850±4.450[e]
30	14.024±1.892[a]	16.247±2.572[b]	23.438±3.848[cd]	25.117±4.774[e]

[a]P <0.05 Mist sauna *vs.* control.

[b]P <0.05 Full immersion bath *vs.* showering.

[c]P <0.05 Mist sauna *vs.* showering.

[d]P <0.05 Full immersion bath *vs.* mist sauna.

eP <0.05 Full immersion bath *vs.* control.

0 to 10 min: Bathing period.

The factor of time was the most marked when it came to fatigue sensations and not the method of bathing as elaborated in Figure 3.2 hereunder (Lee et al, 2012: *This is an Open Access article distributed under the terms of the Creative Commons Attribution License (*http://creativecommons.org/licenses/by/2.0*), which permits unrestricted use, distribution, and reproduction in any medium, provided the original work is properly cited*):

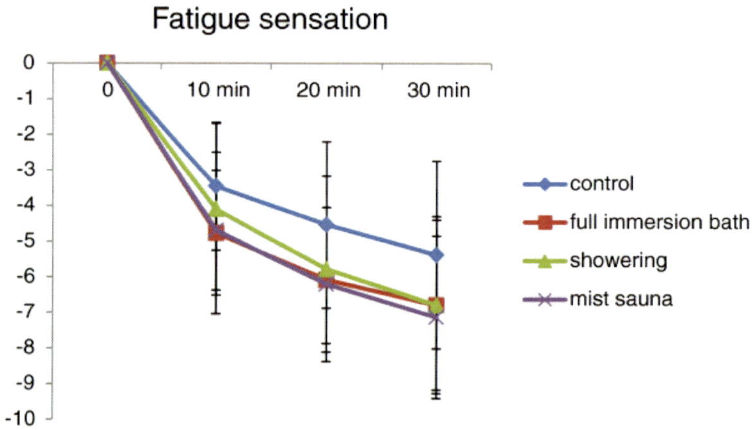

Figure 3.2.

The Mean reading i.e. the ◇MNF of the Electromyography was markedly higher in the full-body immersion intervention when compared with that of no bathing parameter. The implication of this is that the participants made an excellent recovery from muscle-fatigue situations with full-body immersion intervention.

The ΔTre as pertains to the Full-body immersion intervention was noticeably higher than that of the no bathing parameter and the showering parameter. The inference from this is that the full-body immersion condition was far more effective with regard to muscle-fatigue when compared with the no bathing or the showering alternatives.

In the ΔSBF and ΔO2Hb, regards to the mist sauna parameter were much higher in the full-body immersion parameter in comparison. One of the considerations that were taken into factor was the impact of physiological response to the presence or absence of water pressure as a reason. While there may be the hypothesis of enhanced blood flow as a result of the pressure of water in the full-body immersion there was no factor of water pressure in the case of mist sauna. Therefore it is a safe assumption that the enhancement of blood flow was a factor of the venous return.

Moreover, ΔTsk when comparing the full-body immersion intervention with the showering or the no bathing scenario was higher by a full fifty minutes. The conclusion from all this is obvious in that the effects of the mist sauna were nearly identical to that of the full-body immersion bathing when it comes to effectiveness in recovery from muscle fatigue. Moreover, ΔTsk demonstrated a full fifty- five minutes increase after five minutes of bathing, in the mist sauna condition when compared with the other conditions.

Contrastingly◇MNF of the electromyograph during the mist sauna was not noticeably higher than that of the other parameters. While the ◇Tre values in the mist sauna were not higher (1.01°C below) for half the participants the resting participants on the contrary displayed an increase. It is to be noted that the temperature of the bathroom was equal to or lesser than forty degrees centigrade while in the case of the resting participants the temperature was more than forty degrees centigrade. As a result, it was easy to arrive at the finding that the Standard Deviation in the case of ◇Tre in the mist sauna parameter was a bit higher (SD=0.23°C) than that during the showering (SD=0.17°C) and full-body immersion bath (SD=0.19°C). Due to the movement of the participants from the pre-room to the bathroom, the temperature of the sauna was lower than the standardized setting of forty degrees centigrade. It was therefore a consideration factor that some of the participants took a longer time to move from one room to another when compared with others, and there was also the consideration that this difference in bathroom temperature in the case of the mist sauna experiment could affect the recovery rate from muscle-fatigue. The two correlations of ◇Tre and bathroom temperature were studied to analyze these correlations and the ◇MNF of EMG and ◇Tre in the mist sauna parameter. Between ◇MNF and ◇Tre we found a positive correlation. What this result displayed was that higher ◇MNF was achieved with higher bathroom temperature. Therefore, between ◇Tre and the bathroom temperature, a positive correlation has been discovered.

To summarize, if the rectal temperature of a participant and the bathroom temperature was low, then ◇Tre of the mist sauna condition was high and the ◇MNF could be higher. Therefore the effect of the mist sauna would be about the same as the full-body immersion conditioned played out at 40°C. The fact that the actual thermal sensation was the same between the mist sauna and the full-body immersion bathing was also verified.

(Lee et al, 2012)

3.3 STUDY 3: BATH TO PROTECT THE HEART

3.3.1 Objective

One of the leading causes of premature death all over the world is cardiovascular disease and has been directly attributed to personal habits like physical exercise, smoking, drinking, and diet- in much the same way bathing is also another form of a lifestyle behavior and varies across cultures and across countries.

One of the ways that are being examined as a probable means to mitigate the incidence of cardiovascular disease is through exposure to heat through full-body immersion tub bathing. In a report that was published in 1998, with respect to a study wherein eight participants with type 2 diabetes mellitus were made to bathe in a tub for six days in a week for a duration of thirty minutes and for three weeks in order to study the impact of heat on the control of diabetes. The results displayed a ten percentile decrease in mean glycosylated hemoglobin and a twenty percentile decrease in mean fasting plasma glucose among the participants. More prospective studies later displayed more impacts on the parameters of blood pressure and weight control.

Laukkanen *et al* (2018) displayed a noticeable reduction in risk in the cases of sudden cardiac death and fatal Cardio Vascular Disease by constant bathing in a sauna by employing a population-based cohort study over a period of six years. That said, the cardiovascular benefits of tub bathing with respect to cardiovascular health are yet to be studied. The total peripheral vascular resistance is reduced while the stroke volume and the cardiac output are noticeably increased as a result of the water pressure that is provided by tub bathing. Sleep quality and self-rated health is also affected in a positive direction by tub bathing. On the contrary, there are also the indicators for sudden death as a result of accidental drowning, especially among the elderly, and heart attack brought about by a sudden change in temperature as a result of heatstroke with the inability to control body temperature by diaphoresis, especially with regards to sauna bathing, which has not been reported.

However, with regards to the cardiovascular risk, there have been no prospective studies that have studied the long-term effect of tub bathing.

3.3.2 Methods

From the period of 1990 up until 2009, thirty thousand seventy-six participants with the age brackets of forty to fifty-nine years with no recorded history of cardiovascular disease or cancer were followed up. The classification was done by way of the frequency of bathing, zero to two times a week, three to four times a week, and almost every day. After adjusting the traditional cardiovascular disease risk factors along with the selected dietary factors, the Cox Proportional Hazards Model was used to estimate the HRs of incident CVDs.

3.3.3 Results

The 538 373 person-years of follow-up yielded a documented total of 2097 incident cases of CVD, with328 coronary heart diseases (CHDs) (275 myocardial infarctions and 53 sudden cardiac deaths) and 1769 strokes (255 subarachnoid hemorrhages,991 cerebral infarctions, 510 intracerebral hemorrhages, and 13 unclassified strokes). The multivariable HRs (95% CIs) for almost daily or every day versus zero to two times/week were 0.72 (0.62 to 0.84, trend $p<0.001$) for total CVD; 0.65 (0.45 to 0.94, trend $p=0.065$) for CHD; 0.74 (0.62 to 0.87, trend $p=0.005$) for total stroke; 0.77 (0.62 to 0.97, trend $p=0.467$) for cerebral infarction; and 0.54 (0.40 to 0.73, trend $p<0.001$) with regard to intracerebral hemorrhage. There were no relations established between tub bathing frequency, the risk of sudden cardiac death, or subarachnoid hemorrhage.

Thus, it was concluded that in the case of the middle-aged Japanese, the frequency of tub bathing was inversely associated with the CVD risk.

(Ukai et al, 2020)

3.4 FINAL POINTS

"Noble deeds and hot baths are the best cures for depression."— Dodie Smith, I Capture the Castle

In a nutshell, many of us have felt "better" be it from a rough day or a blue mood when we shower or take a dip in water. Research too is bringing out the observations backed by scientific proof that a bath can heal us physically and mentally!

3.5 REFERENCES

1. Goto, Y., Hayasaka, S., Kurihara, S., & Nakamura, Y. (2018). Physical and Mental Effects of Bathing: A Randomized Intervention Study. *Evidence-based complementary and alternative medicine : eCAM*, *2018*, 9521086. https://doi.org/10.1155/2018/9521086.
2. Lee, S., Ishibashi, S., Shimomura, Y., & Katsuura, T. (2012). Physiological functions of the effects of the different bathing method on recovery from local muscle fatigue. *Journal of physiological anthropology*, *31*(1), 26. https://doi.org/10.1186/1880-6805-31-26.
3. Johnston BL, Watt EW, Fletcher GF. Oxygen consumption and hemodynamic and electrocardiographic responses to bathing in recent post-myocardial infraction patients. Heart Lung. 1981;10:666–671.
4. Laukkanen JA, Laukkanen T, Kunutsor SK. Cardiovascular and Other Health Benefits of Sauna Bathing: A Review of the Evidence. Mayo Clin Proc. 2018 Aug;93(8):1111-1121. doi: 10.1016/j.mayocp.2018.04.008. PMID: 30077204.
5. Ukai T, Iso H, Yamagishi K, *et al*. Habitual tub bathing and risks of incident coronary heart disease and stroke. *Heart* 2020;106:732-737.

Chapter 4

Water as the Earth's Buffer and Immune System

Vladyslav V. Goncharuk

A.V. Dumansky Institute of Colloid Chemistry and Chemistry of Water, National Academy of Sciences, 42 Academician Vernadskoho Boulevard, Kyiv 03680, Ukraine

ABSTRACT

A comprehensive review on water as a buffer of the planet and its immune system is presented. The relationship between the quality of drinking water, the level of development of human intelligence, and its health is highlighted. The problems of obtaining cytogenetically and physiologically safe high-quality drinking water are discussed along with the problems of its decontamination. The solutions to maintain stable development of aquatic ecosystems are proposed. It is shown that tap water is the habitat of multiresistant genetically-mutated microorganisms.

Citation: Water as the Earth's Buffer and Immune System Vladyslav V. Goncharuk. Hindawi Publishing Corporation Journal of Chemistry Volume 2013, Article ID 472323, 14 pages http://dx.doi.org/10.1155/2013/472323

WATER RESOURCES OF THE PLANET

The bulk of the planet's water is concentrated in the seas and oceans and constitutes around 1350 million km^3. This is salt water, with an average mineralization of 35.0 g/L, which makes it unsuitable not only for drinking but also for agriculture needs. Fresh water is localized mainly in the North and South poles and in the mountains in the form of glaciers. This resource constitutes 30–50 million km^3. However, this source of fresh water is almost unobtainable for human use.

The earth's rivers and lakes contain about 0.4 million km^3 of fresh water that is accessible for mankind. Subterranean regions of the planet have substantial stocks of water, mainly saline and salt ones. At a depth of 800 m and up to 1600 m, the water resource constitutes around 4 million km^3.

Thus, our planet has only 3% of fresh water of its total amount, with its bulk in the pack ice of the Arctic and Antarctic. The resources of fresh water accessible to human constitute only 0.06%, 0.8 million km^3.

At present, more than 40 countries of the world experience an absolute water deficiency (e.g., the Near East, Africa, Indochina, and Australia). A fifth of the population in Europe and the Americas drinks contaminated water, which does not meet international standard criterion. According to the World Health Organization (WHO) official data, around 80% of human diseases around the globe are related to the consumption of low-quality drinking water [1].

The first attempts to establish priorities for conservation and functioning resilient natural resources were defined in the documents of the United Nations (UN) conference in Rio de Janeiro (June, 1992) and the World Summit on Sustainable Development in Johannesburg "Johannesburg Agreement on Sustainable Development" (August 2002) [2]. "Every inhabitant of the planet has the right for clean drinking water," is the formula declared by the UN [3, 4]. This humane statement is unfortunately not supported by any specific actions. An ill-advised economy, the pursuit of profits without accounting for ecological problems, and the harsh competitive struggle of monopolies have resulted in the majority of surface and underground drinking water supply sources being depleted and heavily polluted. As a result, polluted water penetrates rivers, seas, and oceans. A fourfold increase of the world population within the 20th century alone—from 1.5 to 6 billion people—has dramatically exacerbated this deficiency of fresh water, which is used not only for drinking purposes but also for industrial activity. All of this combined has resulted in the worsening of the global ecological situation.

Artesian groundwater is a source of drinking water that is reliably protected from human impacts. We currently know of more than 150 different types of freshwater and saline drinking water treatments.

The most abundant impurities in underground waters are two-valence iron and manganese, fluorine ions, nitrates, ammonia, hydrogen sulfide, hard salts, and an increased salt content. Normally, these impurities substantially exceed maximum allowable concentrations for drinking water. Their presence is caused by natural factors of a geological nature. At the same time, one has to remember that high-quality water for human health should contain a wide spectrum of micro-impurities, vitally important, biologically active elements, and natural organic compounds. It is these compounds dissolved in water that provide its taste, smell, transparency and physiological properties.

Surface-level fresh waters are the second most common source of the drinking water supply. However, the global nature of human activities in the 20th century in terms of industrial development, progress in agriculture, transportation, urban utilities, the formation of megalopolises, and the growth of cities and settlements resulted in the wide-scale pollution and contamination of surface waters. The composition of wastewaters is constantly changing due to the synthesis of new chemical substances, often possessing toxic, carcinogenic, and mutagenic properties.

One can identify the following most hazardous types of pollutants entering the environment which eventually affect the water supply:(i)chemical pollutants:(1)inorganic compounds:(a)heavy metal ions,(b)salts,(c)toxic, biologically active substances,(2)organic compounds:(a)oil products,(b) phenols,(c)pesticides,(d)surface active substances,(e)chlorine-containing compounds,(f)xenobiotics;(ii)bactericidal and viral pollutants;(iii) radioactive substances of natural and anthropogenic originisotopes of elements;(iv)mutagenic compounds of organic and inorganic origin;(v) mycotic pollutants.

Seas and oceans may also serve as the source of drinking water supply. It is known that the average salt content in water constitutes about 35 g/l. It is natural that such water cannot be used for drinking. There are various approaches intended to desalinate water. In practice, desalinization technologies such as distillation, membrane, and electrochemical methods are used.

Advantages and Drawbacks of Water Desalination by Reverse Osmosis for Drinking Water Purposes

Desalination of sea (ocean) water is one of the fastest growing sectors of the economy. The introduction of reverse osmosis technology has had a big impact on the rapid development of desalination. Reverse osmosis is used for water treatment and purification of drinking water from heavy metals, nitrates, nitrites, surfactants, phenols, hardness, organic and microbiological contaminants, and organochlorine compounds.

At present, this method is one of the most popular methods of purifying drinking water, including artesian well water for the industrial production of bottled drinking water. The reverse osmosis method of water purification produces water which contains virtually no contaminants. This method has more economical operation and maintenance than any other technology. The effectiveness of using reverse osmosis technology for the desalination of sea water is three times more effective in terms of energy consumption and the degree of purification than the other methods.

At the same time, the obvious disadvantage of reverse osmosis is the lack of selectivity of the membranes to remove contaminants. This means that not only toxic impurities and microorganisms are removed, but also vital minerals and microelements. Will this water fulfill physiological requirements? Of course not! After all, purity of drinking water is not the only criteria of water quality. Also important is the extent to which drinking water is the source of elements necessary for normal body functioning. Water obtained by the reverse osmosis method cannot be called drinking water; one gets somewhat distilled water.

The results of studies on biotesting demineralized (distilled) water show that it affects both the physiological and intracellular processes in the body.

The required presence in the optimal drinking water content of calcium and magnesium is confirmed in Figure 1. Regions supplied by water with a low concentration of calcium and magnesium exhibit a significant (30–40%) increase in the incidence of hypertonic disease (see Figure 1).

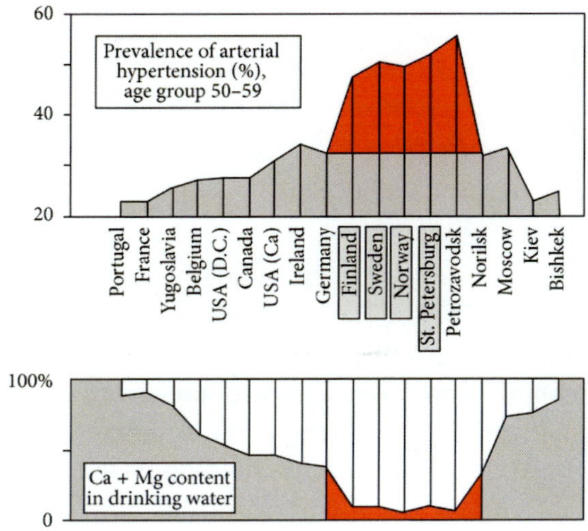

Figure 1: Effect of calcium and magnesium in the drinking water on the incidence of hypertonic disease [5].

It is necessary to do adjustments to the salt composition in water obtained by reverse osmosis [6]. This is achieved by the introduction of salts required for normal human activity: calcium, magnesium, sodium, and potassium. Taking into account this fact, factories using the reverse osmosis method need to provide special mineralizers, in order to restore the optimal mineral content of water. Thus, an artificial formation of water for drinking purposes is created. This is not the best option, but is vitally necessary where there is no other source of fresh water.

The Proof of the Failure of World's Approach to Assessing the Quality of Drinking Water and New Scientifically-Based Approaches to Assessing Its Quality

Life on earth is only possible as a result of the presence of water. The emergence and development of biological diversity on our planet is due to the presence of water, which has unique physical and chemical properties. Almost a hundred years it is known that water—not a simple compound where apart of protium there are also heavy isotope patterns however no one has ever thought about why they exist and what their role.

The unique properties of water is determined by its isotope composition and, first of all, by the ratio of protium and deuterium. The natural concentration of deuterium in the world's oceans, in salt and fresh waters varies within the interval 90–180 ppm. The stable optima, concentration of deuterium in water, constitutes 150 ppm. It is within this concentration interval that our ordinary water possesses maximum biological activity both in sea and in fresh water bodies. A decrease or an increase of the deuterium concentration in water results in a radical change of its physicochemical and biological properties [7–9].

Water is currently viewed from fundamentally new positions. The comprehensive study of water has yielded quite new results [10]. It has been found that if the deuterium concentration in water is adjusted, all ideas about its fundamental properties accepted today are changed drastically. Both the melting point and the freezing point, and other water properties, are substantially changed depending on the relation of the isotope composition of hydrogen. If deuterium is completely removed from water, then it will not exist on our planet in its liquid state. Consequently, water structure, just as all physicochemical parameters, is determined by the presence of deuterium. The results regarding the influence of deuterium on water properties became the basis for the discovery of a fundamentally new method for monitoring water quality.

Here, questions arise. What is drinking water? How can the quality of drinking water be assessed? What are the optimal parameters that water should possess in order to be completely safe according to biological and physiological points of view?

The development and the setting of regulations and standards for drinking water quality in different countries since the late 19th and early 20th centuries have changed from simply regulating the macrocomponents of natural water to more insightful knowledge about the impact of anthropogenic contaminants and toxic microcomponents of water on human organisms [11].

In connection with the worsening of the ecological state of surface and underground sources, the issue of water quality control, used by people for drinking purposes, became more acute [12, 13]. As a result of the low quality of drinking water, real threats to the sanitary-epidemiological situation in various regions of the planet crop out.

An increase of the quality of the matter being controlled in national standards of various countries of the world does not solve the issue of obtaining safe drinking water at centralized water-treatment stations either. All this calls for ever more expensive equipment and the complication of technological processes. Special attention should be paid to the fact that over the last two to three decades, an uncontrolled sharp increase of chemical compounds in the environment took place [11].

While the environment experiences these catastrophic changes, the quality control of dozens of substances in tap water does not guarantee that it really is drinkable and safe for human health. A high level of a technogenic load on water bodies, the use of imperfect technologies of water treatment, and secondary contamination of the water in distribution networks result in the ingress to the drinking water of a substantial amount of inorganic and organic pollutants, whose joint effect on the human organism causes the effect of synergism known in chemistry and biology, threatening human health. In this situation there is no possibility to provide the population of any country and continent with quality drinking water that is safe for human health. In addition, drinking water obtained from surface sources is unsafe due to the presence microtoxins in it of. As a result of our research, we found the presence of toxic micromycetes in surface waters and distribution systems. These micromycetes are not disinfected by current techniques used at centralized water treatment stations, even when chlorinated with high doses.

Hence, classic approaches to the assessment of drinking waters are useless. The notion of "normalizing maximum allowable concentrations in drinking water of different toxicants" purportedly safe for human health, is, as a matter of fact, immoral.

QUALITY OF DRINKING WATER: PROBLEMS AND SOLUTION

The harmonic development of Homosapiens concurrently with the planet's biosphere calls for the immediate prevention of environmental pollution. Our generation has witnessed qualitative changes of natural and drinking waters. For the first time in the history of mankind, the necessity of introducing quality standards for drinking water has appeared. These standards over time were improved, and new ones were introduced. Substances both of natural and anthropogenic origins were hazardous for human health and were strictly controlled. General human and worldwide approaches for drinking water quality evaluation were established. The first countries in the world that developed state standards for drinking water quality were the USA and the USSR. Owing to a high level of bacterial pollution of surface waters, drinking waters were disinfected with chlorine. This was one of the greatest mistakes of mankind. Since water always contains organic compounds, its chlorination inevitably results in the formation of very toxic, mutagenic, and carcinogenic organochlorine compounds. For the first time in the history of mankind, people began to drink chlorinated water hazardous for human health—technogenic water!

At the beginning of the 20th century, the level of surface-water contamination with chemical and bacterial components did not achieve a critical state. Therefore, small doses of chlorine, used for disinfection of water, do not lead to the formation of substantial concentrations of hazardous organochlorine compounds. However, the turbulent development of industry, agriculture, the formation of megapolises, and a dramatic increase of the population size of the world (a fourfold rise over the 20th century) resulted in a catastrophic level of bacterial and chemical contamination of water sources used for conditioning drinking water. It entailed the necessity of using large doses of chlorine for the disinfection of water. As a result, there was an increase in the concentration of organochlorine compounds in drinking water. Given such a dangerous factor, new, more sophisticated technologies of water treatment began to be developed. These technologies include preliminary filtration from suspended particles, primary chlorination, and chemical water treatment by coagulation concurrent with aluminum, iron, and flocculants of organic and inorganic origin, then filtration on sand and carbon filters. For suppressing the development of microorganisms in pipelines, the water again is treated with chlorine of such a concentration that at the outlet of the faucet of every consumer the content of residual activated chlorine is within the range of 0.3–0.5 mg/dm^3.

Schematic chlorine dioxide, possessing a higher oxidizing potential, is used instead of chlorine in a number of countries, according to regulations. Potentials of some oxidants are given in Table 1.

Table 1

Chemical substance and its formula		Oxidizing potential, eV
Chlorine	Cl_2	1.36
Chlorine dioxide	ClO_2	1.57
Hydrogen peroxide	H_2O_2	1.78
Ozone	O_3	2.07
Atomic oxygen	O	2.42
Hydroxyl radical	$HO^{\cdot}$	2.80

The current technology has its advantages and disadvantages. Advantages include the fact that the decontamination process occurs more effectively. However, it also means that chlorine dioxide may produce a broader set of organochlorine compounds.

An especially negative side of the abovementioned technologies is the use of coagulants containing aluminum. On the one hand, this final introduction of many types of contaminants into the initial water is a necessary stage of treatment. On the other hand, it introduces a new, very dangerous contamination of drinking water with residual aluminum compounds. It was known for a long time that aluminum ions contained in drinking water are exceptionally toxic and affect human health. Several publications deal with this research. This is why the World Health Organization from year to year toughens the requirements regarding the concentration of residual aluminum in drinking water.

Based on these two important prerequisites, namely, the presence in drinking water of organochlorine compounds and aluminum, I allow myself to discuss drinking water not as tap water but as technogenic water. Never before has mankind consumed water for drinking purposes which contained very toxic organochlorine compounds and also aluminum ions, which are byproducts of the modern water treatment technology.

Contaminations of the initial water reached such a level that water treatment facilities built in compliance with the effective world standards already are not capable of preventing the ingress of substances to drinking water. In this regard, their joint effect became a real threat to human health [14].

All this results in the necessity of a search for new alternative water treatment technologies to obtain safe water. Here, first and foremost, one should pay attention to natural, ecologically safe oxidants, which could be used in advanced technologies for obtaining high-quality drinking water from the severely fouled sources.

In the last century, for the first time, Russian scientists proposed a fundamentally new approach in water treatment technology—the use of ozone instead of chlorine. The world first ozonizer was developed in Lomonosov Moscow State University, and the first ozonation station for water treatment was built and commissioned in St. Petersburg before World War I. Since then, many years passed before the world community started to treat this technology with due attention.

Below are specified the potential oxidants used today in water treatment, which are the most promising for creating new technologies.

The analysis of the cited data is evidence of the fact that the maximum reactivity is possessed by a hydroxyl radical, which is capable of oxidizing any organic compound into carbon dioxide and water. It is formed in interaction of the ozone with water or hydrogen peroxide with water by the following mechanism:

$$O_3 + H_2O \longrightarrow H_2O_2 + O_2$$

$$H_2O \xrightarrow{h\nu} HO^{\bullet} + HO^{\bullet}$$

$$O_3 + HO^{\bullet} \longrightarrow HO_2^{\bullet} + O_2$$

$$O_3 + HO_2^{\bullet} \longrightarrow HO^{\bullet} + 2O_2$$

$$O_3 + H_2O_2 \xrightarrow{\text{Very slowly}} HO_2^{\bullet} + O_2 + HO^{\bullet}$$

$$O_3 + HO_2^{-} \xrightarrow{\text{Quickly}} HO^{\bullet} + O_2^{\bullet-} + O_2$$

$$O_3 + H_2O_2 \longrightarrow HO_3^{\bullet} + O_2^{\bullet-} + H^{+} \quad (1)$$

In aqueous alkaline solutions ozone is decomposed with the formation of $O_3^{\bullet-}$ according to the following reaction:

$$O_3 + HO^{-} \longrightarrow O_3^{\bullet-} + OH^{\bullet} \quad (2)$$

Then the process follows the schematic with separation of the anion radical of atomic oxygen $O_2^{\bullet-}$ superoxide anion radical $O_2^{\bullet-}$, and then molecular oxygen:

$$HO^{\bullet} + OH^{-} \longrightarrow O^{\bullet-} + H_2O$$

$$2O^{\bullet-} + H_2O \longrightarrow O + 2OH^{-}$$

$$O^{\bullet-} + O \longrightarrow O_2^{\bullet-}$$

$$2O_2^{\bullet-} + H_2O \longrightarrow HO_2^{\bullet-} + OH^{-} + O_2 \quad (3)$$

Acid conjugate to anion radical $O_2^{\bullet-}$ that is, radical $HO_2^{\bullet-}$ is a weaker donor of electrons than $O_2^{\bullet-}$, while in the system as following:

$$O_3^{\bullet-} = O_2 + O^{\bullet-} \quad (4)$$

The most active particle is the anion radical of atomic oxygen. Therefore, many reactions of $O_3^{\bullet-}$ are actually reactions of $O^{\bullet-}$. In an aqueous solution, radicals $O_3^{\bullet-}$, $O_2^{\bullet-}$, and $O^{\bullet-}$ are also formed in photolysis or γ-radiolysis.

According to reactivity the active particles may be arranged in the following series:

$$HO^{\bullet} > O_2^{\bullet-} > O_3^{\bullet-} > O^{-} > HO_2^{\bullet} \quad (5)$$

It has been found that impurities of hydrogen peroxide produce a rather strong catalytic effect on ozone decomposition. Schematically it may be displayed as the following:

$$HO_3^{\bullet} > HO^{\bullet} + O_2$$

$$O_3 + HO_2^{\bullet} > HO^{\bullet} + 2O_2$$

$$O_3 + O_2^{\bullet-} > O_3^{\bullet-} + O_2$$

$$O_3^{\bullet-} + H_2O > HO^{\bullet} + HO^{-} + O_2 \quad (6)$$

The results of our research confirmed that the identified processes are substantially accelerated in the presence of catalysts (both homogenous and heterogeneous) and under ultraviolet radiation.

It is exactly these results that were used for creating fundamentally new water treatment technologies for any types of pollutants in combination with a set of physical-chemical and biological methods. The research of the kinetics and the mechanism of all processes taking place in aqueous systems provided us with the possibility to develop not only new complex technologies, but new equipment for water treatment. These methods may help obtain high quality drinking water from effectively any source of drinking water supply as far as pollution is concerned. They have an exceptionally great significance for solving the issue of water disinfection from any microbiological and chemical source of pollution.

We offer three ways to solve these problems in the selection of the appropriate parameters—concentration, temperature, pH environment, and other factors.

The first is a simple water treatment with ozone or hydrogen peroxide.

The second is photooxidative disinfection:

$$O_3 + H_2O \xrightarrow{h\nu}$$

$$H_2O_2 + H_2O \xrightarrow{h\nu}$$

$$O_3 + H_2O_2 + H_2O \xrightarrow{h\nu} \quad (7)$$

The third is the combinated effect of three factors simultaneously:

- photocatalytic disinfection with ozone

$$O_3 + H_2O \xrightarrow{h\nu + cat} \quad (8)$$

- photocatalytic disinfection with hydrogen peroxide:

$$H_2O_2 + H_2O \xrightarrow{h\nu + cat} \quad (9)$$

- photocatalytic disinfection using the combined effects of ozone and hydrogen peroxide:

$$O_3 + H_2O_2 + H_2O \xrightarrow{h\nu + cat.} \quad (10)$$

- photocatalytic disinfection with chlorine:

$$Cl_2 + H_2O \xrightarrow{h\nu + cat.} \quad (11)$$

To date, we are unaware of more efficient and environmentally friendly methods of water treatment.

WATER AND HUMAN HEALTH

The wide-scale use of chlorine in the technology for centralized drinking water supply stations is determined by the following three main reasons. The use of chlorine ensures:(i)primary decontamination of raw water;(ii) substantial improvement of the coagulation process whereby phyto- and zooplankton and many organic and inorganic substances are removed from water;(iii)prevention of the biological fouling of treatment facility tanks, filtering media, water system networks and equipment.

However, as was indicated above, water treatment with chlorine is accompanied by the formation of a whole range of highly toxic halogen-containing organic compounds such as chlorinated methanes, phenols, aldehydes, ketones, acids and especially hazardous polychlorinated biphenyls up to dioxins—the most toxic matter known on earth. Many years of medical-biological research demonstrated that the long-term use of such drinking water results in an emergence of neurotoxins, cardiovascular diseases, and oncological diseases of liver, kidneys, and the hematopoietic system of human and produces mutagenic effects. It is these compounds that lead to the development of impotence. Therefore, the technology of water ozonation has been used increasingly in water treatment systems all over the world. Ozone as an alternative reagent simultaneously plays both the role of a disinfectant and an oxidant. In addition, as was noted above, it possesses a higher oxidizing potential, especially the products of its interaction with water. Therefore, the rate of its interaction with all classes of organic compounds is very high. It decolorizes water very well, eliminates fetid smells and odors, removes iron and manganese, suppresses the growth of algae, and so forth. At the same time, one should remember that the use of ozone in the technology of water treatment is possible only in conjunction with other physicochemical methods.

Ecology—The Health of Human

Having addressed all the above problems, the world-wide issue of determining the degree of risk in the ecology-health system becomes a high-priority one [4]. The hygienic standards and regulations for the levels of environmental pollutants effective today are mostly of a declarative nature. In each specific country, in every region of the world, one needs to assess the priorities of pollutants by the risk levels to human health.

As sources of the risk I would separate four factors playing the most important part in the life of human:(1)drinking water,(2)air,(3)soil,(4)foods.

It is necessary to carry out a complex monitoring of all environmental aspects. Why is monitoring needed? First of all, for finding priority pollutants in each specific region that produces maximum negative impact on human health. Certainly, one may follow the path of eliminating all types of pollutions, but this requires fantastically large financial costs. The most expedient formula of the ecological policy in every country should be the following: minimum costs → maximum results, starting from that of the highest priority, the most dangerous pollutants to human health.

One should isolate such main priorities of risk levels:(i)chemical pollutants,(ii)bacterial pollutants,(iii)radioactive and isotopic pollutants,(iv) mutagenic pollutants.

One should not forget that complex pollution of the environment with different classes of toxic substances very often results in synergism of their effects, when each toxicant's individual effect is much weaker than a blend of different toxicants. The most glaring among the known examples of the synergetic effect is a joint presence of radioactive matter and asbestos, which strengthen the negative effect of each other sixfolds.

As risk indicators let us point out the following five factors whose presence may help unambiguously assess the state of the environment:(i) genetic violations,(ii)disease incidence,(iii)birth rate,(iv)death rate,(v) average longevity.

Bottled Water: Issue of Disinfection and Preservation

Over the last two decades, the technology of water bottling and its sale through a wide network of shops has become very popular [14, 15]. There are just a few who would think about the enormous scientific and technical issues of preserving the quality of such waters. First of all, the water being subjected to bottling should be immaculate in terms of chemical, biological, and organoleptic indicators and in addition should preserve its qualities in a closed state.

We talked about the issue of water disinfection above, but the issue of its preservation requires a more detailed consideration. Carbon dioxide is the most common preservative, which diluting in water, forms carbonic acid. The taste of such acid water is experienced by everybody who drank aerated water in their lifetime. This water is used for extreme situations of quenching one's thirst rather than for its systematic consumption.

The other effective reagent preserving water is silver or, to be more accurate, silver ions. Since ancient times, the technology of water storage in silverware has been known. Such water can be drunk without worry since the concentration of silver ions is meaningless, but the disinfection process is rather lengthy. It can be amplified a thousand times using the electrochemical technology of silvering water. However, the excessive silver concentration is dangerous. Over time, silver ions are adsorbed on the walls of the vessel and rather quickly lose their disinfecting properties. Under these conditions, secondary bacterial contamination is possible.

Unscrupulous companies may use as preservatives special chemical substances whose functional properties have effects similar to those of antibiotics, which suppress or destroy microorganisms. These substances present enormous hazard for human health. The consumption of such water by drinking leads to the sterilization of the digestive tract causing a known disease—dysbacteriosis. The chemical nature of the containers used is a very serious issue in the process of water bottling. There are a sufficiently large number of investigations, whose results are an evidence of the toxicity of polymer containers. The safest containers are glassware, glass containers, and enameled and ceramic ware.

"GREEN" CHEMISTRY

The contemporary world early in the 20th century found itself on the verge of the global ecological crisis, which threatens the very existence of civilization. Industrial production still remains high-waste. That is why a new science has appeared and continues to develop effectively. This is the science of ecological economy, which tries to unite social, economic, and ecological systems. Ecological economy should take into account ecological costs when assessing the economic performance of production. A modern economy should develop with the account of ecological factor.

It is quite clear that the correction of the current unfavorable situation does not look to be an easy or quickly resolved issue. This results from the impossibility of allocating sufficient investments for the solution of the ecological issues.

In the world science, this area of science and technical investigations has been referred to as "green chemistry".

When we talk about drinking water quality we, first of all, should think and care about the quality of drinking water supply sources, and this is almost the entire water basin of the planet. The solution to the existing issue of protecting the whole ecosystem from the products of mankind's life activity lies in the quest for fundamentally new technical solutions. These solutions aim not only to overcome the consequences, but should prevent the causes, leading to the unfavorable ecological aftermath.

The most widely known water contamination, referred to as water "color," is caused by blue-green algae—cyanobacteria. An important adverse influence on the water basin is related to their ability to synthesize biologically active substances possessing very high toxicity. The ingress to water of such biogenic elements as phosphorus, nitrogen, potassium, sulfur, and also organic matter from industrial, domestic, rainfall, and waste waters of the agricultural production created the prerequisites for intensive development of algae of various systematic groups, which results in the formation of water "color".

The facts of the formation of toxins by algae have been known for a rather long time, as early as the 1960s, while the systematic research of the causes of this phenomenon and factors increasing the toxicity started about 20 years ago. The most serious attention at present has been drawn to the toxicant microcystine, produced by cosmopolite cyanobacteria, causing water "blooming" in water bodies of virtually all countries of the world. In terms of biological activity, toxins of cyanobacteria surpass by many times such known substances as strychnine, curare, a range of fungi toxins, and potassium cyanide. They exceed the toxicity of potassium cyanide by as much as 50–1000 times. Such a strong poison as curare exceeds them by least 10 times. Hence, it is quite obvious that such water presents a serious threat to the life of fish, birds, and humans. Even swimming in "blooming" water pools is dangerous for human health, not to mention the use of this water as a source of drinking water supply.

All these facts indicate the necessity of forming the Global World Program for the Protection of the Planet's Water Basin against Pollution.

THE BIOSPHERE AND CIVILIZATION—THE ISSUE OF COMPATIBILITY

The biosphere in its classical state no longer exists. We are witnessing the apocalypse of the biosphere, which started in the middle of the 20th century. Transforming the environment—the biosphere, whose part is human himself—society underestimated the fact that creating the technogenic environment by its existence violates the basic laws of the biosphere, the laws of the universe.

Nowadays we live in the transition period of the coexistence of two worlds—the biosphere and the noosphere. There is no return to the biosphere. Man as a creature of the biosphere origin should find new ways for his survival and development. The gene pool of the biosphere has started to intensively rearrange itself—the turbulent process of evolution has begun, which has given birth to ecological changes on a global scale. Mass extinction of the existing species and emergence of new ones are observed across the board. Especially intensive is the evolution of infections. This is evidenced by a widely known fact of the adaptation of pathogenic microflora to antibiotics. This is an illustrative manifestation of mutagenesis—the emergence of a new, earlier absent quality—resistance to antibiotics. Complex genomes are created by nature during millions of years, while in modern laboratories they are created within a year. Man has designed a new artificial habitat, the TechnoSphere, kept safe from all biological processes of the biosphere, without connection with which he simply cannot exist.

The issues of nuclear energy and its impact on the environment bear a complex nature. Since the radioactivity phenomena have been discovered, the irreversible processes of dramatic increase of background radiation all over the planet have been seen [16].

The intensive development of the nuclear industry led to the emergence of a considerable number of objects presenting potential radiation danger: nuclear power plants with reactors of various types, chemical mining plants, and plants for processing nuclear fuel. The radionuclide compositions of pollutants emitted to the environment during the operation and also in emergency situations substantially differ, which makes the use of a single flow chart of purification impossible in all mentioned cases.

Among the numerous aspects of protecting the water basin from technogenic pollutants, one of the most important aspects is achieving the effective purification of various types of pollutants, including radioactive pollutants. Achievement of this task is hampered by the general worsening of the ecological situation, accompanied by the appearance in the waters of an ever larger amount of various toxicants of organic and inorganic nature. Together with the necessity of development and the employment of more complex and sophisticated technologies to achieve purification levels to meet the existing standards, this begs the question of the applicable range of the standards themselves. Here the ever increasing restraint, which currently manifests itself despite the sufficient theoretical and practical substantiation at the moment of adoption, is determined by insufficient accounting of the possibility for aggravation of the adverse influence of toxicants on live organism in their synergetic effect. The synergetic effect of pollutants of various types and radionuclides seems to be one of the most dangerous for human health.

The adverse effect of radioactive elements of the environment is of a complex nature. Despite the fact that it is well studied, the issue of the processing and burial of radioactive waste, which today like "a nuclear gene" are either on the sea bottom or in the mines under earth and are waiting for their time, is still outstanding.

There is one more aspect of this issue which has not been sufficiently studied yet. I mean the issue of the impact of nonradioactive isotopes on the development of biological objects and, first of all, hydrogen isotopes, in particular, deuterium. The impact of the isotopic composition of substances on kinetic characteristics of numerous reactions has been known for a long time.

It turns out that deuterium strongly affects the chemical parameters of chemical and biological processes since it forms a stronger bond with oxygen in a water molecule, unlike light hydrogen—protium. It is also known that an increased concentration of deuterium in drinking water results in acceleration of the organism aging process. Deuterium manifests toxic properties with respect to living organisms [17].

Nowadays, one of the most serious issues is increased levels of background radioactivity and an increase of the concentration of heavy isotopes with respect to their light prototypes. First of all, it concerns vitally important elements such as hydrogen, oxygen, and carbon, which are a basis of protein life in our biosphere and affect the mechanism of biological evolution.

Thus, ecology is a science studied, strictly speaking, the state of the biosphere, where human is its indispensable part. However, today, when it is quite obvious that the biosphere is being transformed, in an increasing degree, to the noosphere transforming into the TechnoSphere, one cannot help taking into consideration that the world has really changed. It is necessary to have a clear notion of the place of human in a quickly changing world. Ecology transforms into a new geological phenomenon, into a new branch of science investigating the interaction of three globally coexisting constituent components of the universe: biosphere, noosphere, and TechnoSphere. The decisive role in their interaction is played by human. The future of our civilization is determined by the level of the awareness of the changes that have taken place in the environment.

A NEW APPROACH TO THE "ENVIRONMENT-HUMAN HEALTH" PROBLEM

Nature built up self renewing systems in the process of evolution supporting life activity of all live bodies. Gastrointestinal tract (GI) is the most impressive system of people and animals. Actually this is a flow-through bioreactor inhabited by many microorganisms. Total number of microorganisms exceeds number of macroorganism's own cells and makes up about 10^{14}–10^{15} cells [18]. Total weight of microorganisms amounts to 3-4 kg. Intestinal tract of healthy people contains more than 500 types of microorganisms. In the different gastrointestinal tract segments, number of bacteria is different and correlate with the state of digestive system (Table 2).

Table 2: Composition of the intestinal organisms of the human [19]

Types of bacteria	Average concentration of microorganisms (as per 1 mL or 1 g)			
	Intestine	Small intestine	Twisted intestine	Colon
Total number	$0–10^3$	$0–10^5$	$10^2–10^7$	$10^{10}–10^{12}$
Anaerobes				
Bacteroides	Seldom	$0–10^3$	$10^3–10^7$	$10^{10}–10^{12}$
Bifidobacteria	Seldom	$0–10^4$	$10–10^9$	$10^8–10^{12}$
Enterococci	Seldom	$0–10^3$	$10^2–10^6$	$10^{10}–10^{12}$
Clostridia	Seldom	Seldom	$10^2–10^6$	$10^6–10^3$
Eubacteria	Seldom	Seldom	Seldom	$10^9–10^{12}$
Aerobes				
Enterobacteria	$0–10^2$	$0–10^3$	$10^2–10^7$	$10^4–10^{10}$
Streptococci	$0–10^2$	$0–10^4$	$10^2–10^6$	$10^5–10^{10}$
Staphylococci	$0–10^2$	$0–10^3$	$10^2–10^5$	$10^4–10^9$
Lactobacteria	$0–10^2$	$0–10^4$	$10^2–10^5$	$10^4–10^{10}$
Fungi	$0–10^2$	$0–10^2$	$10^2–10^4$	$10^4–10^6$

Set of genes being a part of bacterial metagenome exceeds approximately 100 times set of genes of human organism [20]. Number of bacteria differs in the various segments of gastrointestinal tract. Number of microorganisms in the mouth cavity in acid environment conditions is little and makes up from 0 to 10^3 CGU (colonies of generatrices units) as per one milliliter of environment. In the lower sections of gastrointestinal tract number of microorganisms is much higher. The main factors determining reproduction of bacteria in the environment in the upper sections of gastrointestinal tract is a fast movement of food, secretion of bile and pancreatic juice. Environmental conditions in the colon are diametrically opposite, that is why number of bacteria in this section of gastrointestinal tract amounts to 10 CGU per one milliliter. *Bifidobacteria* and *Bacteroids* dominate among several hundreds of bacteria in the intestinal tract which share constitutes 25 and 30%, correspondingly, with regard to the total number of anaerobic bacteria. Intestinal tract of a child contains no bacteria before his birth. Fast innidiation of child's intestine with bacteria which are a part of the composition of intestinal and vaginal flora of the mother takes place at the moment of birth. As a consequence it is formed complicated community of microorganisms consisting of *Bifidobacteria*, *Lactobacteria*, *Enterobacteria*, *Clostridia*, and Gram-positive cocci.

Then composition of microbial population is subject to changes due to the influence of several factors of the environment, the most important of which is child's feeding. As early as 1900 Tissier proved that the main components of intestine microflora of the breastfed children are *Bifidobacteria*. Such bifidodominant microflora exercises protective functions and is instrumental in maturation of the mechanisms of child's immune response. On the other hand, artificially fed children have much less *Bifidobacteria* in the colon, and species' composition of intestine microflora is less diversified.

Composition of intestine microflora is fairly individual and is formed from the first days of child's life, nearing indices of an adult person by the end of the 1st-2nd year of life, and undergoes some changes at the elderly age (Table 3).

Table 3: Composition of intestine microflora in various age groups

Name of the group of microorganisms	Children of the first year of life	Adult people	People at the elderly age
Bifidobacteria	10^{10}–10^{11}	10^{9}–10^{10}	10^{9}–10^{10}
Lactobacteria	10^{6}–10^{7}	10^{7}–10^{8}	10^{6}–10^{7}
Bacteroides	10^{7}–10^{9}	10^{9}–10^{10}	10^{10}–10^{11}
Fusobacteria	$<10^{6}$	10^{8}–10^{9}	10^{8}–10^{9}
Veillonella	$<10^{5}$	10^{5}–10^{6}	10^{5}–10^{6}
Eubacteria	10^{6}–10^{7}	10^{9}–10^{10}	10^{9}–10^{10}
Peptostreptococcus	$<10^{5}$	10^{9}–10^{10}	10^{10}
Clostridium (lecitinpositive)	$<10^{5}$	$<10^{5}$	$<10^{6}$
Clostridium (lecitinnegative)	$<10^{7}$	10^{7}–10^{8}	10^{8}–10^{9}
Enterobacteria: *Escherichia coli* (with typical properties)	10^{7}–10^{8}	10^{7}–10^{8}	10^{7}–10^{8}
	$<10^{4}$	$<10^{7}$	$<10^{7}$
Enterococci	10^{6}–10^{7}	10^{7}–10^{8}	10^{6}–10^{7}
Staphylococci: *Staphylococci epidermidis; Staphylococci aureus*	$<10^{5}$	$<10^{4}$	$<10^{3}$
	$<10^{1}$	$<10^{2}$	$<10^{2}$
Aerobic bacteria	$<10^{3}$	$<10^{5}$	$<10^{5}$
Fungi pertaining to *Candida* genus	$<10^{3}$	$<10^{4}$	$<10^{4}$

Numerous research works revealed that there is a connection between composition of intestine microflora and children's immune resistance [21–24]. Distribution of microorganisms in the gastrointestinal tract has quite exact laws. Most microorganisms (about 90%) are presented constantly in one or another section of gastrointestinal tract and represent the main (resistant) microflora, facultative (supernumerary, accompanying) microflora (makes up 10%), and 0.1–0.02% is accounted for accidental (transit or residual) microorganisms. Thus, the basic microflora in the colon is represented by anaerobic microorganisms, while aerobic bacteria make up an accompanying microflora. *Staphylococci*, *Clostridium, Proteus*, and Fungi belong in the colon to the residual microflora (Tables 2 and 3). Besides, there are about 10 intestine viruses and certain representatives of the nonpathogenic protozoa in the intestine. The quantity of strict and facultative anaerobes is always one order more than aerobes in the colon, and obligate anaerobes are immediately adhesived on epithelial cells, facultative anaerobes are located higher, followed by aerobic microorganisms. Thus, anaerobic bacteria (mainly *Bifidobacteria* and *Bacteroides*), which on the whole constitute about 60% of the total number of anaerobic bacteria, represent the most permanent and numerous group of intestine microflora carrying out the basic functions.

The whole complex of microorganisms and macroorganisms creates specific symbiosis. Recently some microbiologists declared a man and his microbes as one whole symbiotic superman. Progress in the development of methods of molecular biology thrust the scientists forward to the new level of understanding of processes of symbiosis of a man and his microflora. There seemed that processes of symbiosis of a man and his microflora were well studied and their further study will not give any specific surprise. However, quick growth of the speed with simultaneous fall of cost of methods of sequence analysis for deoxyribonucleic acid (DNA) (determining its nucleotide sequence), simultaneous growth of PC capacity, and development of Internet made it possible to analyze information about large areas of genomes. New (population) approach appeared in genetics of microorganisms after decoding of chromosomes of hundreds of types of certain bacteria: analysis of genes of all bacteria inhabiting a certain natural area at once. The population of "human bioreactor" is of course one of the most important for the investigation of microbial population. The first study made to look in a new way at the intestinal microbiota was published in 1999 by the group of scientists from the National Institute for Agronomic Research (France) and University of Reading (Great Britain). The authors decided to apply method of sequence analysis of genes of 16S RNA for the investigation of microbial population of intestines (16S RNA is certificate of identity of bacteria). Since Pasteur times the first and obligatory stage of identification of microorganisms was their cultivation in culture media. But a great number of important (useful and pathogenic) microbes do not grow in any media. Due to the development of bioinformatics and appearance of modern methods of molecular biology, it has become possible to study before inaccessible and uncultivated bacteria and to begin to put in order ultraentangled classification of already known prokaryotes [20].

It was established using new methods that only 24% of obtained sequences of 16S RNA belonged to the already known microorganisms. Three fourths of microflora in the intestine of each person avoided attention of the researchers armed by the methods of classic microbiology for more than one-hundred years! The scientists just could not select conditions for cultivation of these bacteria because the most capricious inhabitants of intestine refused to grow in traditional microbiological media. Just the same situation is observed in the chlorinated water of the water supply system. For the time being it was established by means of molecular method that in the microbiota of an adult person are presented 10 out of 70 large bacterial taxons. About 90% of our microbes belong to Firmicutes type (to them belong, e.g., known by everybody as Lactobacteria responsible for milk souring) and Bacteroidetes—strict anaerobes (organisms able to live only in absence of oxygen). The last are often used as indicators of contamination of natural waters by sewage runoff. Remaining populations (10%) are divided between taxa Proteobacteria (among others intestinal bacteria (*Escherichia coli*) belong to them), Actinobacteria (from one of actinomycetes types was obtained antibiotic streptomycin), *Fusobacteria* (usual inhabitants of mouth cavity and are frequent cause of parodontosis), Verrucomicrobia (not long ago a type of these microbes was found in geothermal spring feeding methane which is more than enough in intestine due to the vital functions of other microorganisms), Cyanobacteria (they are often called “blue-green algae”), Spirochaetes, Synergistes, and VadinBE97.

Feeding of microorganisms inhabiting intestine are provided at the cost of so-called nutrients coming from superposed sections of gastrointestinal tract, which are not digested with their own enzymatic systems and are not soaked in the thin intestine. Proteins and carbohydrates not soaked in the thin intestine are exposed to deeper bacterial decomposition in the blind intestinal—mainly by intestinal bacteria and anaerobic microorganisms. These agents are needed to provide energy and flexible requirements of microorganisms. Ability to use nutrients for its vital functions depends on the enzymatic systems of various bacteria. According to this, the following types of bacteria are conditionally marked out:(i)mostly with saccharolytic activities, carbohydrates are basic energy substrates which is characteristic for the saprophyte flora;(ii)mostly with proteolytic activities which are used with proteins for energy goals (representatives of pathogenic and conditionally pathogenic flora);(iii)bacteria with mixed activities.

Accordingly, predominance in food of one or another nutrient and violation of their digestion will stimulate growth of various microorganisms [25, 26]. Carbohydrate nutrients are especially needed for the normal vital functions of intestinal microflora. They were considered as "ballast", but later it was found out their importance for the intestinal microflora and people's health on the whole. According to the modern definition, prebiotics are partly or fully indigestible food which selectively stimulates growth and metabolism of one or several groups of microorganisms, providing normal composition of microbiocenosis [27–30].

Microflora of the gastrointestinal tract performs the following functions [31–35].(i)Morphokinetic and energy effects (energy supply of epithelium, regulation of intestinal peristalsis, organism's thermal supporting, regulation of the differentiation, and regeneration of epithelial fabrics).(ii)Forming of protective barrier of mucous membrane of intestine, suppression of growth of pathogenic microflora.(iii)Immunogenic role (stimulation of immune system, stimulation of tissue immunity including generation of immunoglobulins).(iv) Modulation of functions of cytochromes P450 in the liver and generation of P450-like cytochromes.(v)Detoxication of toxic substances and compounds. (vi)Production of various biologically active compounds, promotion of certain medicaments.(vii)Mutagenic/antimutagenic activity (growth of resistance of increase of epithelial cells to the mutagens (carcinogens), destruction of mutagens).(viii)Regulation of cavities' gas composition.(ix)Regulation of behavioral reaction.(x)Regulation of replication and gene expression of prokaryotic and eukaryotic cells.(xi)Regulation of the programmed death of eukaryotic cells (apoptosis).(xii)Depository of microbial genetic material. (xiii)Participation in etiopathogenesis of the diseases.(xiv)Participation in water-salt metabolism, sustention of organism's ionic homeostasis.(xv) Forming immunological tolerance to the food and microbial antigens.(xvi) Participation in colonization resistance.(xvii)Providing homeostasis of symbiotic relations of prokaryotic and eukaryotic cells.(xviii)Participation in the metabolic activity: metabolism of proteins, lipo (delivery of lipogenesis substrates), and carbohydrates (delivery of gluconeogenesis substrates) and regulation of bile acids, steroids, and other macromolecules.

Thus, gastrointestinal tract bioreactor is represented with sequencing circuit of microorganisms' specific communities each of which performs certain functions and transfers to the following community substances required for its life activity and functioning. Gastrointestinal tract bioreactor is now the most advanced advice providing complete destruction of the wide range of organic and inorganic matters.

Aerotanks and digestors are used in the practice of water treatment where aerobic and aerobic processes take place. However, each reactor works individually and, what is extremely important, a limited number of microorganisms participate in their work.

Creation of communities' chain of microorganisms where conditions necessary for their normal life activity (gastrointestinal tract type) will be will allow in the visible future to develop an universal bioreactor for destruction of the wide spectrum of organic and inorganic matters in the waste waters.

TAP WATER IS THE MEDIUM FOR MULTIRESISTANT MICROORGANISMS—MUTANTS

As it is known, the main task of chlorine treatment of the drinking water is water sterilization against pathogenic microorganisms. However, this task is not fulfilled to the full extent because of high adaptive capacities of single-celled animals (protozoa).

Scientists of the Swiss Federal Institute for water supply problems found out with the new method of microbiologic control based on flow cytometry that the content of microorganisms 100–1000 times higher in the sanitarily pure drinking water than the standard method. What is happening with the chlorinated water in the obsolete water distribution systems of the centralized water supply of the people? Under these at-first-sight aseptic conditions, the conglomerate of various microorganisms multiresistant to chlorine and its compounds is developed. One-celled organisms, having large adaptive flexibility, completely adapt to the habitation in the tap water. They are developing and propagating themselves in the water and educe their metabolic by-products having toxicant properties into aqueous media. Thus, in the water supply lines of the water distribution systems is forming their own "microcosmos" of the communities of multiresistant microorganisms, including pathogenic ones. They die in the dechlorinated aqueous media.

Emergence of adaptation by combinative way sometimes results in the new mutations which being included into system of genotype changes a phenotype. This path is called the adaptation of a combinative. The use of sublethal concentrations of disinfectants in water treatment processes to the emergence of strains of microorganisms, resistant to chlorine and chlorine compounds.

Adaptation emerging in the combinatorial way sometimes leads to the new mutation which, being included in the genotype system, changes the phenotype. Application of sublethal concentrations of disinfecting agents used during drinking-water treatment contributes to the appearance of microorganism strains, resistant to chlorine and its compounds. This phenomenon is proved by the studies of Mokiyenko and colleagues [36, 37], who considered very important medical problem "drinking water—water-caused infections." The authors concern themselves with the nature of bacteria resistance to the biocides effect, including chlorine and its compounds. In this connection they take notice of deterioration of the situation with infectious diseases in Ukraine on account of the drinking water of poor quality [38]; they make a hypothesis of chlorine resistance of bacteria which rests on very complicated process. It consists of two stages: information-spatial interaction of the receptive body and a substrate [39, 40].

According to this hypothesis the problem of bacteria's adaptive multiresistance to chlorine is closely connected with hormesis. This phenomenon is determined with two-phase effect of chemical matters directed to the water disinfection. Small doses produce stimulating effect, and large doses result in the inhibition of biological processes. hormesis connections "dose-effect" are observed in the biota representatives of all levels of organization starting from viruses and bacteria and ending with endotherms and human beings within the wide range of doses [41, 42].

In the papers [43, 44], investigating dynamics of microbial cenosis in the bottled slightly carbonated and strongly carbonated mineral waters, the scientists come to conclusion as to inadequacy of the carbonation methods in question. It was noted substantial increase of saprophyte microorganisms even after the first month of storage of these waters. Saturation, according to the authors, does not guarantee sanitarian safety of the waters. The data is provided which testifies about adaptation of microorganisms to the high concentrations of carbon dioxide.

The mechanism of adaptation of microorganisms to the aggressive environment is based on the process of adaptation development. Process of adaptation development, as any other process, is inconsistent internally. It consists of interaction of two counter-parties: of partial disorganization of already fully formed norm of organism's and species' reaction and its next new organization [45, 46].

In the process of adaptation of microorganisms, takes place a change in the structure and physiological processes under the influence of the environmental factor. Degree of adaptation of microorganisms to the new conditions depends on resistance of this culture to the intensity of impact of external environmental factors. Some species of microorganisms, for example, acid-alcohol- alkali-resistant, diphtheritic group and fungi, are changeable somewhat less than typhous group, round bacteria, and anaerobes. The last adapt easier to the disturbing factors of the environment. Adaptation of the microorganisms is expressed mostly in the attitude to the temperature and availability of oxygen. The better and more perfect adaptation is realized, the slowly and gradually increases impact of the new factors [47].

New conditions can make microorganisms to become less demanding to the environment, limit their physiological requirements or sporulate. Thus, the morphology of microbical cell and its structure sometimes is changes.

Mutations lead to the emergence of bacteria strains with new properties. Mutagenic effect takes place often in case of encountering harmful chemical compounds in the environment where microorganisms usually reproduce themselves; a part of them die and the most stable survive and produce multiresistant strains, sometimes even towards antibiotics and immune serum.

The primary effect of mutagenic agent does not necessary lead to the true mutation. A new phenotype shows up only when changed gene starts to function. There is a possibility to accumulate and to highlight mutants with various defects: with violation of processes of transport or usage of the substrate, with the defects of intermediate metabolism, hypersensirivity to the temperature, and so on [48].

Theoretically, mutations provoked by radiations, chemical matters or other factors could lead to the extinction of bacterial population; however, there are biochemical mechanisms in any living cell that are able to fully or partially restore parent structure of DNA. Complex of ferments catalyzing reactions of corrections of DNA damage make up systems of reparation. They differ fundamentally as to their biochemical mechanism of genome reconstruction. There are three known main mechanisms of correction of DNA defects: immediate reversion from DNA damage to the initial structure, excision (falling out) of damaged initial structure with the further reconstruction, and activation of mechanisms providing immunity to damage.

Thus, there is a high-scale possibility of formation in the tap water of multiresistant populations to the chloride and its compounds. It is largely due to the fact that prokaryotes are the most ancient living organisms on the Earth. They successfully existed during various eras owing to their unique adaptability to the drastic change of physical-chemical factors of media. They are widely presented in these latter days which are described dangerously increasing anthropogenic pollution of the environment.

Microorganisms adapt easily to the aggressive media because they have high mutagenic properties. They are agamic and are reproducing themselves by cell division that is also a positive factor for their adaptive possibilities. Thus, if in the changed media conditions survives at least one cell-mutant, in case of intensive division it forms a new population of microorganisms with changed phenotypic characteristics. Animal unit in such population of microorganisms differ from the natural forms by the appearance, structure of colonies, and physiological characteristics.

An important factor influencing the adaptation of *microbiotas* to biocides as well as to chloride is hormesis. It allows microorganisms to stimulate its biological activity on exposure of small doses of decontaminant. Such "training" increases immunity of the cells to the higher doses and increases number of mutations in the population holding out hope to the survival of microbiota in the aggressive media.

It is necessary to change from time to time decontaminant substances to make the process of adaptation of microorganisms to the tap water more difficult or impossible. For this purpose, it is necessary to use a set of biocides differing as to chemical composition and mechanism of action. But realization of such technological process of decontamination of drinking water at the water treatment works of the centralized water supply is not a trivial task.

TOXIC INFLUENCE OF BACTERIA ON TEST ORGANISMS IN THE DRINKING WATER

The influence of *Escherichia coli* bacteria and products of their metabolism on the toxic level of the drinking water was not studied before. As a result of our last research work [49] it was found that the drinking water including bottled water contains *colibacillus*. Standard research as to the usage of *Escherichia coli* bacteria as an index of sanitary-hygienic condition of the water does not include toxic evaluation of the water bacterized with *Escherichia coli*.

It is established that metabolites of *Escherichia coli* bacteria separated in the aqueous media considerably deteriorate its quality, inhibit roots' growth on the plant body, demonstrate chronic toxicity on hydra, and provoke acute toxicity on *Ceriodaphnia*.

The obtained results on the cellular level showed that deviations from control were observed starting from the concentration of microorganisms in the amount of 10^2. Even in case of finding few bacteria, the inflammatory processes affect hematological indices in the leucocytes of fish blood.

The leucocytes play a very important role for the protection of the organism against bacterial and mycotic infections. Growth of neutrocytosis in the blood is a response of the organism to the bacterial and many other infections. Lymphopenia (decrease of the number of lymphocytes) is typical for the initial stage of infectious-toxic process and is related with their migration from the vessels to the tissues (inflammatory tissue).

In the studied samples of the bottled drinking water with different concentrations of bacteria, significant increase of nucleus rate of the fish tissues is observed. Structural and quantitative changes of cell nucleus were observed in case of bacteria concentration equal to *Escherichia coli* 10^2. Changes in the components of cell nucleus which are carriers of genetic information leads to the mutation of cells. This can be a reason for error exclusion of one or another disease and lead to the proliferation of oncologic cells.

According to WHO acting hygienic requirements to the water quality from the centralized drinking water supply systems, total number of *E. coli* in 1 mL of water should not exceed 3 CGU. Our research proved that existence of bacteria *E. coli* in the drinking water is inadmissible.

REFERENCES

1. World Health Organization, *Guidelines on Drinking Water Quality Control. Recommendation*, vol. 1, World Health Organization, Geneva, Switzerland, 2004.
2. United Nations, *Report of the World Summit on Sustainable Development*, United Nations, New York, NY, USA, 2002.
3. World Water Day, "Clean Water for a Healthy World," 2010, http://www.unepcom.ru/?go=razdel&level=2&cid=66.
4. Global Environment Outlook, *GEO4. Environment For Development. United Nations Environment Programme*, Prog Press, Valletta, Malta, 2007.
5. "How to turn water into the water," Mineral additive, "SEVERYANKA", 2011, http://www.koshcheev.ru/2011/10/20/severyanka/.
6. V. V. Goncharuk, "SOS: drinking water," *Journal of Water Chemistry and Technology*, vol. 32, no. 5, pp. 255–283, 2010.
7. V. N. Lobyshev and L. P. Kalinichenko, *Izotopnyye Effekty D_2O V Biologicheskikh Sistemakh (Isotope Effects of D_2O in Biological Systems)*, Nauka, Moscow, Russia, 1978.
8. G. Somlyai, G. Laskay, T. Berkenyi et al., "The biological effects of deuterium-depleted water, a possible new tool in cancer therapy," *Zeitschrift fur Onkologie*, vol. 30, no. 4, pp. 91–94, 1998.
9. V. V. Goncharuk, A. A. Kavitskaya, I. Y. Romanyukina et al., "Revealing water's secrets: deuterium depleted water," *Chemistry Central Journal*, vol. 7, no. 1, pp. 103–108, 2013.
10. V. V. Goncharuk, V. B. Lapshin, T. N. Burdeinaya et al., "Physicochemical properties and biological activity of the water depleted of heavy isotopes," *Journal of Water Chemistry and Technology*, vol. 33, no. 1, pp. 8–13, 2011.
11. M. J. Figueras and J. J. Borrego, "New perspectives in monitoring drinking water microbial quality," *International Journal of Environmental Research and Public Health*, vol. 7, pp. 4179–4202, 2010.
12. M. A. Shannon, P. W. Bohn, M. Elimelech, J. G. Georgiadis, B. J. Marĩas, and A. M. Mayes, "Science and technology for water purification in the coming decades," *Nature*, vol. 452, no. 7185, pp. 301–310, 2008.

13. C. A. Stedmon, B. Seredyńska-Sobecka, R. Boe-Hansen, N. Le Tallec, C. K. Waul, and E. Arvin, "A potential approach for monitoring drinking water quality from groundwater systems using organic matter fluorescence as an early warning for contamination events," *Water Research*, vol. 45, no. 18, pp. 6030–6038, 2011.
14. S.-L. Loo, A. G. Fane, W. B. Krantz, and T. Lim, "Emergency water supply: a review of potential technologies and selection criteria," *Water Research*, vol. 46, no. 10, pp. 3125–3151, 2012.
15. C. Reimann, M. Ayräs, A. Chkushin et al., *Environmental Geochemical Atlas of the Central Barents Region. Special Publication of the Central Kola Expedition, Geological Survey of Finland and Geological Survey of Norway*, Schweizerbart Science, Stuttgart, Germany, 1998.
16. E. V. Sobotovich and V. V. Dolin, Eds., *Tritij Y Biosferi (Tritium in the Biosphere)*, Nauk Dumka, Kyiv, Ukraine, 2012.
17. In Search of the Fountain of Youth, "Preliminary analysis of deuterium's role in DNA degradation," 2003, http://www.springboard4health.com/notebook/health_deuterium_analysis.
18. V. V. Goncharuk, *Nauka O Vode*, Naukova Dumka, Kiev, Ukraine, 2010.
19. I. A. Meshkov, M. M. Anikhovskaya, M. Y. Yakovleva et al., "Intestinal endotoxin in regulation of hemostasis activity and in pathogenesis of the DIC syndrome," *Human Physiology*, vol. 131, no. 6, pp. 91–96, 2005.
20. G. A. Zavarzin, *Bacteria and Composition of Atmosphere*, Nauka, Moscow, Russia, 1984.
21. S. V. Belmer, T. V. Gasilina, and A. V. Malkoch, "Znacheniye prebiotikov dlya funktsionirovaniya kishechnoi microflory (Importance of prebiotics for functioning of intestinal microflora)," *Russkiy Meditsinskiy Zhurnal*, vol. 3, pp. 151–153, 2008.
22. S. V. Belmer and A. V. Malkoch, "Kishechnaia microflora y znacheniye prebiotikov dlya eyo funktsionirovaniya (Intestinal microflora and importance of prebiotics for its functioning)," *Lechfshchiy Vrach*, vol. 4, pp. 25–29, 2006.
23. V. Poroyko, "Glubokouvazhayemy microb (Dear microbe)," *Populyarnaya Mechanika*, vol. 4, pp. 44–47, 2008.
24. P. L. Stark and A. Lee, "The microbial ecology of the large bowel of breast-fed and formula-fed infants during the first year of life," *Journal*

of Medical Microbiology, vol. 15, no. 2, pp. 189–203, 1982.

25. S. Salminen, C. Bouly, R. M. C. Boutron et al., “Functional food science and gastrointestinalphysiology and function,” *British Journal of Nutrition*, vol. 80, supplement 1, pp. 147–171, 1998.
26. A. I. Khavkin, “Microbiotsenos kishechnika y immunitet (Intestine’s microbiocenosis and immunity),” *Russkiy Meditsinskiy Zhurnal*, vol. 11, no. 3, pp. 1–16, 2003.
27. G. R. Gibson and M. B. Roberfroid, “Dietary modulation of the human colonic microbiota: introducing the concept of prebiotics,” *Journal of Nutrition*, vol. 125, no. 6, pp. 1401–1412, 1995.
28. S. Salminen, R. Tanaka, H. Takayama et al., “Effects of administration of TOS and Bifidobacterium breve 4006 on the human faecal Hoga,” *Bifidobacter Microflora*, vol. 2, pp. 17–24, 1983.
29. B. Björkstén, E. Sepp, K. Julge, T. Voor, and M. Mikelsaar, “Allergy development and the intestinal microflora during the first year of life,” *Journal of Allergy and Clinical Immunology*, vol. 108, no. 4, pp. 516–520, 2001.
30. S. V. Belmer and T. V. Gasilina, “Ratsionalnoye pitaniye y sostav kishechnoy microflory (Balanced feeding and composition of intestine microflora),” *Voprosy Detskoy Diyetologii*, vol. 1, no. 5, pp. 17–20, 2003.
31. A. F. Voronin and B. A. Tenderov, *Funktsionalnoye Pitaniye (Balanced Feeding)*, Koloss-Press, Moscow, Russia, 2002.
32. S. V. Belmer, A. V. Malkoch, and O. N. Minushkin, “Znacheniye prebiotikov dlya funktsionirovaniya kishechnoy mikroflory (Importance of prebiotics for the functioning of intestine microflora),” in *Proceedings of the 13th Congress Detskikh Gastroenterologov*, Actualniye voprosy abdominalnoy patologii u ditey, Moscow, Russia, March 2006 2006.
33. R. D. Rolfe, “Interactions among microorganisms of the indigenous intestinal flora and their influence on the host,” *Reviews of Infectious Diseases*, vol. 6, supplement 1, pp. S73–S79, 1984.
34. D. A. Jenkins, C. W. C. Kendall, and V. Vuksan, “Inulin, oligofructose and intestinal function,” *Journal of Nutrition*, vol. 129, no. 7, pp. 1431–1433, 1999.
35. C. A. Edwards and A. M. Parrett, “Intestinal flora during the first months of life: new perspectives,” *British Journal of Nutrition*, vol.

88, supplement 1, pp. S11–S18, 2002.

36. A. V. Mokiyenko, A. I. Gozhenko, N. F. Petrenko et al., “Water and water caused infections Voda y vodnoobuslovlennyye,” Odessa, ART-B, 2: 288, 2008.
37. A. V. Mokiyenko, A. I. Gozhenko, and N. F. Petrenko, “Water chlorination: disinfectation or adaptiveness, inactivation or stimulation?” *Vesnik NAN Ukrainy*, vol. 1, pp. 32–40, 2012.
38. M. A. Anfriychik, “Infectious diseases and demographic crisis in Ukraine,” *Journal of Academy of Medical Sciences*, vol. 13, no. 3, pp. 533–542, 2007.
39. A. V. Mokiyenko, A. I. Gozhenko, and N. F. Petrenko, “Mechanism of formation of adaptive multiresistance of bacteria to biocides from the point of view of contribution of fundamental basis of supramolecular chemistry,” *Vesnik NAN Ukrainy*, vol. 8, pp. 49–56, 2010.
40. A. V. Mokiyenko, A. I. Gozhenko, and N. F. Petrenko, “Adaptive multiresistance of bacteria to the analysis of contribution to the evolution of epidemiological process,” *Preventive Medicine*, vol. 2, pp. 90–95, 2011.
41. L. M. Shafran, A. I. Gozhenko, N. F. Petrenko et al., “For the substantiation of hormesis as a fundamental medical paradigm (review of literature and results of own research),” *Present Day Problems of Toxicology*, vol. 2, no. 3, pp. 13–23, 2011.
42. A. V. Mokiyenko and N. F. Petrenko, “hormesis as trigger mechanism of regulatory mutations and its role in the formation of multiresistance bacteria,” *Contemporary Problems of Industrial Toxicology*, vol. 5, pp. 47–52, 2011.
43. O. M. Khmelevska, A. V. Mokiyenko, and S. I. Nikolayenko, “Hygienic evaluation of dynamics of microbal coenosis of the bottled slightly carbonated water,” *Medicine Today and Tomorrow*, vol. 4, no. 49, pp. 161–165, 2010.
44. O. M. Khmelevska, A. V. Mokiyenko, and S. I. Nikolayenko, “Hygienic evaluation of illustratory microorganisma from the bottled on-carbonated, slightly carbonated and highly carbonated mineral waters of Ukraine,” *Hygiene of Population Centers*, no. 58, pp. 122–126, 2011.

45. E. V. Ermilova, *Molecular Aspects of Adaptation of Prokaryote*, Khimizdat, Sankt-Peterburg, Russia, 2012.
46. V. L. Petukhov, V. S. Korotkevich, and S. Z. Stambekov, "Genetics. Sem.GPI, Novosibirsk," 2007.
47. L. S. Buzoleva, *Adaptation of Pathogenic Bacteria to the Environmental Abiotic factors*, Press, Vladivostok, Russia, 2001.
48. S. N. Schelkunov, *Genetic Engineering*, Siberian University Edition, Novosibirsk, Russia, 2004.
49. V. V. Goncharuk and M. R. Vergolyas, "Toxic influence of bacteria on test organisms in the drinking water," *Journal of Water Chemistry and Technology*, 2013.

Chapter 5

Thirst and Drinking Paradigms: Evolution from Single Factor Effects to Brainwide Dynamic Networks

Lawrence E. Armstrong [1] and Stavros A. Kavouras [2]

[1] Human Performance Laboratory and Department of Nutritional Sciences, University of Connecticut, Storrs, CT 06269-1110, USA
[2] Arizona State University, College of Health Solutions, Hydration Science Lab, Phoenix, AZ 85004, USA

ABSTRACT

The motivation to seek and consume water is an essential component of human fluid–electrolyte homeostasis, optimal function, and health. This review describes the evolution of concepts regarding thirst and drinking behavior, made possible by magnetic resonance imaging, animal models, and novel laboratory techniques. The earliest thirst paradigms focused on single factors such as dry mouth and loss of water from tissues. By the end

Citation: Thirst and Drinking Paradigms: Evolution from Single Factor Effects to Brainwide Dynamic Networks. Lawrence E. Armstrong, and Stavros A. Kavouras. Nutrients 2019, 11, 2864

of the 19th century, physiologists proposed a thirst center in the brain that was verified in animals 60 years later. During the early- and mid-1900s, the influences of gastric distention, neuroendocrine responses, circulatory properties (i.e., blood pressure, volume, concentration), and the distinct effects of intracellular dehydration and extracellular hypovolemia were recognized. The majority of these studies relied on animal models and laboratory methods such as microinjection or lesioning/oblation of specific brain loci. Following a quarter century (1994–2019) of human brain imaging, current research focuses on networks of networks, with thirst and satiety conceived as hemispheric waves of neuronal activations that traverse the brain in milliseconds. Novel technologies such as chemogenetics, optogenetics, and neuropixel microelectrode arrays reveal the dynamic complexity of human thirst, as well as the roles of motivation and learning in drinking behavior.

Keywords: dehydration, vasopressin, magnetic resonance imaging, neural network, motivation

INTRODUCTION

Water comprises over 80% of human brain, cardiac, skeletal muscle, kidney and gastrointestinal tissues [1]; it also is the medium in which metabolism, excretion, absorption, secretion and diffusion occur. Because water is essential for the survival of humans, selective pressures have forged mechanisms that regulate total body water (TBW) volume during periods of abundance, perturbation and insufficiency. When euhydrated, human TBW is regulated daily to within ±0.22% of body mass (±0.17 L) in a mild environment, and to within ±0.48% of body mass (± 0.38 L) during exercise-heat stress [2,3]. When assessed once per week, the average biological variability of TBW is 2.0% (0.9 L in a 74 kg healthy male) [4]. This vital stability, in a TBW pool of 44 L (i.e., 60% of 75 kg body mass), is accomplished via a complex, dynamic network of sensory nerves, brain integration, and neuroendocrine responses. Figure 1 and Figure 2 consolidate information from multiple publications [5,6,7,8,9,10,11] and represent the dynamic complexity of human fluid–electrolyte regulation. Figure 1 summarizes the varied homeostatic responses that occur (e.g., the sensation of thirst) in response to osmotically driven intracellular dehydration (left side) and extracellular hypovolemia (right side) that includes reduced circulating blood plasma and blood pressure. The latter state is a more immediate threat to life than cellular dehydration, thus interstitial fluid (i.e., part of the extracellular fluid between cells) serves as a buffer which can be mobilized as needed. When extracellular fluid volume depletion is extreme (e.g., >10% loss of body mass), physiological compensation includes vigorous drinking [12] and increased sodium consumption [13]. Figure 2 illustrates that a reduced motivation to seek and consume water (final state) ultimately results from normalization of intracellular hydration, body fluid osmolality, blood pressure, and extracellular volume.

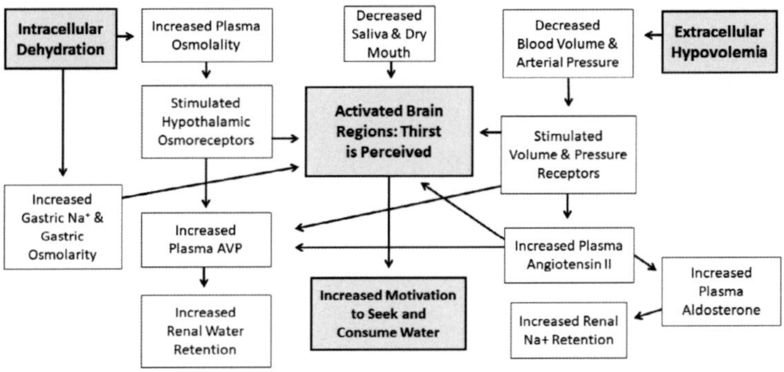

Figure 1: The thirst drive and motivation to seek/consume water are vital aspects of the homeostatic regulation of total body water volume and tonicity, in response to intracellular dehydration, increased plasma osmolality, decreased plasma volume, decreased blood pressure, and extracellular hypovolemia. Abbreviation: Na^{+}, sodium.

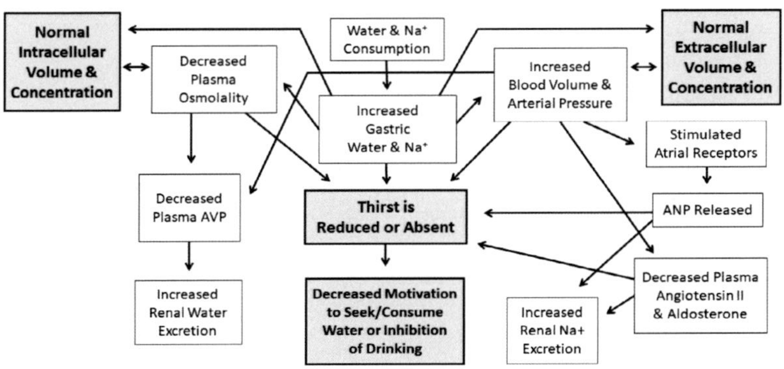

Figure 2: Homeostatic normalization of intracellular hydration, plasma osmolality, blood pressure, and extracellular volume (i.e., due to water and food intake), which result from a persistent, strong motivation to drink. These responses result in reduced thirst and decreased motivation to seek/consume water.

THIRST AND DRINKING BEHAVIOR

The neuroendocrine aspects of TBW regulation are mostly autonomic and subconscious. Other than the clinical signs and symptoms which occur at severe levels of dehydration (e.g., headache, difficulty concentrating, or collapse [14]), thirst and the motivation to drink are among the few fluid-relevant sensations that humans perceive when mildly dehydrated, beginning at the level of 1%–2% body mass loss. As such, thirst is an integral, conscious aspect of TBW regulation and, during normal sedentary daily activities, is an adequate stimulus for total fluid replacement [15]. Although vitally important to optimal health and physiological functions, the definition, component parts, and mechanisms of thirst have evolved and have been debated since the 19th century [16,17], as presented in Table 1. The following paragraphs describe the ways that research findings, methodological/technological advances, and animal research have influenced these paradigm shifts.

Table 1: Evolution of concepts and biological techniques regarding the nature and mechanisms of thirst and drinking behavior

Observations, Perspectives and Paradigms [a]	Publications [b]
When cholera patients were treated with intravenous saline and sodium bicarbonate, their intense thirst resolved.	[18]
Thirst is a local sensation and originates in the mouth and throat. Peripheral sensations (e.g., dry mouth, stomach contraction) caused by a water deficit become the instigating and sustaining conditions that motivate water seeking and drinking.	[19,20]
Thirst is a generalized sensation arising from the loss of water from tissues as dehydration progresses.	[21,22,23,24]
Water consumed orally, but not entering the stomach (i.e., exiting via esophageal fistula), does not stop the act of drinking. However, water injected directly into the stomach of an animal causes water intake to end.	[21]
Specific brain functions have precise anatomical localizations. Thirst arises from stimulation of a thirst center in the brain.	[25,26,27,28]
Anesthesia of the pharynx (i.e., bilateral section of nerves) does not affect thirst.	[22]
"True" thirst arises from lack of water in tissues whereas "false" thirst arises from dryness of mouth and throat. Moistening this region without restoring tissue water affords only temporary relief from thirst. A dry mouth is neither necessary nor sufficient to account for drinking in all circumstances.	[24]
Hypothalamic-pituitary neuroendocrine responses (unspecified) influence thirst.	[29,30,31]
Cellular dehydration stimulates thirst, with no change of cellular osmotic pressure.	[32,33]

Non-osmotic oropharyngeal sensations influence thirst and drinking behavior.	[34,35,36,37]
Electrical stimulation of specific brain areas induces drinking in unanesthetized animals, suggesting a localized thirst center.	[38,39,40,41,42]
Gastric distension satiates thirst.	[43,44]
Increasing the extracellular fluid (plasma) osmolality to varying levels allows determination of the plasma concentration at which thirst appears (i.e., the thirst threshold). The central drinking threshold decreases in response to intravenous administration of arginine vasopressin.	[45,46]
Hypothalamic damage alters thirst. Regulatory drinking is abolished by bilateral lesions in the lateral hypothalamus.	[47,48,49]
Rather than a single stimulus to drinking, multiple factors (osmotic pressure, sodium and chloride concentrations, intracellular and intracellular volumes) stimulate drinking.	[35,50]
Administration of active substances (i.e., saline, cholinergic agents) to specific brain loci via implanted brain cannulas induced drinking.	[41,51,52,53]
Hypovolemia is an independent and potent stimulus of thirst.	[33,54,55,56]
Cardiovascular feedback to the brain (i.e., volume, pressure, osmolality) modulates thirst.	[33,57,58,59,60,61,62,63]
A control model of thirst was developed on the basis of physiological research and was simulated using a digital computer.	[64,65]
The renin-angiotensin system mediates thirst and stimulates a search for water. Angiotensin II (Ang II) stimulates drinking. Injection of an Ang II antagonist (saralasin) directly into cerebral spinal fluid blocks drinking. Ang II also drives salt appetite and consumption of salt.	[10,61,66,67,68,69,70]
Loss of intracellular water and extracellular water stimulate drinking independently and, when loss of body water affects both, the signals to drink are additive. This phenomenon is named the "double depletion hypothesis of thirst".	[71,72,73]
The kidneys may influence drinking behavior by affecting the: (1) volume of water in the body (e.g., a diuretic may cause thirst secondary to urinary water loss without influencing thirst directly); (2) amount of solute in the body (e.g., while leisurely consuming a hypertonic beverage, some of the salt is excreted); and (3) release of angiotensin II, which stimulates drinking at specific brain loci (i.e., subfornical organ, organum vasculosum of lamina terminalis).	[6]
Two thirst states exist. The first is induced by a state of physiological need (i.e., reestablishes homeostasis of volume and concentration after dysequilibrium occurs; it is regulatory), and the second is not primarily regulatory (i.e., non-homeostatic).	[7]
Multiple variables stimulate thirst and drinking behavior of rats: intragastric sodium chloride, intragastric water, increased or decreased arterial blood pressure, decreased plasma volume, increased or decreased plasma osmolality, and increased plasma or local angiotensin II concentration.	[8]

The anterior cingulate cortex is recognized as a thirst center in the brain and is associated with consciousness of thirst and the pleasantness of drinking.	[74,75]
Utilizing optogenetic laboratory techniques, neuroscientists can elucidate neuron dynamics during thirst and drinking behavior, as well as the downsteam pathways by which neurons transmit information to other brain regions. See text for methodological details. [c]	[76,77,78]
Optogenetic techniques have allowed identification of specific rodent brain loci that (a) establish a persistent aversive state when the animal is dehydrated; (b) regulate motivation for water intake, thirst, and their circadian influences; (c) induce and reduce water consumption; (d) distinguish selective water and salt intake; and (e) detect water at the tongue via taste sensations.	[79,80,81,82,83,84,85,86,87]
Utilizing chemogenetic laboratory techniques, neuroscientists have discovered relationships between brain activity, brain neural circuits, thirst, and drinking behavior in freely moving animals. See text for methodological details. [c]	[88]
Combining chemogenetic and optogenetic methods, investigators have identified rodent brain loci that (a) induce drinking while consuming food (i.e., prandial drinking), (b) suppress food intake when water is unavailable (dehydration anorexia), (c) stimulate thirst-quenching signals that lead to rapid satiety following consumption of fluids but not solids, (d) are influenced by circulating hormones (i.e., Ang II), and (e) interpret environmental cues associated with water intake.	[89,90,91,92]
Two distinct neural populations in the brain trigger or suppress thirst. This suggests an innate brain circuit that initiates and stops animal water-drinking behavior, and likely functions as a center for thirst control in the brains of mammals.	[80]
Thirst-promoting neurons in the brain respond to inputs from the oral cavity during eating and drinking, which they then integrate with information about blood composition. This suggests a neural mechanism to explain behaviors such as the prevalence of drinking during meals, the rapid satiation of thirst, and the thirst-quenching influence of oral cooling.	[93]
The aversive quality of thirst, and the motivation to drink, drive the desire to quench thirst. Localized brain neuron activity is proportional to the strength of this aversive emotion (i.e., negative valence).	[82,83,93]
Sensory neurons that perceive extracellular osmolality, volume and blood perfusion pressure (thereby producing the sensation of thirst) converge on the same brain region as the neurons that release arginine vasopressin (AVP, antidiuretic hormone). As such, elevated extracellular fluid osmolality stimulates the sensation of thirst to promote water intake, and the release of vasopressin that enhances water reabsorption in the kidney.	[63]
Several hormones associated with eating and satiety have been proposed to modulate thirst neurons and vasopressin release; these include amylin, cholecystokinin, ghrelin, histamines, insulin, and leptin.	[63]

The water and salt content of the gastrointestinal tract are precisely measured and communicated to the brain, to control the drinking behavior of mice. This osmosensory signal (a) involves the vagus nerve, (b) is integrated with oropharyngeal and blood-borne signals, and (c) is transmitted from the gut to forebrain neurons that control thirst and vasopressin release.	[11]
In the rodent brain, activation of approximately 24,000 neurons in 34 brain loci revealed a global brainwide representation of a thirst-motivated state. This state appears to moderate the propagation of sensory information and its transformation into behavioral output.	[94]

[a] most of the above observations, perspectives, and paradigms arise from research involving rodents; [b] publications prior to 1920 are reviewed thoroughly by Fitzsimons (1973); [c] see Section 5.1 (below) Optogenetics and Chemogenetics.

Across intellectual disciplines, the definitions and perspectives of thirst are quite different. For example, the physiologist views thirst as a sensation induced by homeostatic imbalance (i.e., intracellular or extracellular water deficit, circulatory hypotension; Figure 1), which acts to counteract water deficiency [71]. Ironically, a few physiologists have published comprehensive reviews of water and electrolyte metabolism [95], regulation of cellular volume [96], and local renin–angiotensin systems mediating diverse physiological functions [97], with little or no mention of thirst and drinking. In contrast, social and behavioral psychologists often view thirst from an environmental perspective associated with meal times and cultural dietary norms, or factors such as motivation, incentive, and learning. This type of drinking ordinarily involves large inter-individual variability [98]. Human factors psychologists observe thirst from the perspectives of fluid availability, climate, and the sensory properties of fluids (i.e., temperature, palatability), with the goal of optimizing human performance, health and safety by designing equipment (e.g., water bottles), altering the environment (e.g., water availability), and modifying work tasks [99]. Cognitive neuroscientists observe activated brain regions, using the sophisticated imaging instruments described below, while test participants perform various tasks (e.g., drinking, swallowing, tasting). Finally, present-day chemogeneticists and optogeneticists study brain loci and communication pathways by observing chemically and genetically modified neurons in real time, to understand their roles in dehydration and thirst-motivated behaviors. Considering these specific content-area perspectives, the present review acknowledges that the thirst drive (i.e., measured via subjective rating scale [100,101]) is complex, dynamic, and distinct from drinking behavior—which is measured as the volume, concentration, temperature, palatability, and ingredients of fluid consumed [102,103].

Investigators have created models to facilitate understanding and communicate information about thirst and drinking behavior. One physiological paradigm [104] acknowledges that thirst arises from integration of physiological, psychological and environmental inputs to the brain (Figure 3). Physiological responses (Figure 1 and Figure 2) involving multiple organ systems maintain set points of osmolality, blood volume and blood pressure; this type of thirst has been named *homeostatic*. Published reviews of homeostatic thirst have presented a historical record of the advances resulting from both human and animal studies. These include the monograph published by A.V. Wolf [16] nearly 60 years ago, and the comprehensive review papers published by J.T. Fitzsimons. An initial Fitzsimons review [17] described early research (1817–1918), three classic theories of thirst, drinking behavior and satiety, the pharmacology of drinking, the rhythmicity of drinking and feeding schedules, disease-related dysfunctions of thirst, and the influences of electrical stimulation, intracellular dehydration, or extracellular dehydration on thirst. Subsequently, he focused on the stimuli that induce thirst in animals (e.g., dipsogens), as well as the hormonal and renal responses to thirst [6]. In 1998, Fitzsimons published a remarkable review of angiotensin, thirst and sodium appetite [7]; topics included renin- and angiotensin-induced drinking, the influences of peptides and neurotransmitters, and experimentally stimulated drinking. Even today, it is difficult to expand these classic, voluminous physiological publications.

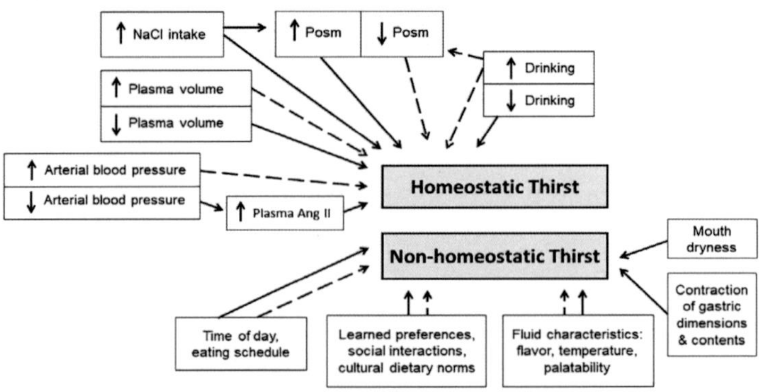

Figure 3: Factors that influence thirst are categorized as either homeostatic (i.e., volume, pressure, concentration) or relatively rapid non-homeostatic (anticipatory) inputs. Solid and dashed arrows represent, respectively, factors that increase and decrease thirst. Abbreviations: NaCl, sodium chloride; Posm, plasma osmolality; Ang II, angiotensin II.

In contrast to homeostatic thirst, non-homeostatic thirst (Figure 3) is influenced by the taste and temperature of a fluid (i.e., alliesthesia [102,103]), mouth dryness, gastric distention, meal contents, meal timing [105], idiosyncratic learned, preferences, occupational schedules, social interactions, and cultural norms [75,98]. Both physiologists and cognitive neuroscientists have recognized that dehydrated humans drink to satiation rapidly across 3–10 min [74,106], causing decreased thirst and motivation to drink, well before elevated plasma osmolality and sodium return to normal concentrations (i.e., a process that may require 15–50 min [107,108,109], and before the consumed fluid is absorbed from the gut [110,111,112]. Thus, human homeostatic thirst is modulated by non-homeostatic oropharyngeal neural signals which rapidly reduce and limit total fluid intake in the presence of a persistent, strong motivation to drink [107]; these signals modulate satiety and oppose over-drinking [113].

HUMAN BRAIN IMAGING TECHNOLOGIES

Development of positron emission tomography (PET) imaging technology involved the noteworthy efforts of scientists, physicians and physicists who moved this technology from invention to research more than 30 years ago, then to clinical acceptance as an imaging modality during the years 1985–1995 [114]. During the initial years of development, PET scans were used almost exclusively to provide (a) an absolute measure of increased tissue metabolism (i.e., focusing on glucose utilization), especially in the diagnosis and progress of cancerous tumors [115,116], and (b) maps of human brain and heart function in three dimensions. Regional blood flow and metabolism also are determined via Fluorine-18-2-fluoro-2-deoxy-d-glucose and Oxygen-15-water PET imaging [116]. As a nuclear medicine technique, PET requires a cyclotron, specialized professional staff (e.g., a cyclotron engineer and nuclear medical technologist) and a positron-emitting radioisotope on a daily basis [117]. Today, new technologies allow noninvasive cell tracking via PET imaging of radiolabeled cells, with or without reporter genes. This technique involves three steps: stable expression of a reporter gene, ex vivo radiolabeling of co-incubated target cells, and transfer of cells for in vivo PET imaging [118]. Functional magnetic resonance imaging (fMRI) expanded enormously at the beginning of the 21st century and was dominated by basic neuroscience inquiry [119]. Published fMRI research spans a wide range of cognitive functions (e.g., taste, vision, hearing, memory, reward prediction, language, motor performance) and clinical disorders (e.g., multiple sclerosis, synesthesia, schizophrenia, psychopathology, post-traumatic stress disorder, migraine headaches, diabetes insipidus) [120]. Sequential fMRI images detect excess blood supply and the calculated oxygen delivery to a brain structure or region (i.e., a method that is named blood oxygen level-dependent responses, BOLD [119]); this method is an indirect measure of brain activity. As such, fMRI represents neural activations and metabolic activity, whereas its predecessor magnetic resonance imaging assessed only anatomical structures. In some research studies, fMRI is complimented by other measurements of brain physiology such as electroencephalography, transcranial magnetic stimulation, and near-infrared spectroscopy [120].

Imaging of Human Thirst and Drinking Behavior

Both fMRI and PET imaging have advanced our understanding of the human brain regions that are activated during thirst (i.e., induced by intravenous injection, fluid restriction, or exercise); mouth irrigation, drinking, satiation; tongue protrusion, touching the tongue, swallowing; and taste stimuli. All of these processes are represented in Table 2, Table 3, Table 4 and Table 5. The ability to assess subjective sensations (i.e., using rating scales) during human imaging offers a great advantage over animal experiments, because the intensity of thirst, taste and other sensations can be correlated with changes of regional blood flow in the brain [92].

Table 2: Activation (A) of human brain regions and loci in response to experimentally induced thirst

Method of Inducing Thirst	Frontal Lobe			Parietal Lobe			Temporal Lobe	Occipital Lobe		Insular Cortex	Striatum		References
	OF	FG	MC	IL	PCG	SC	TG	LG	CU		PT	CN	
T_{IV}	A	A			A		A					A	[110] [a]
T_{IV}	A	A			A	A	A	A	A	A		A	[74] [a]
T_{IV}								A	A				[111] [a,b]
T_{IV}		A			A		A	A	A	A		A	[108] [a,c]
T_{IV}		A	A	A			A	A	A	A			[121] [a]
T_{IV}	A	A	A	A		A	A		A				[112] [a]
T_{FR}	A			A						A			[122] [c]
T_{FR}							A			A			[123] [c]
T_{30}	A	A		A			A				A		[124] [d]
T_{60}	A	A		A			A		A	A	A		[75] [c]

Method of Inducing Thirst	Cingulate Cortex			Limbic System						Brain Stem		Cerebellum [e]					
	ACC	MCC	PCC	HI	PG	T	HY	LT	A	P	PAG	I–IV	V	VI–VII	VIII–IX	DN	
T_{IV}	A			A	A	A					A		A	A			[110] [a]

T_{IV}	A	A	A	A	A	A			A	A							[74] [a]
T_{IV}		A	A		A	A						A	A	A	A		[111] [a,b]
T_{IV}	A		A		A	A	A	A		A		A	A	A	A		[108] [a,c]
T_{IV}	A	A															[121] [a]
T_{IV}	A	A				A											[112] [a]
T_{FR}	A										A						[122] [c]
T_{FR}									A								[123] [c]
T_{30}	A		A					A									[124] [d]
T_{60}	A	A				A				A	A	A	A	A	A	A	[75] [c]

[a] positron emission tomography; [b] magnetic resonance imaging; [c] functional magnetic resonance imaging; [d] fMRI with pulsed arterial spin labeling; [e] I–IV, anterior hemisphere, lingula, central; V, anterior quadrangulate lobule, culmen; VI–VII, posterior quadrangulate lobule, declive, superior semilunar lobule; VIII–X, pyramis, uvula, nodulus, biventral lobule, tonsillar; DN, dentate nucleus. Abbreviations: T_{IV}, hyperosmolar thirst induced via hypertonic saline; T_{FR}, thirst induced by 6-8 h fluid restriction; T_{30}, thirst induced by 30 min of cycling exercise plus 4.5 h fluid restriction; T_{60}, thirst induced by 60 min of cycling exercise with no fluid intake; OF, orbitofrontal cortex; FG, frontal gyrus; MC, primary motor cortex; TG temporal gyrus; IL, inferior parietal lobule; PCG, postcentral gyrus; SC, somatosensory cortex; TG, temporal gyrus; LG, lingual gyrus; CU, cuneus; PT, putamen; CN, caudate nucleus; ACC, anterior cingulate cortex; MCC, midcingulate cortex; PCC, posterior cingulate cortex; HI, hippocampus; PG, parahippocampal gyrus; T, thalamus; HY, hypothalamus including the organum vasculosum of the lamina terminalis (OVLT); LT, lamina terminalis; A, amygdala; P, pons; PAG periaqueductal gray matter.

Table 3: Deactivation (D) of human brain regions in response to mouth irrigation and drinking to satiation

Deactivation Stimulus	Frontal Lobe			Parietal Lobe			Temporal Lobe	Occipital Lobe		Insular Cortex	Striatum		References
	OF	FG	MC	SL	IL	PCG	TG	LG	CU		PT	CN	
MI, S	D	D					D					D	[110] [a]
MI, S										D			[74] [a]
MI, S	D									D			[122] [b]
MI, S		D		D				D	D	D			[123] [b]
S													[111] [a]
S		D										D	[108] [a,b]
S													[121] [a]
S	D	D											[112] [a]
S													[124] [c]
S													[75] [b,d]

Deactivation Stimulus	Cingulate Cortex			Limbic System						Brain Stem		Cerebellum				References
	ACC	MCC	PCC	HI	PG	T	HY	LT	A	P	PAG	I–IV	V	VI–VII	VIII–X	
MI, S				D	D	D										[110] [a]
MI, S	D	D	D		D	D			D							[74] [a]
MI, S											D					[122] [b]
MI, S		D														[123] [b]
S												D	D	D	D	[111] [a]

S	D		D							D					D	[108] [a,b]
S		D														[121] [a]
S		D				D										[112] [a]
S	D															[124] [c]
S																[75] [b,d]

[a] positron emission tomography; [b] functional magnetic resonance imaging; [c] fMRI with pulsed arterial spin; [d] no deactivations were considered or reported. Abbreviations: MI, mouth irrigation; S, drinking to satiation; D, reduced brain image signal strength, below a predetermined threshold; SL, superior parietal lobule; additional brain loci abbreviations appear in Table 2 footnote.

Table 4: Human brain region activations (A) in response to tongue movements, touching the tongue, water held in the mouth, and swallowing

Experimental Condition	**Frontal Lobe**					**Parietal Lobe**				**Occipital Lobe**				**Insular Cortex**	**Striatum**		**References**
	OF	**FG**	**MC**	**RA**	**PMC**	**PN**	**IL**	**PCG**	**SC**	**CU**	**OO**	**LG**	**VC**		**PT**	**CN**	
T					A			A	A								[125] [a,b]
TE		A			A	A		A	A						A		[126] [d]
TE	A	A	A		A	A	A		A	A				A			[127] [d]
VS							A	A	A					A	A		[128] [a,b]
VS		A	A					A	A					A			[129] [d]
VS		A	A		A	A		A	A	A	A			A			[126] [d]

VS, R			A		A				A				A				[130] [d]
W	A	A					A	A	A					A		A	[131] [a,b]
SW	A				A			A						A			[132] [a,b,c]
SW		A	A	A	A			A	A			A		A			[131] [a,b]
SW		A	A		A	A	A	A	A	A				A			[127] [d]
Experimental Condition	**Temporal Lobe**				**Cingulate Cortex**			**Limbic System**		**Brain Stem**			**Cerebellum**				**References**
	TG	**FG**		**AA**	**ACC**	**PCC**		**T**	**A**	**DB**	**P**	**MB**	**I – IV**	**V**	**V I – VII**	**VIII–X**	
T	A																[125] [a,b]
TE					A			A									[126] [d]
TE	A							A		A		A					[127] [d]
VS	A	A			A			A							A	A	[128] [a,b]
VS																	[129] [d]
VS	A				A												[126] [d]
VS, R						A											[130] [d]
W	A							A			A			A	A	A	[131] [a,b]
SW				A	A				A	A					A		[132] [a,b,c]
SW												A		A	A	A	[131] [a,b]
SW								A				A		A		A	[127] [d]

[a] positron emission tomography; [b] magnetic resonance imaging; [c] transcranial magnetic stimulation mapping; [d] functional magnetic resonance imaging. Abbreviations: T, voluntary tongue protrusion and bilateral touching the tongue; TE, voluntary tongue elevation inside the mouth; VS, voluntarily swallowing own saliva on cue, no injection; R, reflexive swallowing induced by injecting water into the pharynx; W, voluntarily held injected deionized water in mouth; SW,

voluntarily swallowed injected deionized water on cue; RA, Rolandic area; PMC, premotor and supplementary motor cortex; PN, precuneus; OO, occipital operculum; VC, visual cortex; FG, fusiform gyrus; AA, auditory association cortex; DB, dorsal brainstem; MB, midbrain; additional brain appear loci abbreviations in footnotes of tables above.

Table 5: Activation (A) of human brain regions and loci in response to taste stimuli

Taste Stimuli	Application Method	Frontal Lobe			Parietal Lobe					Occipital Lobe	Insular Cortex	Striatum	References
		OF	**FG**	**LG**	**IL**	**PCG**	**SC**	**SG**	**AG**	**LG**		**CN**	
D, Na	I										A	A	[133] [a]
D, F	P	A	A	A						A	A		[134] [a,b]
Na, Sac	I	A		A						A	A	A	[135] [b,d]
F	S		A	A			A	A	A		A		[136] [c]
F	I	A	A								A		[137] [c]

Taste Stimuli	Temporal Lobe	Cingulate Cortex		Limbic System			Brain Stem	Cerebellum				Reference
	TG	**ACC**	**MCC**	**HI**	**PG**	**T**	**P**	**I–IV**	**V**	**VI–VII**	**VIII–X**	
D, Na	A	A		A	A	A						[133] [a]
D, F												[134] [a,b]
Na, Sac	A		A	A	A	A						[135] [b,d]
F	A											[136] [c]
F		A										[137] [c]

[a] positron emission tomography; [b] magnetic resonance imaging; [c] functional magnetic resonance imaging; [d] magnetoencephalography. Abbreviations: D, deionized water; Na, sodium chloride solution; I, injected into mouth via tube; F, 4–8 different taste stimuli were applied to the tongue; P, filter paper placed on tongue; Sac, saccharine solution; S, sip and spit protocol; SG, supramarginal gyrus; AG, angular gyrus; additional brain loci abbreviations are defined in previous table footnotes.

Table 2 summarizes the findings of 10 human laboratory studies in which investigators induced thirst. Because the experimental methods varied (i.e., thirst subsequent to intravenous hypertonic saline, fluid restriction, low-intensity exercise, and combinations of these methods), considerable differences exist in the brain regions that were activated. This suggests that the human brain processes different forms of dehydration (e.g., extracellular or intracellular) in unique ways, depending on the homeostatic disruptions of tonicity and volume experienced. This is reasonable, given that (a) the brain constantly monitors and regulates intravascular volume and plasma osmolality, and (b) fluid restriction and exercise-induced dehydration both decrease plasma volume but intravenous infusion expands plasma volume [124]. Table 2 also demonstrates that certain brain loci were activated in at least 50% of experiments, regardless of the method employed by investigators. These loci include the orbitofrontal cortex, frontal gyrus, inferior parietal lobe, temporal gyrus, cuneus, insular cortex (i.e., insula), anterior cingulate cortex, midcingulate cortex, and the thalamus. However, it is important to note that tongue movements, swallowing, and oropharyngeal sensations were not controlled, but concurrently activated specific brain loci during these experiments.

Table 3 summarizes the same 10 human imaging studies presented in Table 2, but from the perspective of deactivation (i.e., determined on the basis of reduced PET and fMRI signal strength, below a pre-established threshold) of brain regions in response to mouth irrigation (n = 4) and subsequent oral drinking to satiation (n = 10). Investigators reported that insular cortex deactivations were observed in three out of four mouth irrigation experiments, but in no drinking-to-satiation trials. This may mean that the insular cortex integrates taste or temperature (i.e., from cool water) information but not satiety. In addition to these deactivated brain loci, it appears that the human brain responds to drinking and dilution of the extracellular fluid by activating inhibitory neural networks that generate an inhibitory emotion ([75]; see Figure 2); these inhibitory effects are described below.

Recognizing that the act of drinking to satiation, as presented in Table 3, involves neuromuscular and sensory actions, Table 4 presents the findings of investigations which focused on tongue protrusion, touching the tongue, and swallowing. Activations were observed in at least 50% of studies, in these eight brain loci: frontal gyrus, premotor cortex and primary motor cortex of the frontal lobe; somatosensory cortex and postcentral gyrus of the parietal lobe; insular cortex (observed in 73% of cited studies); temporal gyrus; and thalamus. As noted in Table 2 above, five of these loci were activated during dehydration and thirst; however, none of the publications summarized in Table 2 reported controlling for tongue movements, tongue sensations, or the act of swallowing.

Table 5 summarizes the findings of five human studies that observed brain activations induced by applying solutions with distinct taste characteristics (e.g., salty, sweet, bitter) onto the tongue or into the mouth. The brain loci that were activated in at least 60% of studies included the orbitofrontal gyrus, frontal gyrus, and lingual gyrus of the frontal lobe, insular cortex, and the temporal gyrus. Of these, it is interesting that three were identified in both Table 2 (i.e., in response to dehydration and thirst) and Table 4 (i.e., in response to tongue movements, tongue touch, and swallowing). Thus, Table 5 suggests that taste is a sensory component of thirst. In support of this concept, both human and animal investigations have shown that ingestive appetite (i.e., fluids and food) is influenced by sensory characteristics [138], especially the desire to drink or not drink a fluid after it has been initially tasted [99,139].

Because little is known about human neural networks that produce the thirst sensation and motivate humans to drink, it is reasonable to ask, "What are the functions of activated brain loci in Table 2, Table 3, Table 4 and Table 5?" In response to this question, Table 6 presents a variety of evidence-based functions and sensations that have been associated with swallowing, thirst, taste, smell, and somatosensory information (e.g., temperature, touch, pressure, osmolality, stretch, movement). Although concurrently activated brain loci may represent neural circuits, this cannot be verified with fMRI and PET brain imaging methods. Human brain neural network algorithms have been developed [140] but have not been implemented in any of the studies presented in the present review. Table 6 also supports previous human fMRI studies which determined that multiple parallel interhemispheric neural pathways are involved in volitional swallowing [129,141], and are located in several spatially discrete cortical and subcortical loci, including a brainstem integrative network [126]. Volitional swallowing requires a complex sequence of carefully timed muscular contractions [132]. These events have been described in four stages: oral preparatory, oral transport, pharyngeal, and esophageal [127]. Numerous muscle actions and brain region activations are involved, as represented by the functions numbered 11–18 in Table 6 footnotes. In contrast, reflexive swallowing of water injected into the pharynx has been associated [130] only with the primary motor cortex (see function 17 in Table 6 footnotes).

Table 6: Proposed functions of human neural networks and concurrently activated brain regions and loci. [a,b.]

Frontal Lobe					Parietal Lobe					Temporal Lobe	Occipital Lobe			Insular Cortex	Striatum		References
RA	OF	FG	MC	PMC	PN	SG	IL	PCG	SC	TG	OO	LG	CU		PT	CN	
									1,4					1,4			[121]
									1,4					1,4			[112]
	4	1,4			1					1,4				1		4	[124]
		1,4												1,4			[108]
	2	2												2			[122]
	5													6			[75]
	2														11,12	11,12	[110]
	4													4			[74]
														5			[123]
5,10	5,10	5,10												5,10	5,10		[142]
	7,9a	7										9a					[134]
										8		8		7,8		8	[133]
	9a	9a												9a			[137]
			13		15	15		15	19				15	15			[126]
								15		15				15	15		[128]

				15					15	15				15				[132]
	15		15						15						15	15		[109]
		16							15					16				[129]
16											16			16				[131]
		15	17	15	15								15	15				[130]
		18		15,18	18		15,18	15,18	15,18	18			18	15,18	18			[127]
Cingulate Cortex					**Limbic System**							**Brain Stem**		**Cerebellum**				
ACC	**ACG**	**MCC**		**PCC**	**HI**	**PG**	**T**	**HY**	**LT**	**A**	**MB**	**P**	**PAG**	**I–IV**	**V**	**VI–VII**	**VIII–IX**	
1,4				1,4		1,4	1,4	1,4							1,4	1,4		[121]
							1,4	1,4						1,4	1,4	1,4	1,4	[112]
1		1,4		1,4		1,4	1,4		1							4	4	[124]
	1,4						1,4							1,4	1,4	1,4	1,4	[108]
																		[122]
		6		5						6	6		6					[75]
2,11,12									2			12	12					[110]
3,4,14				3,4,14	4,14	4	3,4,14	3		3								[74]
5																		[123]
5,10							5,10			5,10								[142]
	9a							9a		7,9a								[134]
	8					8	8											[133]
9a																		[137]
13																		[126]
							15									15	15	[128]
										15						15		[132]
							15											[109]

																	[129]
																	[131]
15																	[130]
15,18						15,18				15,18				15	15		[127]

[a] these evidence-based functions are described in the Introduction and Discussion sections of each publication; [b] concurrently-activated brain regions and neural networks are suggested by identical numbers within each study (row). Fluid homeostatic functions: 1, responds to changes of plasma osmolality and/or sodium concentration. Non-homeostatic thirst, taste, and sensory functions: 2, responds to water intake or mouth irrigation; 3, dry mouth sensation; 4, thirst sensation; 5, discriminates fluid pleasantness/unpleasantness; 6, discriminates fluid pleasantness/unpleasantness during overdrinking when satiated; 7, integrates taste and olfactory sensations, recognized as flavor; 8, taste sensation specific to salt; 9a, integrates multiple taste stimuli; 9b, integrates multiple thirst-related stimuli; 10, processes the intensity of fluid characteristics. Non-homeostatic tongue movement, swallowing, and facial functions: 11, muscular motor activity; 12, coordinates motor outputs related to appetitive reward (e.g., thirst); 13, plans and executes movements; 14, processes spatial memory and visual association; 15, regulates volitional swallowing; 16, processes somatosensory or motor information during repetitive swallowing; 17, regulates reflexive swallowing; 18, integrates voluntary tongue elevation inside the mouth; 19, facial sensations. Abbreviations: ACG, anterior cingulate gyrus; other brain loci are defined in the text and Table 2 footnote.

Figure 4 is an idealistic illustration drawn by the present authors, of the regions in the human brain that have been associated with thirst and drinking behavior, and based on the majority of brain imaging studies shown in Table 2, Table 4 and Table 5. Although our understanding of the neural pathways between these human brain loci is incomplete at present, it is widely accepted that some of these brain regions integrate a variety of sensory and motor inputs [129,137], and that complicated behavioral states (e.g., thirsty or satiated) are controlled by discrete clusters of neurons, some composed of only 1000–2000 individual cells in mice [143]. As examples, the following loci in Figure 4 are recognized as integrative hubs: orbitofrontal cortex (i.e., taste, small, visual, auditory, visceral inputs [137]), thalamus (i.e., mediates motivation and emotional drive [144]), insular cortex (i.e., processes sensory, emotions, movement, thirst, taste, touch inputs [129]), and anterior cingulate cortex (i.e., thirst, emotions, cognition, motor actions [42,145]).

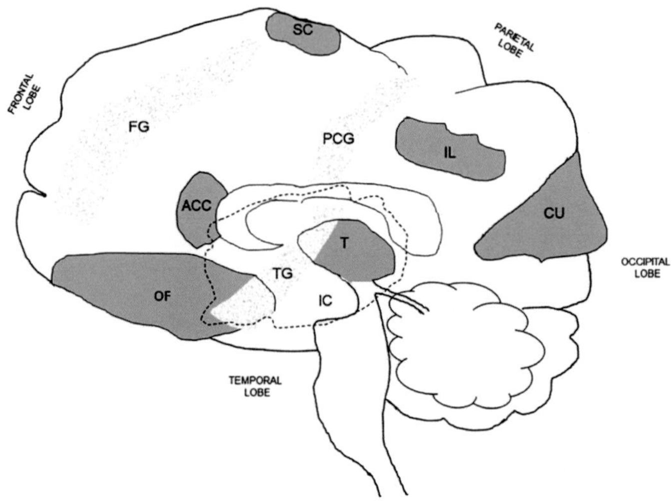

Figure 4: Idealistic drawing of human brain loci (dark shading) which have been associated with thirst, tongue movement, touching the tongue, swallowing, and taste (Table 2, Table 4 and Table 5). Three ridges (gyri) on the cerebral cortex surface are shaded lightly. The dashed region represents the insular cortex (IC) which lies deep within the lateral surface of the brain. Brain loci abbreviations are defined in the Table 2 footnote.

Limitations of Human Brain Imaging

Limitations exist in the conduct and interpretation of brain imaging studies. In Table 2, for example, the experimental protocols which induced thirst (i.e., some spanning >1 h) also altered intracellular and extracellular osmolality, water movements, as well as kidney, endocrine, and cardiovascular responses. Although these 10 studies (Table 2) focus on thirst, it is impossible to know the extent to which complex physiological responses (Figure 1 and Figure 2) influenced PET and fMRI findings. Similarly, Table 3 summarizes studies that focused on mouth irrigation and drinking to satiation, but the duration of these protocols ranged from 0.3 to 3.0 h. Within these studies, numerous digestive actions (e.g., peristalsis, swallowing) and orofacial movements (e.g., lips, tongue) may have occurred, introducing variability into the imaging data (Table 2 and Table 3). Further, with few exceptions [111], the investigators who studied thirst and drinking seldom distinguished or acknowledged the influence of subtle oropharyngeal movements, touch sensations (Table 4) or taste responses (Table 5) as coexisting variables. This is important because six brain regions that were activated in response to taste stimuli (Table 5; orbitofrontal gyrus, frontal gyrus, insular cortex, caudate nucleus, temporal gyrus, thalamus) also were activated in response to thirst (Table 2) and swallowing (Table 4). Due to this overlap, brain activation patterns by themselves do not allow unique identification of a task or the specific characteristics of a given task [146]. Further, some brain regions (e.g., anterior cingulate cortex, insula, thalamus) are highly integrative and serve as relay points for peripheral cardiovascular signals likely via midbrain nuclei. This makes it exceptionally difficult to delineate specific mechanisms for brain activations.

Human factors also may limit the interpretation of brain imaging data (Table 2, Table 3, Table 4 and Table 5) in three ways. First, inter-individual differences may be large, making it difficult to draw inferences about brain regions and networks. Large inter-individual variance has been acknowledged in studies of thirst and hunger [98], volitional swallowing [132], left versus right hemispheric dominance during swallowing [129], tongue movements [126], mental rotation, perception and memory [147,148]. Specifically, individual variance is due to differences in the ways that humans use, prioritize and integrate visceral, motivational, affective, and cognitive information; these differences arise from genetic influences, childhood developmental changes, depression, anxiety, eating disorders and subtypes of obesity [98]. Second, the selective attention which test participants devote to experimental interventions likely influence brain activations. A research group led by van Rinj [142] recorded fMRI scans of 27 women while they tasted water, fruit juice and tomato juice; additionally, these women were instructed to focus on different aspects of these fluids: pleasantness, taste intensity, and caloric contents. As a result of differential selective attention, brain region activations were different when women focused on the intensity and pleasantness of these fluids. This factor was not considered in the publications cited in Table 2, Table 3, Table 4 and Table 5; as such, the influence of selective attention is unknown and difficult to control during experiments. Third, the induction and satiation of thirst is difficult to replicate within a single session and across different days, mostly due to the many homeostatic and non-homeostatic factors [8,10,35,50,105] that influence thirst (Figure 1, Figure 2 and Figure 3) and the dynamic complexity of brain integration.

NEURAL NETWORKS

As described above, Table 1 illustrates the evolution of concepts regarding the nature and mechanism of thirst. Beginning with single factors (i.e., dry mouth, loss of water from tissues) in the 19th century, paradigms of thirst have progressed greatly due to investigations of brain loci and neural networks. However, neural networks provide four advantages over discrete brain loci, for the homeostatic regulation of body water and fluid osmolality, and initiation of thirst [149]. First, the organization of neural networks allows visceral somatosensory information to be broadly distributed from several relatively small loci which act as neural and endocrine ports of access into the brain. Second, the inherent adaptability, plasticity and information storage capability of neural networks are much greater than a single brain locus. Third, complimentary parallel neural networks provide redundancy so that damage or disease within a portion of the system is less likely to be catastrophic. Fourth, interconnected brain networks cooperate in three dimensions to accomplish functions efficiently [150].

ANIMAL RESEARCH COMPLIMENTS HUMAN BRAIN IMAGING

The rodent model of thirst and drinking behavior [63,93,151] has allowed identification of neurons that relay information regarding the status of plasma volume, vascular perfusion pressure, angiotensin II, ingested water passing through the mouth and throat, gastrointestinal water, extracellular sodium concentration and osmolality. Figure 5 illustrates and summarizes the findings of five rodent model publications which focused on thirst, thirst-related motivation, drinking behavior, and downstream signals to other brain loci. Although the authors of these investigations do not agree on all activated neural circuits and brain loci functions, they consistently agree that the lamina terminalis (a set of interconnected brain structures that coordinate the homeostatic responses to fluid imbalance; see dashed line in Figure 5) detects extracellular osmolality, angiotensin II, and fluid consumption [82,90,92,151]. The lamina terminalis amalgamates many input signals (e.g., plasma osmolality; examples appear in Figure 1, Figure 2 and Figure 3) and relays them to the anterior cingulate cortex and the insular cortex, to produce the conscious perception of thirst [92,124,151,152] and arginine vasopressin (AVP) release [93]. In addition, the thalamic paraventricular nucleus relays thirst-related signals to the median preoptic nucleus of the anterior hypothalamus (MnPO; a part of the lamina terminalis), as evidenced by the immediate, copious drinking that occurs when the paraventricular nucleus is photostimulated (see Section 5.1 below) [92]. Similarly, the nucleus of the solitary tract (NTS) and parabrachial nucleus (PBN) transmit anticipatory (non-homeostatic) oropharyngeal, baroreceptor, plasma sodium concentration, and upper gastrointestinal tract information via the vagus nerve to the MnPO [90,92]. As shown in Figure 5, the paraventricular nucleus of the hypothalamus (PVH) and the supraoptic nucleus (SON) are important downstream targets of the lamina terminalis that control release of AVP from the posterior pituitary into the circulation. Thus, PVH signaling influences not only urine production and blood pressure, but also the autonomic responses of heart rate and natriuresis [11].

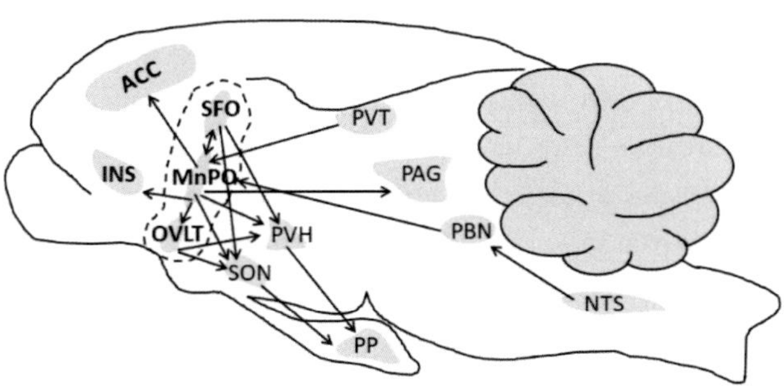

Figure 5: Rodent brain loci that have been associated (2016–2019) with dehydration-induced drinking [82,90,91,92,94,113,151]. In this idealistic illustration, arrows denote evidence-based neural circuits that integrate intracellular or

extracellular signals and modulate thirst, drinking, rapid satiety, overdrinking, and aversive motivational drives. Abbreviations: IC, insular cortex (insula); SFO, subfornical organ; PVT, paraventricular thalamic nucleus; PP, posterior pituitary; additional brain loci are defined in the text and Table 2 footnote.

Prior to the year 2000, the vast majority (91%) of the findings described in the preceding paragraph were discovered with the aid of two animal research methods: (a) lesioning/oblation of specific brain regions, or (b) microinjection/microinfusion. The former approach was utilized in more than 30 studies spanning 50 years, whereas the latter was employed in more than half of 86 publications (i.e., involving the awareness of thirst, drinking behavior, the functions of specific brain regions, and neural circuits) that were reviewed by the present author. The laboratory techniques of arterial or venous ligation, and electrical stimulation of specific brain loci, were employed in only 9% of these publications.

Optogenetics and Chemogenetics

Since the turn of this century, investigators have emphasized that it is not enough to know how individual brain loci and isolated networks function; studying how networks interact with each other is equally important. Today, the frontier is not brain network science, it is the science of *networks of networks* and the ways that networks rapidly connect and disconnect across the entire brain mass [153]. The development of two new laboratory techniques has uniquely enhanced our understanding of rodent brain regions and neural networks that influence thirst, motivation to seek water, and drinking (Table 1 and Table 7). The first is optogenetics, which allows researchers to visualize genetically targeted neurons in living animals, and to track electrical and biochemical events within living neural circuits [154]. Optogenetics utilizes light to control cells in living neurons that have been genetically modified to activate the membrane ion channels in light-sensitive proteins. This technology has given scientists the ability to stimulate or inhibit genetically defined populations of neurons with temporal resolution that is millisecond-precise [143]; in comparison, an fMRI brain scan requires several seconds to accomplish. These characteristics explain why optogenetics has been recognized as one of the most significant new methods of the present century, across all fields of science and engineering [155]. The second technique, named chemogenetics or pharmacogenetics, is similar to optogenetics but it utilizes chemically engineered molecules instead of light and light-sensitive membrane channels [156]. One form of chemogenetics employs directed molecular evolution. Known as Designer Receptors Exclusively Activated by Designer Drugs (DREADD), this tool is used by neuroscientists to activate specific neuronal circuits within the brain that influence thirst and body fluid homeostasis [143,157]. The combination of chemogenetics with imaging techniques in freely moving animals now makes it possible to analyze the complex whole-brain networks that are fundamental to behavior.

Table 7: The importance of incentive, motivation and learning in thirst and the drinking behavior of rodents, primates, and humans

Paradigms	Authors
Peripheral sensations (e.g., dry mouth, stomach contraction) caused by a water deficit become the instigating and sustaining conditions that motivate water seeking and drinking.	[19,20]
We will not fully understand thirst until we acknowledge motivation and the elusive neurological questions it poses: how does the urge to drink arise in the brain, what compels animals to anticipate water consumption, what creates the hedonic state of thirst, and what gives rise to the excitement of water need or the pleasure of consumption?	[72]
Animals can learn to drink in specific situations via classical conditioning. For example, neutral stimuli (i.e., having no effect on behavior) elicit drinking following their repeated association with thirst-inducing treatments (e.g., water deprivation, hypovolemia, hypertonic saline injections). Conditioned physiological changes do not elicit this drinking behavior. [a]	[158]
Animals can learn to associate specific tastes with the effects that result after fluid consumption; this learning influences subsequent experiences with those fluids. [a]	[159]
Palatability of a fluid (i.e., pleasant flavor, aversive taste, temperature) can override the homeostatic control (i.e., plasma hyperosmolality, extracellular volume depletion) of fluid balance in humans.	[138]
Small animal drinking is behaviorally complex. A small animal drinks in connection with eating, in anticipation of thirst, and because it learns the location of a water source. It paces drinking across a diurnal cycle. It will forego drinking if too much searching is required. [a]	[160]

Motivation of animals can be determined by measuring: (a) how hard they work (e.g., number of times they will press a lever, distance they will travel in a maze) to obtain water, rather than how much water is consumed before satiety mechanisms terminate drinking; and (b) the extent to which they will tolerate aversive consequences (e.g., water containing different concentrations of bitter tasting quinine, electrical shock) in order to obtain water. [b]	[104]
Motivation-specific response systems exist in the primate brain. That is, some neurons respond to the sight or taste of food but not water, and other neurons respond to the sight and/or taste of water but not food. Such motivational specificity is required of a system that guides and controls hunger- or thirst-motivated behavior.	[161]
Behavioral responses depend partly on physiology and vice versa. In some situations, normal drinking in rats may be largely due to non-physiological factors.	[162]
The arousal of a motive has no necessary connection with homeostatic regulation. Some motives arise without needs, and some needs arise without motives.	[163]
Much physiology-induced human behavior is learned. Virtually all ingestive appetite is acquired and is influenced by sensory characteristics (i.e., taste, smell, texture) or the cooling (i.e., pleasurable) effects of water. This is true for babies, young children and adults.	[139]
The initiation or termination of drinking could result from past personal experiences. A variety of associations with the consequences or outcomes of drinking in specific situations may leave a desire (i.e., to subsequently drink or not to drink certain beverages), when those situations are next encountered.	[139]
Basic needs and motives such as thirst cause a heightened perceptual readiness to environmental cues that are instrumental in satisfying these needs (e.g., recognizing a water source that otherwise might go unnoticed).	[164]
Utilizing chemogenetic and optogenetic laboratory techniques, neuroscientists activate specific brain neurons to identify the circuitry and cellular signals that influence/generate behaviors, innate drives, memories, learning, and motor functions. See text for methodological details. [c]	[88,151,157, 165]

The theory of incentive motivation states that the power of external stimuli (i.e., visual recognition) is calibrated dynamically, based on the current body state of the organism (e.g., motivation is high when the body is dehydrated and low when normally hydrated).	[166]
The activity of dehydration-activated neurons (i.e., in specific brain regions) establishes a scalable, persistent, and aversive internal state that dynamically controls thirst-motivated behavior. This state motivates drinking to quench thirst. Localized brain neuron activity is proportional to the strength of this aversive emotion (i.e., negative valence). [a]	[79,81,82,83,93]
Rodent brain loci have been identified that control motivational processes such as water seeking, drinking, and cessation of drinking.	[90,92]
The motivated behavior of rodents to seek and consume water is regulated by thirst; this regulation involves modulation of brain-wide neural population dynamics. [c]	[94]

[a] researchers cannot know if an animal is thirsty; when referring to non-humans, the acts of seeking and consuming water are assumed to indicate that a thirst-like drive exists; [b] these measurements will not always correlate with the amount of water consumed during free access (*ad libitum* drinking) because the amount consumed reflects the amount of water required to terminate drinking, rather than the initial motivation of an animal to obtain water; [c] see section (above) 5.1 Optogenetics and Chemogenetics.

In 2019, W.E. Allen and colleagues [94] provided arguably the most enlightening thirst-related paradigm advance of the 21st century. Rather than utilizing fMRI scans of brain activations or neural circuits, this research team recorded high-density microelectrode array activations of approximately 24,000 individual neurons throughout 34 rodent brain loci, during several hundred thirst-related tasks (i.e., in a water restricted state, animals were trained to receive water as a reward for correct choices). Investigators specifically targeted thalamic and hypothalamic nuclei directly downstream of the MnPO, as well as diverse second-order regions (e.g., insular cortex) that were anatomically connected by axonal pathways to these downstream regions. As mice gradually consumed water, more than half of the recorded neurons responded to a water-predicting olfactory cue and sustained water acquisition activity. These procedures revealed a global representation of the thirst-motivational neural network, as depicted in Figure 6 [94]. Once satiated, this wave of brainwide activations (i.e., flow of information) was restricted to only a transient change of activity, with no motivated behavioral response. Subsequently, while mice were sated, localized optogenetic activation of hypothalamic thirst-sensing neurons rapidly converted brainwide activity to the pre-satiation state. The complexity of these brainwide neural network phenomena confirms research conducted more than a decade before. Bourque and colleagues [167] described the dynamic complexity of relationships among plasma osmolality, osmoreceptors, ingestive behaviors, sympathetic outflow, renal function, extracellular tonicity/volume, cardiovascular balance, and thermoregulation.

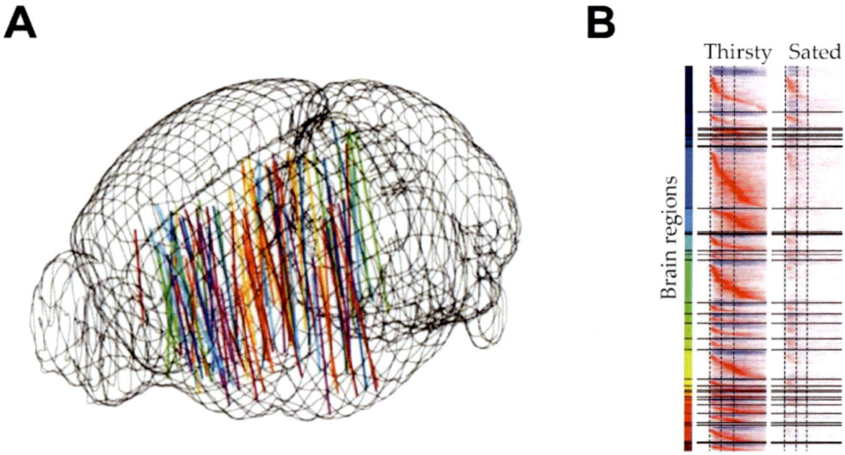

Figure 6: (A) Tracks of neuropixel electrode activations, recorded in mouse forebrain and midbrain regions during head-fixed, thirst-motivated choice behavior. (B) Brainwide activity dynamics of individual neurons from 31 brain loci, in response to an olfactory cue, while a mouse was both thirsty and sated. Reproduced with the permission of the publisher [94].

When compared to static brain imaging research (Table 2, Table 3, Table 4 and Table 5), the work of Allen et al. [94] is one of several rodent studies that exemplify the new behavioral insights and thirst-related paradigms which optogenetic and chemogenetic methods provide. For example, optogenetic or combined optogenetic–chemogenetic techniques have allowed the following discoveries to be made: (a) an innate brain circuit involving subfornical organ and organum vasculosum of the lamina terminalis neurons—from which signals are transferred downstream to the MnPO [84,91]—that can turn on and off an animal's water-drinking behavior [80,82]; (b) thirst-promoting neurons (i.e., subfornical organ) respond to oropharyngeal inputs during eating and drinking as well as the composition of blood, to anticipate how food and water consumption will alter fluid balance in the future, and then adjust drinking behavior preemptively [93]; (c) thirst and salt appetite are driven by distinct groups of angiotensin II receptor neurons in the subfornical organ [85]. Far from the 19th century observations of thirst that involved single factors (Table 1), these techniques associate behaviors (i.e., motivation, incentive, seeking water or salt) with specific brain regions and neural circuits.

Limitations of Animal Models

The fluid–electrolyte balance of all vertebrates is subject to stringent homeostatic controls which maintain intracellular and extracellular ionic and osmotic conditions that are critical for normal cell functions [168]. Mammals, fish, amphibians, reptiles, birds, rodents, and humans share the common needs of maintaining osmolality, total body water, extracellular volume, and blood pressure. However, vertebrates obtain and conserve water and essential electrolytes via a wide range of taxonomic-specific evolutionary adaptations, including sodium appetite, restricted water loss from the body surface, regulation of urine contents, and water storage [60]. These vertebrate mechanisms of fluid–electrolyte balance are necessarily diverse, due to differences of environmental conditions (e.g., land, water, air, temperature, solar radiation, water availability) and life activities (e.g., avoiding predators, seeking food and water, migration) [60]. As a result, large species-specific differences of water consumption exist (i.e., expressed as % of body weight/24 h): man, 3%; dog, 5%; cattle, 6%, rabbit, 11%, and rat, 16% [3]. Even when comparing different strains of rats, divergent patterns of water intake have been reported during food deprivation (e.g., ranging from little change to an 80% decrease, relative to ad libitum baseline intake), prompting the authors to suggest that generalizations regarding rodent drinking behavior must be made with caution [169].

The human brain weighs 1.5 kg and consists of 86 billion neurons, whereas the mouse brain weighs 0.4 g and contains 70 million neurons [170]. Thus, it is relevant to ask, "Is it valid to generalize rodent thirst, drinking behavior, brain circuits, and neural networks to humans?" The answer to this question is not universally accepted; some research teams believe that neural and endocrine bases of thirst in rodents are similar to those of humans [9,42,92,93,124,171,172], whereas others emphasize rodent-human differences [124,172]. As noted above, the anterior cingulate cortex (ACC, acting in concert with the insular cortex) is essential to the conscious perception of thirst. However, the exact location, size, structure and connectivity of the ACC in nonhuman primates is not agreed upon by neurophysiologists, and it is not always obvious which areas of the rodent frontal cortex should be considered as equivalent to the human ACC [144]. Similar questions also arise in neuroscience fields other than thirst, where some researchers emphasize mouse–human anatomical and physiological similarities (e.g., Parkinson's disease [173]; the neuroprotective benefits of exercise to counteract effects of aging [174]), some investigators acknowledge obvious mouse–human differences (e.g., size and complexity of the cerebral and cerebellar cortex, hemispheric dominance, hemispheric specialization [145,170,175], whereas others describe both similarities and differences in mouse and human brains (e.g., neural network organization in Alzheimer disease pathways [176]). Thus, after more than 125 years of experimental neuroscience, mouse and rat experiments may or may not have strong generalizability to humans, especially considering the fact that the mammalian cerebral cortex has proven to be far more variable across species than was believed two or three decades ago [175].

A CONTEMPORARY PUBLIC HEALTH PROBLEM: LOW DAILY WATER CONSUMPTION

Public health surveys spanning decades have studies nutritional status and its association with health promotion and disease prevention; these surveys have revealed a remarkable variability in the 24 h total water intake (TWI) of apparently healthy individuals. For instance, in the third National Health and Nutrition Examination Survey, the lowest to highest decile of TWI was 1.69 to 7.93 L/24 h for men and 1.25 to 6.16 L/24 h for women [177]. Interestingly, the plasma osmolality of the lowest and highest water consumers were nearly identical for males (279 and 280 mOsm/kg) and females (277 and 277 mOsm/kg). In a cross-sectional study, investigators examined hydration biomarkers of healthy adults with low (LOW; 0.74 L/24 h) and high (HIGH; 2.70 L/24 h) water intakes [178]. They reported significantly elevated urinary hydration biomarkers (e.g., urine osmolality), plasma AVP, and plasma cortisol in LOW, without a LOW versus HIGH difference of plasma osmolality or thirst (i.e., this state has been described as underhydration [179]). In a subsequent intervention study involving women, Johnson and colleagues [180] evaluated the effects of water intake modification. At baseline, thirst and blood concentration were not different between groups but plasma AVP and urine osmolality were significantly elevated in LOW. Next, controlled water intake was increased in LOW and reduced in HIGH. Interestingly, serum osmolality was not different between groups at baseline, and did not change significantly during the 4 d water intervention for either group, giving the appearance of similar health states.

The principal public health question is whether chronically elevated plasma AVP (i.e., and concentrated urine) could increase the risk of chronic diseases, as previously had been described for kidney disorders. Multiple epidemiological studies from Sweden [181,182,183], United Kingdom [184], and France [185,186,187] have reported that elevated AVP, assessed by its surrogate marker copeptin, is positively associated with diabetes. In a study involving rats, AVP administration for four weeks induced glucose intolerance and hepatic steatosis [188]. Recently, a clinical trial observed 60 healthy males and females; osmotically elevated plasma AVP acutely impaired glucose regulation [189]. Similarly, several studies have linked LOW to the development of diabetes [185,186,187,190,191,192]. External to these effects on glucose regulation and diabetes, the pharmacologic blockade of AVP membrane receptors is currently under investigation in patients with rapidly progressing renal disease [186,187,188,189,190,191,192,193]. The above findings indicate that a low daily water intake activates homeostatic mechanisms involving AVP to maintain water and osmotic balance, and exposes LOW to an increased risk of chronic diseases. This raises important questions. Why are some people perfectly content to consume a very small TWI each day? Considering that only subtle physiological differences distinguish LOW from HIGH, what aspects of thirst are not homeostatic but rather involve trial and error, new information, or a strong dislike originating from a previous experience?

LEARNING, MOTIVATION AND AVERSION

Thirst and drinking behavior are distinctly influenced by learning and motivation, in ways that classical physiologists have seldom considered or measured (Figure 1 and Figure 2 [162,163,166]) when they investigated homeostatic thirst. Table 7 describes several of these influences, as observed in both human and animal studies that span 165 years. The following are especially relevant: (a) drinking is behaviorally complex [160]; (b) hypovolemia elicits a true motivational state of thirst rather than mere reflexive drinking [56]; (c) the thirst drive and its related motivation to drink cause a heightened perceptual readiness to respond to environmental cues that may meet this need [164,166]; (d) specific human brain loci have been associated with perceptions, emotions, innate drives, memories, learning, and motor activities [88,157,165].

Although drinking is a fundamental behavior, the means by which the human brain transforms a need for water into a specific motivational drive is only superficially understood [84]. As a potential site of this transformation, the anterior cingulate cortex interests neuroscientists greatly. Heilbronner and Hayden [145] propose that ACC neurons (Figure 4 and Figure 5) link contexts with strategies by integrating diverse task-relevant information, to create a complex representation of the task environment and exert abstract control over decisions and actions. The ACC also has strong connectivity to motor centers [194]. These observations are consistent with the concept of a neural hub that integrates information from multiple homeostatic and non-homeostatic factors (Figure 3), then elicits thirst and motivates the host to seek and consume water [90,92,93,107,124,151].

A consistent and substantial body of evidence [152] describes the roles of the lamina terminalis, the ACC and the insular cortex (see Figure 4) as a neural pathway that generates thirst, motivation to seek water, and goal-directed drinking behavior. Homeostatic (i.e., plasma osmotic and endocrine) signals (Figure 3) are integrated by the lamina terminalis [124] and forwarded to the ACC, which translates these signals into emotional responses, arousal and affect by assigning cost–benefit values to behavioral options, considering alternative choices in an unpredictable or changing environment, and guiding decisions so that the most appropriate action is taken [152,195]. Anatomically, the ACC is uniquely located at the interface of the frontal cortex, the motor system and subcortical structures, allowing it to integrate multiple signals during development of motivation [195]. After attempting to acquire water (i.e., either successfully or unsuccessfully), the ACC purportedly updates host beliefs and internal models of decision making [196]. The insular cortex serves functions similar to those of the ACC [152,194] and also integrates neural signals from different loci in the lamina terminalis (e.g., subfornical organ, organum vasculosum of the lamina terminalis, median preoptic nucleus). Indeed, both the ACC and the IN are deactivated (i.e., fMRI signal strength) when drinking results in satiety; this state generates an aversive emotion that decreases motivation to drink and protects the host from over-drinking [123,194]. Thus, both the unpleasant effect of thirst and the pleasant effect of drinking regulate water consumption [82] and ultimately total body water balance (Figure 2).

Drinking behavior also is influenced by learning that involves past experiences, environmental characteristics, and pleasant or unpleasant sensations [139]. According to the classical "drive-reduction" hypothesis, animals learn specific behaviors that reduce the level of an aversive drive state (e.g., thirst). By reducing negative affect signals, a preference for cues associated with lessening of physiological need states (i.e., seeking and consuming water) can be learned [89]. It also is possible that humans learn to associate subtle cues (e.g., the time of day or a large sweat loss during prolonged exercise [197]) with impending dehydration and thus drink in anticipation of plasma volume and osmolality changes [13]. Interestingly, the ACC has been theoretically associated with learning in both rats and primates [145]. However, learning requires more than change in a single brain region; it requires multiple interconnected networks that reconfigure their connections during the learning process. Even when a person learns a relatively simple task, large portions of the brain become involved [198].

Two additional aspects of thirst and drinking behavior have been reported by investigators during the past five years. The first is based on the recognition that brain neurons within the MnPO encode an aversive motivational drive; these nerves represent a distinct homeostatic neuronal cell type with a distinct biological function that opposes thirst in mice [82,89]. The second involves fMRI evidence and extensive arguments regarding a human mechanism that opposes swallowing after rehydration and satiation of thirst. This inhibitory mechanism generates sensations of unpleasantness that are associated with the insular cortex, midcingulate cortex, amygdala, and periaqueductal gray matter. Previous research has implicated these regions in discomfort and aversion [75]. Subsequently, this research group conducted experiments involving ratings of swallowing effort and regional brain imaging [109], as participants prepared to swallow small volumes of liquid while they were thirsty and after they had overdrunk. After overdrinking, regional brain activations occurred in the motor cortex, prefrontal cortices, posterior parietal cortex, striatum, and thalamus when participants prepared to swallow. Effort ratings provided unequivocal support for swallowing inhibition, with a threefold increase in effort after overdrinking; these effort ratings were correlated with activations in the right prefrontal cortex and pontine regions of the brainstem. The authors concluded that swallowing inhibition in humans likely is a "hard-wired" process which helps the body to avoid the detrimental effects of overdrinking that could result in water intoxication and/or death [109]. Although not widely appreciated, both the aversive motivational drive and swallowing inhibition are part of the complex, dynamic regulation of fluid–electrolyte balance.

SUMMARY AND IMPLICATIONS FOR FUTURE RESEARCH

The preceding paragraphs and Figure 3 emphasize that central nervous system homeostatic regulation of thirst and fluid intake integrates osmotic, ionic, hormonal, intracellular, and extracellular signals; concurrently, non-homeostatic regulation integrates oropharyngeal, environmental, social and cultural factors, fluid characteristics, and learned preferences. The motivation to seek and consume water arises from the integration and transfer of these signals to specific brain loci, and their conversion to decisions and actions via mechanisms that are not fully understood. Thus, thirst and drinking behavior have attracted, across more than 180 years (Table 1), the interest of investigators from numerous specialized research areas (e.g., physiology, medicine, behavioral psychology, cognitive neuroscience, optogenetics); these specialists measure different variables, use different methods and instruments, develop different concepts of thirst and drinking behavior, and as a general rule seldom communicate directly. Thus, most current paradigms and models are incomplete.

Most current concepts of thirst and drinking behavior have arisen from rodent studies, which may or may not be applicable to humans, and human brain imaging which has identified numerous active brain regions (e.g., concurrent with thirst or consuming water) but cannot causally ascribe specific motives or actions to each. Specifically, the following brain loci provide interesting promise (Table 1, Table 2 and Table 3) for future brain imaging studies of thirst and drinking: anterior cingulate cortex, insular cortex, orbitofrontal cortex, frontal gyrus, posterior cingulate cortex, thalamus, and cerebellum. In addition, Table 4 distinguishes the brain regions that apparently function in the muscular actions of swallowing or drinking (e.g., postcentral gyrus, primary motor cortex, premotor and supplementary motor cortex) but not in the awareness of thirst; importantly, the insular cortex was identified in 73% of Table 4 publications and in 100% of those cited in Table 5. However, considering only individual brain loci and localized neural pathways results in incomplete paradigms of thirst and drinking behavior. Today, the focus is on networks of networks, with thirst and satiety conceived in terms of hemispheric waves of neuronal activations [94] that move across the brain in milliseconds.

The future of thirst research lies in the newly developed techniques of chemogenetics, optogenetics, and neuropixel microelectrode arrays, which reveal important aspects of the dynamic complexity of human thirst, water seeking, and drinking. Although these recently developed invasive techniques (i.e., brain surgery, implanted electrodes, genetic manipulations) are limited to animal models and are not ethically possible in human research, it is our expectation that noninvasive methods someday will provide detailed paradigms that describe the conversion of numerous afferent signals into motivation, decisions, and actions that counteract perturbations of body water volume and concentration.

AUTHOR CONTRIBUTIONS

L.E.A. and S.A.K. wrote, reviewed, and edited manuscript drafts, and approved the final manuscript; conceptualization, L.E.A.; design and creating of tables and figures, L.E.A.

REFERENCES

1. Wang Z., Deurenberg P., Wang W., Pietrobelli A., Baumgartner R.N., Heymsfield S.B. Hydration of fat-free body mass: Review and critique of a classic body-composition constant. *Am. J. Clin. Nutr.* 1999;69:833–841. doi: 10.1093/ajcn/69.5.833.
2. Adolph E.F., Dill D.B. Observations on water metabolism in the desert. *Am. J. Physiol.* 1938;123:369–378. doi: 10.1152/ajplegacy.1938.123.2.369.
3. Adolph E.F. *Physiological Regulations.* Jacques Cattell Press; Lancaster, PA, USA: 1943.
4. Bartoli W.P., Davis J.M., Pate R.R., Ward D.S., Watson P.D. Weekly variability in total body water using 2H2O dilution in college-age males. *Med. Sci. Sports Exerc.* 1993;25:1422–1428. doi: 10.1249/00005768-199312000-00017.
5. Epstein A.N. Fourth JAF Stevenson Memorial Lecture: The physiology of thirst. *Can. J. Physiol. Pharmacol.* 1976;54:639–649. doi: 10.1139/y76-090.
6. Fitzsimons J.T. The physiological basis of thirst. *Kidney Int.* 1976;10:3–11. doi: 10.1038/ki.1976.74.
7. Fitzsimons J.T. Angiotensin, thirst, and sodium appetite. *Physiol. Rev.* 1998;78:583–686. doi: 10.1152/physrev.1998.78.3.583.
8. Stricker E.M., Sved A.F. Thirst. *Nutrition.* 2000;16:821–826. doi: 10.1016/S0899-9007(00)00412-3.
9. McKinley M.J., Johnson A.K. The physiological regulation of thirst and fluid intake. *News Physiol. Sci.* 2004;19:1–6. doi: 10.1152/nips.01470.2003.
10. Thornton S.N. Thirst and hydration: Physiology and consequences of dysfunction. *Physiol. Behav.* 2010;100:15–21. doi: 10.1016/j.physbeh.2010.02.026.
11. Zimmerman C.A., Huey E.L., Ahn J.S., Beutler L.R., Tan C.L., Kosar S., Bai L., Chen Y., Corpuz T.V., Madisen L., et al. A gut-to-brain signal of fluid osmolarity controls thirst satiation. *Nature.* 2019;568:98–102. doi: 10.1038/s41586-019-1066-x. [PMC free article]
12. Cheuvront S.N., Kenefick R.W. Dehydration: Physiology, assessment, and performance effects. *Compr. Physiol.* 2011;4:257–285. [PubMed]

13. Rolls B.J. Physiological determinants of fluid intake in humans. In: Ramsay D.J., Booth D., editors. *Thirst Physiological and Psychological Aspects*. Springer; London, UK: 1991. pp. 391–399.
14. Greenleaf J.E., Harrison M.H. Water and electrolytes. In: Layman D.K., editor. *Nutrition and Aerobic Exercise.* American Chemical Society; Washington, DC, USA: 1986. pp. 107–124.
15. Greenleaf J.E. Problem: Thirst, drinking behavior, and involuntary dehydration. *Med. Sci. Sports Exerc.* 1992;24:645–656. doi: 10.1249/00005768-199206000-00007.
16. Wolf A.V. *Thirst: Physiology of the Urge to Drink and Problems of Water Lack.* Charles C. Thomas; Springfield, IL, USA: 1958.
17. Fitzsimons J.T. Thirst. *Physiol. Rev.* 1972;52:468–561. doi: 10.1152/physrev.1972.52.2.468.
18. Latta T. Letter from Dr. Latta to the Secretary of the Central Board of Health, London, affording a view of the rationale and results of his practice in the treatment of cholera by aqueous and saline injections. *Lancet.* 1832;2:274–277. [PubMed]
19. Colin G. *Traité de Physiologie Comparée des Animaux Domestiques.* Volume 1. JB Baillière; Paris, France: 1854. p. 434.
20. Cannon W.B. The physiological basis of thirst. *Proc. R. Soc. Lond. B.* 1918;90:283–301. doi: 10.1098/rspb.1918.0017.
21. Bernard C. *Leçons de Physiologie Expérimentale Appliquée à la Médecine: Faites au Collège de France.* JB Baillière et Fils; Paris, France: 1856. pp. 50–51.
22. Schiff M. *Leçons Sur la Physiologie de la Digestion: Faites au Muséum D'histoire Naturelle de Florence.* Volume 1. Hermann Loescher; Florence, Italy: 1867. pp. 41–42.
23. Mayer A. Variations de la tension osmotique du sang chez les animaux prives de liquides. *C. R. Seances Soc. Biol. Fil.* 1900;52:153–155.
24. Wettendorff H. *Ph.D. Thesis.* Free University of Brussels, Faculty of Medicine; Brussels, Belgium: 1900. Modifications du Sang Sous L'influence de la Privation D'eau: Contribution à L'étude de la Soif.
25. Nothnagel H. Durst und Polydipsie. *Virchows Arch.* 1881;86:435–447. doi: 10.1007/BF01915727.
26. Kahler O. Die dauernde Polyurie als cerebrales Herdsymptom. *Z. Heilk.* 1886;7:105–220.

27. Paget S. On cases of voracious hunger and thirst from injury or disease of the brain. *Tr. Clin. Soc. Lond.* 1897;30:113–119.
28. Leschke E. Ueber die Durstempfinding. *Eur. Arch. Psychiatry Clin. Neurosci.* 1918;59:773–781. doi: 10.1007/BF02251858.
29. Bailey P., Bremer F. Experimental diabetes insipidus. *Arch. Intern. Med.* 1921;28:773–803. doi: 10.1001/archinte.1921.00100180091006.
30. Fisher C., Ingram W.R., Ranson S.W. Diabetes Insipidus and the Neurohormonal Control of water balance: A contribution to the structure and function of the hypothalamico-hypophyseal system. *Am. J. Med. Sci.* 1938;196:729. doi: 10.1097/00000441-193811000-00031.
31. Leib D.E., Zimmerman C.A., Knight Z.A. Thirst. *Curr. Biol.* 2016;26:R1260–R1265. doi: 10.1016/j.cub.2016.11.019.
32. Gilman A. The relation between blood osmotic pressure, fluid distribution and voluntary water intake. *Am. J. Physiol.* 1937;120:323–328. doi: 10.1152/ajplegacy.1937.120.2.323.
33. Fitzsimons J.T. Drinking by rats depleted of body fluid without increase in osmotic pressure. *J. Physiol.* 1961;159:297–309. doi: 10.1113/jphysiol.1961.sp006809.
34. Bellows R.T. Time factors in water drinking in dogs. *Am. J. Physiol.* 1938;125:87–97. doi: 10.1152/ajplegacy.1938.125.1.87.
35. Adolph E.F., Barker J.P., Hoy P.A. Multiple factors in thirst. *Am. J. Physiol.* 1954;178:538–562. doi: 10.1152/ajplegacy.1954.178.3.538.
36. Oatley K. *Neuropsychology of Thirst: New Findings and Advances in Concepts.* Winston & Sons; Washington, DC, USA: 1973. Simulation and theory of thirst; pp. 199–228.
37. Stricker E.M., Hoffmann M.L. Presystemic signals in the control of thirst, salt appetite, and vasopressin secretion. *Physiol. Behav.* 2007;91:404–412. doi: 10.1016/j.physbeh.2007.04.007.
38. Brugger M. Fresstrieb als hypothalamisches Symptom. *Helv. Physiol. Pharmacol. Acta.* 1943;1:183–193.
39. Andersson B. Polydipsia caused by intrahypothalamic injections of hypertonic NaCl-solutions. *Experientia.* 1952;8:157–158. doi: 10.1007/BF02170230.
40. Greer M.A. Suggestive evidence of a primary "drinking center" in hypothalamus of the rat. *Proc. Soc. Exp. Biol. Med.* 1955;89:59–62. doi: 10.3181/00379727-89-21716.

41. Andersson B., McCann S.M. A farther study of polydipsia evoked by hypothalamic stimulation in the gloat. *Acta Physiol. Scand.* 1955;33:333–346. doi: 10.1111/j.1748-1716.1955.tb01213.x.
42. Sewards T.V., Sewards M.A. The awareness of thirst: Proposed neural correlates. *Conscious. Cogn.* 2000;9:463–487. doi: 10.1006/ccog.2000.0462.
43. Towbin E.J. Gastric distension as a factor in the satiation of thirst in oesophagastomised dogs. *Am. J. Physiol.* 1949;159:533–541. doi: 10.1152/ajplegacy.1949.159.3.533.
44. Paintal A.S. A study of gastric stretch receptors. Their role in the peripheral mechanism of satiation of hunger and thirst. *J. Physiol.* 1954;126:255–270. doi: 10.1113/jphysiol.1954.sp005207.
45. Wolf A.V. Osmometric analysis of thirst in man and dog. *Am. J. Physiol.* 1950;161:75–86. doi: 10.1152/ajplegacy.1950.161.1.75.
46. Szczepanska-Sadowska E., Kozlowski S., Sobocinska J. Blood antidiuretic hormone level and osmotic reactivity of thirst mechanism in dogs. *Am. J. Physiol.* 1974;227:766–770. doi: 10.1152/ajplegacy.1974.227.4.766.
47. Stevenson J.A., Welt L.G., Orloff J. Abnormalities of water and electrolyte metabolism in rats with hypothalamic lesions. *Am. J. Physiol.* 1950;161:35–39. doi: 10.1152/ajplegacy.1950.161.1.35.
48. Teitelbaum P., Stellar E. Recovery from the failure to eat produced by hypothalamic lesions. *Science.* 1954;120:894–895. doi: 10.1126/science.120.3126.894.
49. Teitelbaum P., Epstein A.N. The lateral hypothalamic syndrome: Recovery of feeding and drinking after lateral hypothalamic lesions. *Psychol. Rev.* 1962;69:74–90. doi: 10.1037/h0039285.
50. Stellar E. The physiology of motivation. *Psychol. Rev.* 1954;61:5–22. doi: 10.1037/h0060347.
51. Grossman S.P. Eating or drinking elicited by direct adrenergic or cholinergic stimulation of hypothalamus. *Science.* 1960;132:301–302. doi: 10.1126/science.132.3422.301.
52. Epstein A.N. Water intake without the act of drinking. *Science.* 1960;131:497–498. doi: 10.1126/science.131.3399.497.
53. Ungerstedt U. Is Interruption of the Nigro-Striatal Dopamine System Producing the “Lateral Hypothalamus Syndrome”? *Acta Physiol.*

Scand. 1970;80:35A–36A. doi: 10.1111/j.1748-1716.1970.tb04858.x.

54. Cizek L.J., Semple R.E., Huang K.C., Gregersen M.I. Effect of extracellular electrolyte depletion on water intake in dogs. *Am. J. Physiol.* 1951;164:415–422. doi: 10.1152/ajplegacy.1951.164.2.415.
55. Stricker E.M. Extracellular fluid volume and thirst. *Am. J. Physiol.* 1966;211:232–238. doi: 10.1152/ajplegacy.1966.211.1.232.
56. Stricker E.M. Some physiological and motivational properties of the hypovolemic stimulus for thirst. *Physiol. Behav.* 1968;3:379–385. doi: 10.1016/0031-9384(68)90066-8.
57. Oatley K. Changes of blood volume and osmotic pressure in the production of thirst. *Nature.* 1964;202:1341–1342. doi: 10.1038/2021341a0.
58. Kaufman S. Role of right atrial receptors in the control of drinking in the rat. *J. Physiol.* 1984;349:389–396. doi: 10.1113/jphysiol.1984.sp015162.
59. Stricker E.M. Central control of water and sodium chloride intake in rats during hypovolaemia. In: Ramsay D.J., Booth D., editors. *Thirst Physiological and Psychological Aspects.* Springer; London, UK: 1991. pp. 194–206.
60. Fitzsimons J.T. Evolution of physiological and behavioural mechanisms in vertebrate body fluid homeostasis. In: Ramsay D.J., Booth D., editors. *Thirst Physiological and Psychological Aspects.* Springer; London, UK: 1991. pp. 3–22.
61. Sagawa S., Miki K., Tajima F., Tanaka H., Choi J.K., Keil L.C., Shiraki K., Greenleaf J.E. Effect of dehydration on thirst and drinking during immersion in men. *J. Appl. Physiol.* 1992;72:128–134. doi: 10.1152/jappl.1992.72.1.128.
62. Wada F., Sagawa S., Miki K., Nagaya K., Nakamitsu S., Shiraki K., Greenleaf J.E. Mechanism of thirst attenuation during head-out water immersion in men. *Am. J. Physiol. Regul. Integr. Comp. Physiol.* 1995;268:R583–R589. doi: 10.1152/ajpregu.1995.268.3.R583.
63. Bichet D.G. Vasopressin and the regulation of thirst. *Ann. Nutr. Metab.* 2018;72:3–7. doi: 10.1159/000488233.
64. Oatley K. A control model of the physiological basis of thirst. *Med. Biol. Eng.* 1967;5:225–237. doi: 10.1007/BF02474532.
65. Toates F.M., Oatley K. Computer simulation of thirst and water balance. *Med. Biol. Eng.* 1970;8:71–87. doi: 10.1007/BF02551751.

66. Fitzsimons J.T., Simons B.J. The effect on drinking in the rat of intravenous infusion of angiotensin, given alone or in combination with other stimuli of thirst. *J. Physiol.* 1969;203:45–57. doi: 10.1113/jphysiol.1969.sp008848.

67. Epstein A.N., Fitzsimons J.T., Rolls B.J. Drinking induced by injection of angiotensin into the brain of the rat. *J. Physiol.* 1970;210:457–474. doi: 10.1113/jphysiol.1970.sp009220.

68. Fischer-Ferraro C., Nahmod V.E., Goldstein D.J., Finkielman S. Angiotensin and renin in rat and dog brain. *J. Exp. Med.* 1971;133:353–361. doi: 10.1084/jem.133.2.353.

69. Ganten D., Minnich J.L., Grenger P., Hayduk K., Brecht H.M., Barbeau A., Boucher R., Genest J. Angiotensin-forming enzyme in brain tissue. *Science.* 1971;173:64–65. doi: 10.1126/science.173.3991.64.

70. Lee M.C., Thrasher T.N., Ramsay D.J. Is angiotensin essential in drinking induced by water deprivation and caval ligation? *Am. J. Physiol. Regul. Integr. Comp. Physiol.* 1981;240:R75–R80. doi: 10.1152/ajpregu.1981.240.1.R75.

71. Fitzsimons J.T. Some historical perspectives in the physiology of thirst. In: Epstein A.N., Kissileff H.R., Stellar E., editors. *Neuropsychology of Thirst: New Findings and Advances in Concepts.* V.H. Winston & Sons; Washington, DC, USA: 1973. pp. 3–33.

72. Epstein A.N. *Neuropsychology of Thirst: New Findings and Advances in Concepts.* Winston & Sons; Washington, DC, USA: 1973. Epilogue: Retrospect and prognosis; pp. 315–332.

73. Ramsay D.J., Rolls B.J., Wood R.J. Body fluid changes which influence drinking in the water deprived rat. *J. Physiol.* 1977;266:453–469. doi: 10.1113/jphysiol.1977.sp011777.

74. Denton D., Shade R., Zamarippa F., Egan G., Blair-West J., McKinley M., Lancaster J., Fox P. Neuroimaging of genesis and satiation of thirst and an interoceptor-driven theory of origins of primary consciousness. *Proc. Natl. Acad. Sci. USA.* 1999;96:5304–5309. doi: 10.1073/pnas.96.9.5304.

75. Saker P., Farrell M.J., Adib F.R., Egan G.F., McKinley M.J., Denton D.A. Regional brain responses associated with drinking water during thirst and after its satiation. *Proc. Natl. Acad. Sci. USA.* 2014;111:5379–5384. doi: 10.1073/pnas.1403382111.

76. Li X., Gutierrez D.V., Hanson M.G., Han J., Mark M.D., Chiel H., Hegemann P., Landmesser L.T., Herlitze S. Fast noninvasive activation and inhibition of neural and network activity by vertebrate rhodopsin and green algae channelrhodopsin. *Proc. Natl. Acad. Sci. USA*. 2005;102:17816–17821. doi: 10.1073/pnas.0509030102.

77. Boyden E.S., Zhang F., Bamberg E., Nagel G., Deisseroth K. Millisecond-timescale, genetically targeted optical control of neural activity. *Nat. Neurosci.* 2005;8:1263–1268. doi: 10.1038/nn1525.

78. Deisseroth K. Optogenetics: 10 years of microbial opsins in neuroscience. *Nat. Neurosci.* 2015;18:1213–1225. doi: 10.1038/nn.4091.

79. Robinson M.J., Warlow S.M., Berridge K.C. Optogenetic excitation of central amygdala amplifies and narrows incentive motivation to pursue one reward above another. *J. Neurosci.* 2014;34:16567–16580. doi: 10.1523/JNEUROSCI.2013-14.2014.

80. Oka Y., Ye M., Zuker C.S. Thirst driving and suppressing signals encoded by distinct neural populations in the brain. *Nature.* 2015;520:349. doi: 10.1038/nature14108.

81. Abbott S.B., Machado N.L., Geerling J.C., Saper C.B. Reciprocal control of drinking behavior by median preoptic neurons in mice. *J. Neurosci.* 2016;36:8228–8237. doi: 10.1523/JNEUROSCI.1244-16.2016.

82. Allen W.E., DeNardo L.A., Chen M.Z., Liu C.D., Loh K.M., Fenno L.E., Ramakrishnan C., Deisseroth K., Fenno L.E., Luo L. Thirst-associated preoptic neurons encode an aversive motivational drive. *Science.* 2017;357:1149–1155. doi: 10.1126/science.aan6747.

83. Gizowski C., Bourque C.W. Neurons that drive and quench thirst. *Science.* 2017;357:1092–1093. doi: 10.1126/science.aao5574.

84. Leib D.E., Zimmerman C.A., Poormoghaddam A., Huey E.L., Ahn J.S., Lin Y.-C., Tan C.L., Chen Y., Knight Z.A. The forebrain thirst circuit drives drinking through negative reinforcement. *Neuron.* 2017;96:1272–1281. doi: 10.1016/j.neuron.2017.11.041.

85. Matsuda T., Hiyama T.Y., Niimura F., Matsusaka T., Fukamizu A., Kobayashi K., Noda M. Distinct neural mechanisms for the control of thirst and salt appetite in the subfornical organ. *Nat. Neurosci.* 2017;20:230–241. doi: 10.1038/nn.4463.

86. Zocchi D., Wennemuth G., Oka Y. The cellular mechanism for water detection in the mammalian taste system. *Nat. Neurosci.* 2017;20:927–934. doi: 10.1038/nn.4575.

87. Gizowski C., Zaelzer C., Bourque C.W. Activation of organum vasculosum neurones and water intake in mice by vasopressin neurones in the suprachiasmatic nucleus. *J. Neuroendocrinol.* 2018;30:e12577. doi: 10.1111/jne.12577.

88. Atasoy D., Sternson S.M. Chemogenetic tools for causal cellular and neuronal biology. *Physiol. Rev.* 2018;98:391–418. doi: 10.1152/physrev.00009.2017.

89. Betley J.N., Xu S., Cao Z.F.H., Gong R., Magnus C.J., Yu Y., Sternson S.M. Neurons for hunger and thirst transmit a negative-valence teaching signal. *Nature.* 2015;521:180–185. doi: 10.1038/nature14416.

90. Zimmerman C.A., Leib D.E., Knight Z.A. Neural circuits underlying thirst and fluid homeostasis. *Nat. Rev. Neurosci.* 2017;18:459–469. doi: 10.1038/nrn.2017.71.

91. Augustine V., Gokce S.K., Lee S., Wang B., Davidson T.J., Reimann F., Gribble F., Deisseroth K., Lois C., Oka Y. Hierarchical neural architecture underlying thirst regulation. *Nature.* 2018;555:204–209. doi: 10.1038/nature25488.

92. McKinley M.J., Denton D.A., Ryan P.J., Yao S.T., Stefanidis A., Oldfield B.J. From sensory circumventricular organs to cerebral cortex: Neural pathways controlling thirst and hunger. *J. Neuroendocrinol.* 2019;31:e12689. doi: 10.1111/jne.12689.

93. Zimmerman C.A., Lin Y.C., Leib D.E., Guo L., Huey E.L., Daly G.E., Chen Y., Knight Z.A. Thirst neurons anticipate the homeostatic consequences of eating and drinking. *Nature.* 2016;537:680–684. doi: 10.1038/nature18950.

94. Allen W.E., Chen M.Z., Pichamoorthy N., Tien R.H., Pachitariu M., Luo L., Deisseroth K. Thirst regulates motivated behavior through modulation of brainwide neural population dynamics. *Science.* 2019;364:253. [PubMed]

95. Manery J.F. Water and electrolyte metabolism. *Physiol. Rev.* 1954;34:334–417. doi: 10.1152/physrev.1954.34.2.334.

96. Macknight A.D.C., Leaf A. Regulation of cellular volume. *Physiol. Rev.* 1977;57:510–573. doi: 10.1152/physrev.1977.57.3.510.

97. Paul M., Mehr A.P., Kreutz R. Physiology of local renin-angiotensin systems. *Physiol. Rev.* 2006;86:747–803. doi: 10.1152/physrev.00036.2005.

98. Stevenson R.J., Mahmut M., Rooney K. Individual differences in the interoceptive states of hunger, fullness and thirst. *Appetite.* 2015;95:44–57. doi: 10.1016/j.appet.2015.06.008.

99. Engell D., Hirsch E. Environmental and sensory modulation of fluid intake in humans. In: Ramsay D.J., Booth D., editors. *Thirst Physiological and Psychological Aspects.* Springer; London, UK: 1991. pp. 382–390.

100. Engell D.B., Maller O., Sawka M.N., Francesconi R.N., Drolet L., Young A.J. Thirst and fluid intake following graded hypohydration levels in humans. *Physiol. Behav.* 1987;40:229–236. doi: 10.1016/0031-9384(87)90212-5.

101. Armstrong L.E., Ganio M.S., Klau J.F., Johnson E.C., Casa D.J., Maresh C.M. Novel hydration assessment techniques employing thirst and a water intake challenge in healthy men. *Appl. Physiol. Nutr. Metab.* 2013;39:138–144. doi: 10.1139/apnm-2012-0369.

102. Hubbard R.W., Szlyk P.C., Armstrong L.E. *Fluid Homeostasis During Exercise.* Benchmark Press; Carmel, IN, USA: 1990. Influence of thirst and fluid palatability on fluid ingestion during exercise; pp. 39–95.

103. Szlyk P.C., Sils I.V., Francesconi R.P., Hubbard R.W., Armstrong L.E. Effects of water temperature and flavoring on voluntary dehydration in men. *Physiol. Behav.* 1989;45:639–647. doi: 10.1016/0031-9384(89)90085-1.

104. Rolls B.J., Rolls E.T. *Thirst.* Cambridge University Press; Cambridge, UK: 1982. The control of normal drinking; pp. 152–166.

105. Kissileff H.R. *Neuropsychology of Thirst: New Findings and Advances in Concepts.* Winston & Sons; Washington, DC, USA: 1973. Nonhomeostatic controls of drinking; pp. 163–198.

106. Denton D.A. *The Hunger for Salt: An Anthropological, Physiological and Medical Analysis.* Springer; Berlin, Germany: 1982.

107. Figaro M.K., Mack G.W. Regulation of fluid intake in dehydrated humans: Role of oropharyngeal stimulation. *Am. J. Physiol. Regul. Integr. Comp. Physiol.* 1997;272:R1740–R1746. doi: 10.1152/ajpregu.1997.272.6.R1740.

108. Egan G., Silk T., Zamarripa F., Williams J., Federico P., Cunnington R., Carabott L., Blair-West J., Shade R., McKinley M., et al. Neural correlates of the emergence of consciousness of thirst. *Proc. Natl. Acad. Sci. USA*. 2003;100:15241–15246. doi: 10.1073/pnas.2136650100.

109. Saker P., Farrell M.J., Egan G.F., McKinley M.J., Denton D.A. Overdrinking, swallowing inhibition, and regional brain responses prior to swallowing. *Proc. Natl. Acad. Sci. USA*. 2016;113:12274–12279. doi: 10.1073/pnas.1613929113.

110. Denton D., Shade R., Zamarippa F., Egan G., Blair-West J., McKinley M., Fox P. Correlation of regional cerebral blood flow and change of plasma sodium concentration during genesis and satiation of thirst. *Proc. Natl. Acad. Sci. USA*. 1999;96:2532–2537. doi: 10.1073/pnas.96.5.2532.

111. Parsons L.M., Denton D., Egan G., McKinley M., Shade R., Lancaster J., Fox P.T. Neuroimaging evidence implicating cerebellum in support of sensory/cognitive processes associated with thirst. *Proc. Natl. Acad. Sci. USA*. 2000;97:2332–2336. doi: 10.1073/pnas.040555497.

112. Farrell M.J., Zamarripa F., Shade R., Phillips P.A., McKinley M., Fox P.T., Blair-West J., Denton D.A., Egan G.F. Effect of aging on regional cerebral blood flow responses associated with osmotic thirst and its satiation by water drinking: A PET study. *Proc. Natl. Acad. Sci. USA*. 2008;105:382–387. doi: 10.1073/pnas.0710572105.

113. Bichet D.G. Vasopressin at central levels and consequences of dehydration. *Ann. Nutr. Metab*. 2016;68:19–23. doi: 10.1159/000446200.

114. Nutt R. The history of positron emission tomography. *Mol. Imaging Biol*. 2002;4:11–26. doi: 10.1016/S1095-0397(00)00051-0.

115. Di Chiro G., DeLaPaz R.L., Brooks R.A., Sokoloff L., Kornblith P.L., Smith B.H., Patronas N.J., Kufta C.V., Kessler R.M., Johnston G.S., et al. Glucose utilization of cerebral gliomas measured by [^{18}F] fluorodeoxyglucose and positron emission tomography. *Neurology*. 1982;32:1323–1329. doi: 10.1212/WNL.32.12.1323.

116. Dimitrakopoulou-Strauss A., Strauss L.G., Burger C. Quantitative PET studies in pretreated melanoma patients: A comparison of 6-[^{18}F] fluoro-$_{L}$-dopa with ^{18}F-FDG and ^{15}O-water using compartment and noncompartment analysis. *J. Nucl. Med*. 2001;42:248–256. [PubMed]

117. Muehllehner G., Karp J.S. Positron emission tomography. *Phys. Med. Biol.* 2006;51:R117–R137. doi: 10.1088/0031-9155/51/13/R08.

118. Moroz M.A., Zanzonico P., Lee J.T., Ponomarev V. Ex vivo radiolabeling and in vivo PET imaging of T cells expressing nuclear reporter genes. In: Dubey P., editor. *Methods in Molecular Biology.* Humana Press; Clifton, NJ, USA: 2018. pp. 153–164. [PubMed]

119. Buxton R.B. *Introduction to Functional Magnetic Resonance Imaging: Principles and Techniques.* Cambridge University Press; Cambridge, UK: 2009. pp. 65–116, 339–424.

120. Logothetis N.K. The underpinnings of the BOLD functional magnetic resonance imaging signal. *J. Neurosci.* 2003;23:3963–3971. doi: 10.1523/JNEUROSCI.23-10-03963.2003.

121. Farrell M.J., Egan G.F., Zamarripa F., Shade R., Blair-West J., Fox P., Denton D.A. Unique, common, and interacting cortical correlates of thirst and pain. *Proc. Natl. Acad. Sci. USA.* 2006;103:2416–2421. doi: 10.1073/pnas.0511019103.

122. De Araujo I.E., Kringelbach M.L., Rolls E.T., McGlone F. Human cortical responses to water in the mouth, and the effects of thirst. *J. Neurophysiol.* 2003;90:1865–1876. doi: 10.1152/jn.00297.2003.

123. Becker C.A., Flaisch T., Renner B., Schupp H.T. From thirst to satiety: The anterior mid-cingulate cortex and right posterior insula indicate dynamic changes in incentive value. *Front. Hum. Neurosci.* 2017;11:1–9. doi: 10.3389/fnhum.2017.00234.

124. Farrell M.J., Bowala T.K., Gavrilescu M., Phillips P.A., McKinley M.J., McAllen R.M., Denton D.A., Egan G.F. Cortical activation and lamina terminalis functional connectivity during thirst and drinking in humans. *Am. J. Physiol. Regul. Integr. Comp. Physiol.* 2011;301:R623–R631. doi: 10.1152/ajpregu.00817.2010.

125. Pardo J.V., Wood T.D., Costello P.A., Pardo P.J., Lee J.T. PET study of the localization and laterality of lingual somatosensory processing in humans. *Neurosci. Lett.* 1997;234:23–26. doi: 10.1016/S0304-3940(97)00650-2.

126. Martin R.E., MacIntosh B.J., Smith R.C., Barr A.M., Stevens T.K., Gati J.S., Menon R.S. Cerebral areas processing swallowing and tongue movement are overlapping but distinct: A functional magnetic resonance imaging study. *J. Neurophysiol.* 2004;92:2428–2493. doi: 10.1152/jn.01144.2003.

127. Malandraki G.A., Sutton B.P., Perlman A.L., Karampinos D.C., Conway C. Neural activation of swallowing and swallowing-related tasks in healthy young adults: An attempt to separate the components of deglutition. *Hum. Brain Mapp.* 2009;30:3209–3226. doi: 10.1002/hbm.20743.

128. Zald D.H., Pardo J.V. The functional neuroanatomy of voluntary swallowing. *Ann. Neurol.* 1999;46:281–286. doi: 10.1002/1531-8249(199909)46:3<281::AID-ANA2>3.0.CO;2-L.

129. Lowell S.Y., Reynolds R.C., Chen G., Horwitz B., Ludlow C.L. Functional connectivity and laterality of the motor and sensory components in the volitional swallowing network. *Exp. Brain Res.* 2012;219:85–96. doi: 10.1007/s00221-012-3069-9.

130. Kern M.K., Jaradeh S., Arndorfer R.C., Shaker R. Cerebral cortical representation of reflexive and volitional swallowing in humans. *Am. J. Physiol. Gastrointest. Liver Physiol.* 2001;280:G354–G360. doi: 10.1152/ajpgi.2001.280.3.G354.

131. Zald D.H., Pardo J.V. Cortical activation induced by intraoral stimulation with water in humans. *Chem. Senses.* 2000;25:267–275. doi: 10.1093/chemse/25.3.267.

132. Hamdy S., Rothwell J.C., Brooks D.J., Bailey D., Aziz Q., Thompson D.G. Identification of the cerebral loci processing human swallowing with $H_2^{15}O$ PET activation. *J. Neurophysiol.* 1999;81:1917–1926. doi: 10.1152/jn.1999.81.4.1917.

133. Kinomura S., Kawashima R., Yamada K., Ono S., Itoh M., Yoshioka S., Yamaguchi T., Matsui H., Miyazawa H., Itoh H., et al. Functional anatomy of taste perception in the human brain studied with positron emission tomography. *Brain Res.* 1994;659:263–266. doi: 10.1016/0006-8993(94)90890-7.

134. Small D.M., Jones-Gotman M., Zatorre R.J., Petrides M., Evans A.C. Flavor processing: More than the sum of its parts. *Neuroreport.* 1997;8:3913–3917. doi: 10.1097/00001756-199712220-00014.

135. Kobayakawa T., Ogawa H., Kaneda H., Ayabe-Kanamura S., Saito S. Spatio-temporal analysis of cortical activity evoked by gustatory stimulation in humans. *Chem. Senses.* 1999;24:201–209. doi: 10.1093/chemse/24.2.201.

136. Cerf-Ducastel B., Van de Moortele P.F., MacLeod P., Le Bihan D., Faurion A. Interaction of gustatory and lingual somatosensory perceptions at the cortical level in the human: A functional magnetic resonance imaging study. *Chem. Senses.* 2001;26:371–383. doi: 10.1093/chemse/26.4.371.

137. Kringelbach M.L., de Araujo I.E., Rolls E.T. Taste-related activity in the human dorsolateral prefrontal cortex. *Neuroimage.* 2004;21:781–788. doi: 10.1016/j.neuroimage.2003.09.063.

138. Rolls B.J., Wood R.J., Stevens R.M. Effects of palatability on body fluid homeostasis. *Physiol. Behav.* 1978;20:15–19. doi: 10.1016/0031-9384(78)90196-8.

139. Booth D.A. Influences on human fluid consumption. In: Ramsay D.J., Booth D., editors. *Thirst Physiological and Psychological Aspects.* Springer; London, UK: 1991. pp. 53–75.

140. Sporns O., Tononi G., Kötter R. The human connectome: A structural description of the human brain. *PLoS Comput. Biol.* 2005;1:e42. doi: 10.1371/journal.pcbi.0010042.

141. Mosier K., Bereznaya I. Parallel cortical networks for volitional control of swallowing in humans. *Exp. Brain Res.* 2001;140:280–289. doi: 10.1007/s002210100813.

142. van Rijn I., de Graaf C., Smeets P.A. It's in the eye of the beholder: Selective attention to drink properties during tasting influences brain activation in gustatory and reward regions. *Brain Imaging Behav.* 2018;12:425–436. doi: 10.1007/s11682-017-9710-2.

143. Graebner A.K., Iyer M., Carter M.E. Understanding how discrete populations of hypothalamic neurons orchestrate complicated behavioral states. *Front. Syst. Neurosci.* 2015;9:111. doi: 10.3389/fnsys.2015.00111.

144. Haber S.N., Calzavara R. The cortico-basal ganglia integrative network: The role of the thalamus. *Brain Res. Bull.* 2009;78:69–74. doi: 10.1016/j.brainresbull.2008.09.013.

145. Heilbronner S.R., Hayden B.Y. Dorsal anterior cingulate cortex: A bottom-up view. *Annu. Rev. Neurosci.* 2016;39:149–170. doi: 10.1146/annurev-neuro-070815-013952.

146. Tagaris G.A., Richter W., Kim S.G., Pellizzer G., Andersen P., U-urbil K., Georgopoulos A.P. Functional magnetic resonance imaging of mental rotation and memory scanning: A multidimensional scaling

analysis of brain activation patterns. *Brain Res. Rev.* 1998;26:106–112. doi: 10.1016/S0165-0173(97)00060-X.

147. Goense J., Bohraus Y., Logothetis N.K. fMRI at high spatial resolution: Implications for BOLD-models. *Front. Comput. Neurosci.* 2016;10:66–79. doi: 10.3389/fncom.2016.00066.
148. Cabeza R., Nyberg L. Imaging cognition II: An empirical review of 275 PET and fMRI studies. *J. Cogn. Neurosci.* 2000;12:1–47. doi: 10.1162/08989290051137585.
149. Johnson A.K., Thunhorst R.L. The neuroendocrinology of thirst and salt appetite: Visceral sensory signals and mechanisms of central integration. *Front. Neuroendocrinol.* 1997;18:292–353. doi: 10.1006/frne.1997.0153.
150. Leech R., Kamourieh S., Beckmann C.F., Sharp D.J. Fractionating the default mode network: Distinct contributions of the ventral and dorsal posterior cingulate cortex to cognitive control. *J. Neurosci.* 2011;31:3217–3224. doi: 10.1523/JNEUROSCI.5626-10.2011.
151. Gizowski C., Bourque C.W. Hypothalamic neurons controlling water homeostasis: it's about time. *Curr. Opin. Physiol.* 2018;5:45–50. doi: 10.1016/j.cophys.2018.06.006.
152. Hollis J.H., McKinley M.J., D'Souza M., Kampe J., Oldfield B.J. The trajectory of sensory pathways from the lamina terminalis to the insular and cingulate cortex: A neuroanatomical framework for the generation of thirst. *Am. J. Physiol. Regul. Integr. Comp. Physiol.* 2008;294:R1390–R1401. doi: 10.1152/ajpregu.00869.2007.
153. Quill E. When networks network: Once studied solo, systems display surprising behavior when they interact. *Sci. News.* 2012;182:18–25. doi: 10.1002/scin.5591820619.
154. Deisseroth K., Feng G., Majewska A.K., Miesenböck G., Ting A., Schnitzer M.J. Next-generation optical technologies for illuminating genetically targeted brain circuits. *J. Neurosci.* 2006;26:10380–10386. doi: 10.1523/JNEUROSCI.3863-06.2006.
155. Pastrana E. Optogenetics: Controlling cell function with light. *Nat. Methods.* 2010;8:24. doi: 10.1038/nmeth.f.323.
156. Whissell P.D., Tohyama S., Martin L.J. The use of DREADDs to deconstruct behavior. *Front. Genet.* 2016;7:70. doi: 10.3389/fgene.2016.00070.

157. Nation H.L., Nicoleau M., Kinsman B.J., Browning K.N., Stocker S.D. DREADD-induced activation of subfornical organ neurons stimulates thirst and salt appetite. *Am. J. Physiol. Heart Circ. Physiol.* 2016;115:3123–3129. doi: 10.1152/jn.00149.2016.

158. Weisinger R.S. *Control Mechanisms of Drinking.* Springer; Berlin, Germany: 1975. Conditioned and pseudoconditioned thirst and sodium appetite; pp. 148–154.

159. Barker L.M., Best M.R., Domjan M.P., editors. *Learning Mechanisms in Food Selection.* Baylor University Press; Waco, TX, USA: 1977.

160. Bolles R.C. Toy rats and real rats: Nonhomeostatic plasticity in drinking. *Behav. Brain Sci.* 1979;2:103. doi: 10.1017/S0140525X00061070.

161. Rolls E.T., Rolls B.J., Rowe E.A. Sensory-specific and motivation-specific satiety for the sight and taste of food and water in man. *Physiol. Behav.* 1983;30:185–192. doi: 10.1016/0031-9384(83)90003-3.

162. Rowland N. *Analysis of Motivational Processes.* Academic Press; London, UK: 1980. Drinking behavior: Physiological, neurological and environmental factors; pp. 39–59.

163. Bolles R.C. *Analysis of Motivational Processes.* Academic Press; London, UK: 1980. Some functionalistic thoughts about regulation; pp. 63–75.

164. Aarts H., Dijksterhuis A., De Vries P. On the psychology of drinking: Being thirsty and perceptually ready. *Br. J. Psychol.* 2001;92:631–642. doi: 10.1348/000712601162383.

165. Armbruster B.N., Li X., Pausch M.H., Herlitze S., Roth B.L. Evolving the lock to fit the key to create a family of G protein-coupled receptors potently activated by an inert ligand. *Proc. Natl. Acad. Sci. USA.* 2007;104:5163–5168. doi: 10.1073/pnas.0700293104.

166. Becker C.A., Schmälzle R., Flaisch T., Renner B., Schupp H.T. Thirst and the state-dependent representation of incentive stimulus value in human motive circuitry. *Soc. Cogn. Affect. Neurosci.* 2015;10:1722–1729. doi: 10.1093/scan/nsv063.

167. Bourque C.W. Central mechanisms of osmosensation and systemic osmoregulation. *Nat. Rev. Neurosci.* 2008;9:519–531. doi: 10.1038/nrn2400.

168. McCormick S.D., Bradshaw D. Hormonal control of salt and water balance in vertebrates. *Gen. Comp. Endocrinol.* 2006;147:3–8. doi: 10.1016/j.ygcen.2005.12.009.

169. Walsh L.L. Differences in food, water, and food-deprivation water intake in 16 strains of rats. *J. Comp. Physiol. Psychol.* 1980;94:775. doi: 10.1037/h0077698.

170. Van Essen D.C., Donahue C.J., Glasser M.F. Development and evolution of cerebral and cerebellar cortex. *Brain Behav. Evol.* 2018;91:158–169. doi: 10.1159/000489943.

171. McKinley M.J., Denton D.A., Oldfield B.J., De Oliveira L.B., Mathai M.L. Water intake and the neural correlates of the consciousness of thirst. *Semin. Nephrol.* 2006;26:249–257. doi: 10.1016/j.semnephrol.2006.02.001.

172. Epstein A.N. Thirst and salt intake: A personal review and some suggestions. In: Ramsay D.J., Booth D., editors. *Thirst Physiological and Psychological Aspects.* Springer; London, UK: 1991. pp. 481–501.

173. Cova L., Armentero M.T. IntechOpen.com.; 2011. [(accessed on 17 October 2019)]. 1980–2011: Parkinson's Disease and Advance in Stem Cell Research. Towards New Therapies for Parkinson's Disease. Available online: https://www.intechopen.com/books/towards-new-therapies-for-parkinson-s-disease/1980-2011-parkinson-s-disease-and-advance-in-stem-cell-research.

174. Vecchio L.M., Meng Y., Xhima K., Lipsman N., Hamani C., Aubert I. The neuroprotective effects of exercise: Maintaining a healthy brain throughout aging. *Brain Plast.* 2018;4:17–52. doi: 10.3233/BPL-180069.

175. Preuss T.M. Who's afraid of Homo sapiens? *J. Biomed. Discov. Collab.* 2006;1:17. doi: 10.1186/1747-5333-1-17.

176. Miller J.A., Horvath S., Geschwind D.H. Divergence of human and mouse brain transcriptome highlights Alzheimer disease pathways. *Proc. Natl. Acad. Sci. USA.* 2010;107:12698–12703. doi: 10.1073/pnas.0914257107.

177. Institute of Medicine, Panel on Dietary Reference Intakes for Electrolytes and Water . *DRI, Dietary Reference Intakes for Water, Potassium, Sodium, Chloride, and Sulfate.* National Academy Press; Washington, DC, USA: 2005.

178. Perrier E., Vergne S., Klein A., Poupin M., Rondeau P., Le Bellego L., Armstrong L.E., Lang F., Stookey J., Tack I. Hydration biomarkers in free-living adults with different levels of habitual fluid consumption. *Br. J. Nutr.* 2013;109:1678–1687. doi: 10.1017/S0007114512003601.

179. Kavouras S.A. Hydration, dehydration, underhydration, optimal hydration: Are we barking up the wrong tree? *Eur. J. Nutr.* 2019;58:471–473. doi: 10.1007/s00394-018-01889-z.

180. Johnson E.C., Munoz C.X., Jimenez L., Le Bellego L., Kupchak B.R., Kraemer W.J., Casa D.J., Maresh C.M., Armstrong L.E. Hormonal and Thirst Modulated Maintenance of Fluid Balance in Young Women with Different Levels of Habitual Fluid Consumption. *Nutrients.* 2016;8:302. doi: 10.3390/nu8050302.

181. Enhörning S., Bankir L., Bouby N., Struck J., Hedblad B., Persson M., Morgenthaler N.G., Nilsson P.M., Melander O. Copeptin, a marker of vasopressin, in abdominal obesity, diabetes and microalbuminuria: The prospective Malmö Diet and Cancer Study cardiovascular cohort. *Int. J. Obes.* 2013;37:598–603. doi: 10.1038/ijo.2012.88.

182. Enhörning S., Hedblad B., Nilsson P.M., Engström G., Melander O. Copeptin is an independent predictor of diabetic heart disease and death. *Am. Heart J.* 2015;169:549–556.e1. doi: 10.1016/j.ahj.2014.11.020.

183. Enhörning S., Wang T.J., Nilsson P.M., Almgren P., Hedblad B., Berglund G., Struck J., Morgenthaler N.G., Bergmann A., Lindholm E., et al. Plasma copeptin and the risk of diabetes mellitus. *Circulation.* 2010;121:2102–2108. doi: 10.1161/CIRCULATIONAHA.109.909663.

184. Wannamethee S.G., Welsh P., Papacosta O., Lennon L., Whincup P.H., Sattar N. Copeptin, insulin resistance and risk of incident diabetes in older men. *J. Clin. Endocrinol. Metab.* 2015;100:3332–3339. doi: 10.1210/JC.2015-2362.

185. Roussel R., Fezeu L., Bouby N., Balkau B., Lantieri O., Alhenc-Gelas F., Marre M., Bankir L. Low water intake and risk for new-onset hyperglycemia. *Diabetes Care.* 2011;34:2551–2554. doi: 10.2337/dc11-0652.

186. Roussel R., Velho G., Bankir L. Vasopressin and diabetic nephropathy. *Curr. Opin. Nephrol. Hypertens.* 2017;26:1–318. doi: 10.1097/MNH.0000000000000335.

187. Roussel R., El Boustany R., Bouby N., Potier L., Fumeron F., Mohammedi K., Balkau B., Tichet J., Bankir L., Marre M., et al. Plasma copeptin, AVP gene variants, and incidence of type 2 diabetes in a cohort from the community. *J. Clin. Endocrinol. Metab.* 2016;101:2432–2439. doi: 10.1210/jc.2016-1113.

188. Taveau C., Chollet C., Waeckel L., Desposito D., Bichet D.G., Arthus M.F., Magnan C., Philippe E., Paradis V., Foufelle F., et al. Vasopressin and hydration play a major role in the development of glucose intolerance and hepatic steatosis in obese rats. *Diabetologia.* 2015;58:1081–1090. doi: 10.1007/s00125-015-3496-9.

189. Jansen L.T., Suh H., Adams J.D., Sprong C.A., Seal A.D., Scott D.M., Butts C.L., Melander O., Kirkland T.W., Vanhaecke T., et al. Osmotic stimulation of vasopressin acutely impairs glucose regulation: A counterbalanced, crossover trial. *Am. J. Clin. Nutr.* 2019 doi: 10.1093/ajcn/nqz236.

190. Carroll H.A., Betts J.A., Johnson L. An investigation into the relationship between plain water intake and glycated Hb (HbA1c): A sex-stratified, cross-sectional analysis of the UK National Diet and Nutrition Survey (2008–2012) *Br. J. Nutr.* 2016;116:1–11. doi: 10.1017/S0007114516003688.

191. Carroll H.A., Davis M.G., Papadaki A. Higher plain water intake is associated with lower type 2 diabetes risk: A cross-sectional study in humans. *Nutr. Res.* 2015;35:865–872. doi: 10.1016/j.nutres.2015.06.015.

192. Johnson E.C., Bardis C.N., Jansen L.T., Adams J.D., Kirkland T.W., Kavouras S.A. Reduced water intake deteriorates glucose regulation in patients with type 2 diabetes. *Nutr. Res.* 2017;43:25–32. doi: 10.1016/j.nutres.2017.05.004.

193. Clark W.F., Devuyst O., Roussel R. The vasopressin system: New insights for patients with kidney diseases. *J. Intern. Med.* 2017;282:310–321. doi: 10.1111/joim.12654.

194. Shackman A.J., Salomons T.V., Slagter H.A., Fox A.S., Winter J.J., Davidson R.J. The integration of negative affect, pain and cognitive control in the cingulate cortex. *Nat. Rev. Neurosci.* 2011;12:154. doi: 10.1038/nrn2994.

195. Walton M.E., Mars R.B. Probing human and monkey anterior cingulate cortex in variable environments. *Cogn. Affect. Behav. Neurosci.* 2007;7:413–422. doi: 10.3758/CABN.7.4.413.

196. Kolling N., Behrens T.E.J., Wittmann M.K., Rushworth M.F.S. Multiple signals in anterior cingulate cortex. *Curr. Opin. Neurobiol.* 2016;37:36–43. doi: 10.1016/j.conb.2015.12.007.

197. Armstrong L.E., Casa D.J., Emmanuel H., Ganio M.S., Klau J.F., Lee E.C., Maresh C.M., McDermott B.P., Stearns R.L., Vingren J.L., et al. Nutritional, physiological, and perceptual responses during a summer ultraendurance cycling event. *J. Strength Cond. Res.* 2012;26:307–318. doi: 10.1519/JSC.0b013e318240f677.

198. Sanders L. Flex time: The brain's ability to shift connections might ease learning. *Sci. News.* 2017;192:22–25.

Chapter 6

Water Quality and Brain Function

Stephen C. Bondy[1] and Arezoo Campbell[2]

[1]Center for Occupational and Environmental Health, Department of Medicine, University of California, Irvine, CA 92617-1830, USA
[2]Department of Pharmaceutical Sciences, Western University of Health Sciences, Pomona, CA 91766-1854, USA

ABSTRACT

In the United States, regulations are in place to ensure the quality of drinking water. Such precautions are intended to safeguard the health of the population. However, regulatory guidelines may at times fail to achieve their purpose. This may be due to lack of sufficient data regarding the health hazards of chronic low dose exposure to contaminants or the introduction of new substances that pose a health hazard risk that has yet to be identified.

Citation: Water Quality and Brain Function. Stephen C Bondy, and Arezoo Campbell. Int. J. Environ. Res. Public Health 2018, 15, 2; doi:10.3390/ijerph15010002.

In this review, examples of different sources of contaminants in drinking water will be discussed, followed by an evaluation of some select individual toxicants with known adverse neurological impact. The ability of mixtures to potentially cause additive, synergistic, or antagonistic neurotoxic responses will be briefly addressed. The last section of the review will provide examples of select mechanisms by which different classes of contaminants may lead to neurological impairments. The main objective of this review is to bring to light the importance of considering trace amounts of chemicals in the drinking water and potential brain abnormalities. There is continued need for toxicology studies to better understand negative consequences of trace amounts of toxins and although it is beyond the scope of this brief overview it is hoped that the review will underscore the paucity of studies focused on determining how long-term exposure to minute levels of contaminants in drinking water may pose a significant health hazard.

Keywords: drinking water quality, reservoirs, neurotoxic metals, neurotoxic organics, public health, brain function

INTRODUCTION

The purity of drinking water forms a critical basis for overall considerations of public health since consumption involves the whole population. Although stringent regulations are in place to ensure that drinking water does not contain harmful chemicals, there are trace amounts of impurities. The overall quality of the drinking water is dependent on the source and purification processes used. Contamination of water supplies rarely entails acute evidence of poisoning but rather encompasses a gradual and progressive impairment of health. This is due to chronic low-dose exposure that leads to bioaccumulation of water-soluble amphiphilic environmental toxins that can also be concentrated in lipid deposits. Once a threshold concentration is reached, cellular dysfunction may ensue. Thus, exposure to trace amounts of water contaminants may pose a risk for increased incidence of slow progressing diseases that are generally associated with aging or impaired development. Many of these disorders involve the central nervous system (CNS), which is especially susceptible to damage during development and then with acceleration of neurodegenerative changes during aging.

The adverse consequences of environmental contaminants that are present in drinking water are based on the developmental phase of organisms. For instance, in utero exposures that cause impaired mitochondrial function can lead to insufficient energy production. Such disruption of normal cellular activity may in turn interrupt proper CNS growth. Another early consequence of exposure to contaminants may be abnormalities in regulated gene expression necessary for appropriate development. Genetic imprinting may also occur during early CNS growth that do not immediately manifest changes in function. However, later in the lifespan of an individual, when aging processes work in concert with these early genetic modifications, predisposition to disease may be enhanced. Exposure during later stages of lifespan may compound age-associated deterioration of CNS performance. For instance, an increase in both oxidative stress and inflammation is observed in senescence. As reviewed in this article, many of the contaminants present in drinking water increase one or both of these processes, which are thought to be key contributors to neurodegenerative disorders. The focus of this review is mainly on adverse consequences in the adult brain rather than direct changes to the fetal CNS.

SOURCES OF CONTAMINATION OF DRINKING WATER

Industrial Waste

Many industrial processes use large quantities of water, and the runoff from these has the potential to pollute residential water supplies. Chemicals reaching water supplies in this manner include both inorganic salts and organic compounds. Recently, hydraulic fracturing (fracking), a method used to extract shale gas, has had a major impact on the quality of water leaving sites of this activity. Following drilling, there is a potential that hydrocarbons such as methane, ethane, and propane may leak and contaminate shallow groundwater. Surface water contamination can also occur from the injected hydraulic fracturing fluid, which contains a variety of chemicals including acids, surfactants, acrylic polymers, and borate compounds, amongst other factors [1]. The presence of neurotoxic acrylamide in water supplies, due to breakdown of polyacrylamide grouting agents, is generally overshadowed by the much greater amounts found in cooked foods and coffee [2] and thus is not discussed here.

Agricultural Runoff

Runoff of materials applied to agricultural crops that often end up in water sources, include pesticide and fertilizer residues. Irrigation with polluted water containing a range of agrochemicals are dominant sources of metals and organic compounds in agricultural soils [3,4]. In addition, livestock farming can add animal wastes and bacteria to the water effluent from fields. Finally, run-off from garden chemicals can make a contribution to the water burden of contaminants. One example of an agricultural contaminant in water sources is nitrate. Nitrate is commonly found because of the prevalent use of nitrogen fertilizers [5]. Neural tube defects have been shown to be four times greater in the progeny of women whose public water supply contained nitrate above the US maximum contaminant level [6].

A large number of pesticides have neurotoxic properties [7]. These are often present in combinations with unknown synergistic or antagonistic consequences. Maximum water concentration levels (MCVs) and regulatory guidance levels (RGVs) of many pesticides vary by several orders of magnitude. Furthermore, these values often exceed the calculated upper values for human health risk allowing for uncertainty. Many implied dose limits are above tolerable daily intake values. Despite the fact that worldwide jurisdictions are making efforts to regulate pesticide levels in water and also in other sources, as yet, current worldwide regulatory mechanisms related to drinking water do not provide safe standards in a manner so as to protect public health. Much improvement in this area is clearly needed [8,9].

Water Treatment

Another threat to the quality of water stores ironically comes from procedures intended to improve the potability of water. Formation of a biofilm in drinking water distribution systems allows microbial growth [10]. One method of disinfection is to use of chlorine and its derivatives. Interaction of these agents with trace amounts of organic compounds in water can lead to formation of stable and potentially harmful chlorinated organics. Chlorinated wastewater has been shown to be cytotoxic to mammalian cells [11]. A combination of chlorine and ammonia, used for disinfecting water causes erosion of copper piping that may increase the content of the metal in the drinking water. The impact of chlorine on corroding copper pipes is pH dependent [12]. Another potentially harmful material added to water in order to precipitate organic matter and thus to clarify water involves the application of alumina as a means of coagulating particulate contaminants. This can increase the aluminum content of water [13]. The increasing prevalence of acid rain leads to aluminum leaching from rocks and can thus further elevate the aluminum burden of water resources. Epidemiological evidence reveals that aluminum levels in drinking water are related to the incidence of Alzheimer's disease (AD) [14]. This is supported by laboratory data with experimental animals exposed to levels of aluminum in drinking water that parallel those found in some residential supplies [15,16]. However, the issue of a causal relation of Al in water and the promotion of AD remains controversial [17].

A recent study evaluated how exposure to disinfection by-products found in tap water during pregnancy relate to adverse neurodevelopmental consequences of the child. Although there was a positive correlation between exposure to disinfection by-products and mental score for girls at one year of age, the difference did not persist when the evaluation was conducted at 4–5 years of age [18]. Thus, more studies are needed to better understand if and how trace amounts of disinfectants in the tap water may have neurodevelopmental or neurodegenerative consequences.

Water Conduits

Water delivery through metal-lined conduits can be another factor that can impair the quality of the final product emerging from household taps. While the use of lead piping has been greatly curtailed in recent years, lead soldering is still a popular means of annealing piping. The evidence that even low levels of lead is a developmental hazard to human populations is unambiguous [19]. In addition, the use of copper tubing is currently widespread. Newly emerging data involving both epidemiological reports and studies on experimental animals are increasingly indicating that water-borne copper can also be a source of neurotoxicity [20,21,22,23].

Consumer Products

Pharmaceutical agents can enter the aquatic environment and eventually the drinking water [24,25]. According to the Center for Disease Control and Prevention (CDC), 48.7% of the population uses at least one prescription drug every 30 days and some of these chemicals gradually get into the water sources. Water from public drinking water supply wells on Cape Cod frequently contained levels of the antibiotic sulfamethoxazole, the anticonvulsant phenytoin, and the surfactant perfluorooctane sulfonate at high levels [26].

In the aquatic environment, pharmacological agents may have adverse effects. Environmentally relevant doses of the anxiolytic drug oxazepam can alter the behavior of fish [27]. Thus, the presence of pharmacological agents in water supplies can lead to adverse consequences to aquatic life and eventually to human health. Metals can enter the water supply as a result of the improper disposal of lead containing batteries and of brake pads, which are high in zinc and copper. A number of metals such as cadmium, zinc, lead, and copper, present in brakes and tires, can also contaminate storm water runoff from roadways [28].

CONSIDERATION OF NEUROTOXIC POTENTIAL OF INDIVIDUAL CONTAMINANTS

Metals

Lead (Pb)

The seepage of lead from industrial wastes into drinking water is an old problem. This is compounded by the existence of significant amounts of lead in some waters, especially those from low-lying areas [4,29]. Whereas lead toxicity was once believed to involve only a small population of industrially exposed workers, the neurotoxicity of very low levels of lead that can be relevant to large populations, has been widely recognized for the last 20 years. This adverse effect is especially pronounced in the developing fetal and neonatal brain. Low-dose exposure to lead has been associated with behavioral and cognitive impairments [30].

One of the more recent incidents involving lead exposure through the drinking water, is from Flint, Michigan. In an effort to cut costs, the city of Flint switched its water source from the Detroit Water and Sewage Department to Flint River water treated at Flint Water Service Center [31]. Many water samples tested subsequently contained lead above the safe level [32]. Blood lead levels in children were significantly increased [33]. In California, contamination from a lead battery recycling plant has led to community concerns regarding potential neurological impacts [34]. An analysis of blood lead levels in children less than six years old revealed that residential proximity to the plant had an impact on these levels [35].

The potential for lead to cause brain abnormalities is not limited to neurodevelopment. A hallmark of Alzheimer's disease is senile plaques composed of amyloid beta (Aβ) protein derived from the amyloid precursor protein (APP). Prenatal lead exposure may have delayed long-term consequences in that it can influence Aβ-related biological pathways that have been implicated in Alzheimer's disease [36]. Early gene imprinting by environmental lead exposure may subsequently enhance expression of genes associated with neurodegenerative disorders [37]. In monkeys, exposure to Pb during an early phase of brain development caused an increase in APP expression and amyloid deposition in later life [38,39].

Aluminum (Al)

The prevalence of Al in drinking water varies considerably [40]. The metal is often added to water as a coagulant of organic matter and the residual levels that are inadvertently solubilized are considered innocuous. However, increasing evidence from both epidemiological and from laboratory studies, suggests that levels of Al found in some drinking water may be harmful. Chronic exposure to such levels can cause neuroinflammation and oxidative stress in experimental animals [15,21].

Al exposure may potentiate the progression of neurodegenerative disorders such as Alzheimer's disease. This is strengthened by findings showing elevated Al in postmortem AD brain tissue [41,42,43]. Al is found in the cerebral arteries of AD patients [44] suggesting that the metal may disrupt aspects of the blood brain barrier [45]. Epidemiological studies have been able to strengthen this connection between Al ingestion and neurodegenerative disease. The prevalence of AD dose-dependently increased in areas where Al concentrations in the drinking water supply are high [46]. A meta-analysis of nine reports associated elevated Al concentrations with diminished cognitive performance [47]. A complicating factor in evaluation of the effects of Al in drinking water, is evidence from human and animal studies showing that the simultaneous presence of silicic acid appears to be protective against Al toxicity, presumably due to formation of an inert aluminosilicate [48,49]. It is difficult to link one specific exposure to AD since it is a pleiotropic disease with many causes. The positive and negative evidence for Al being a factor in AD has recently been evaluated and summarized [50].

Copper (Cu)

Low concentrations of copper salts in water that were previously considered harmless may also be related to adverse neurological effects. Unlike lead and aluminum, Cu is a trace element that has a biological role as a cofactor in many enzymes. However, because the free forms of Cu are toxic, there are many mechanisms that are in place to keep the metal bound to proteins.

Two hereditary disorders that disrupt Cu homeostasis (Menkes' and Wilson's diseases) illustrate the neurotoxic potential of Cu [51].

Cu in the drinking water has been shown to be more toxic than corresponding concentrations of the metal that are present in food [52]. The levels of the metal in drinking water are governed by the pH and the types of conduits that are used. Households that use Cu tubing in the plumbing system have higher concentrations of the metal due to gradual corrosion. However, the amounts of Cu in tap water are much lower if the water is allowed to run before use [53]. The establishment of regulations concerning copper content of residential water set the maximum standard for copper as 1.3 ppm [54]. Levels of up to 7.8 mg/L can be reached if water remains stagnant in corroding copper pipes [55].

Increased copper ingestion from drinking water can be associated with gastrointestinal symptoms [56]. There are concerns that the metal may also play a causative role in neurodegenerative disorders such as Alzheimer's disease and Parkinson's disease (PD). These concerns are based on the findings that the non-ceruloplasmin bound component of Cu is increased in the serum of AD patients [57] and cerebrospinal fluid of patients with PD [58]. There is evidence from a transgenic mouse model of AD that chronic copper exposure accelerates disease-associated pathology [22,23] and occupational exposure to Cu has been linked to an increased risk of PD [59]. Free Cu has been associated with cognitive decline [60] and this observation strengthens its potential role in AD [61].

Specific Cu-binding sites are present on the amyloid precursor protein [62] as well as the amyloid peptide [63] where the metal attaches [64]. Upon binding, Cu undergoes redox cycling which leads to formation of reactive oxygen intermediates [65,66]. Low-dose exposure to Cu in the drinking water significantly increased markers of oxidative stress in the brain of exposed animals [21] and increased activation of the transcription factor, AP-1 [67]. Since AP-1 is activated by redox status, the mechanism by which Cu modulates the transcription factor is likely related to its stimulation of oxidative processes. Inability of microglia to sequester Cu bound to amyloid plaques may enhance inflammatory events known to be exacerbated in AD [68]. Most recently, levels of Cu as low as 2 μM (a level only 10% of maximal levels of Cu recommended by the EPA) in the drinking water of aged mice, have been found to inhibit the actions of low-density lipoprotein receptor-related protein 1 (LRP1), which is involved in the transport of β-amyloid out of cells. This leads to accumulation of β-amyloid and neuroinflammation within the brain [69]. Overall, this concordance of laboratory and studies of human populations suggests a major, largely unacknowledged hazard.

Thus, Cu in drinking water may have the ability to cause adverse neurological effects through several separate but intertwined mechanisms.

Arsenic and Cadmium

The presence of inorganic arsenic or cadmium in drinking water can present a serious hazard. Chronic arsenic poisoning is found in large parts of Bangladesh and adjoining parts of India, where ground water is severely contaminated with heavy metals [70]. One study of nearly a million subjects in a part of West Bengal reported a prevalence rate of arsenicosis in over 15% of the inhabitants. The highest level of arsenic found in drinking water was over 1300 μg/L and values of over 100 μg/L were common. Peripheral neuropathy was present in 16% of cases [71].

Arsenic exposure has also been linked to neurodevelopmental abnormalities. For example, in a case control study conducted in Bangladesh, mothers with higher arsenic exposure and folate deficiency were found to be at a higher risk of giving birth to an infant with neural tube defects. This was associated with histone modifications, which were taken to imply epigenetic effects [72]. Furthermore, cognitive [73,74] and motor [75] function is lower in children that consume arsenic contaminated water for a prolonged period. This metal may also enhance biological processes associated with neurodegenerative disorders. For example, arsenic promotes accumulation of α-synuclein, a neuronal protein that plays an important role in Parkinson's disease [76]. An epidemiological relation between arsenic exposure and a higher incidence of neurodegenerative disorders has been reported [77]. This may be attributable to the ability of arsenic to promote oxidative stress and inflammation, both of which are associated with neurodegeneration [78].

The hazards of arsenic in drinking water are not confined to third world countries. Some parts of the US, notably Texas and the Great Lakes Basin, have arsenic levels of over 50 μg/L in groundwater [79].

While the neurotoxic aspects of arsenic are overshadowed by its potential as a carcinogen, damage to the developing nervous system poses a grave long-term risk. Similarly, while cadmium exposure is associated with nephrotoxicity, low doses of cadmium in drinking water can promote excess free radical related oxidative events. The brain appears to be more sensitive to such changes than other organs [80]. It is likely that exposure to neonates is more harmful compared to adults [81]. Cadmium levels in drinking water have been found to exceed the permissible limits of the World Health Organization in both Egypt and Iran [82,83]. The hazard posed by seemingly low-level exposures to several heavy metals, such as cadmium, mercury and lead is exacerbated by the tendency of these metals to accumulate in tissues over time.

Organic Materials

Halogenated Residues

Organochlorine pesticides owe their effectiveness to their stability rather than to their reactivity. This durability allows prolonged allosteric interaction with key receptor sites and ion channels. Their non-reactiveness is the same quality that can permit their persistence in aqueous media. Another class of stable organohalogen compounds that has diffused into the environment, includes polychlorinated biphenyls (used as electrical insulators) and polybrominated biphenyls (used as fire retardants). The inadvertently produced dioxins are also in this group. Due to their exceptionally low reactivity, trace amounts of these materials can persist for extended periods in water supplies. Despite their inertness, such compounds can selectively target specific biological sites and impede their function for long periods by allosteric means. Their neurotoxicity is well established [84].

A further presence of organohalogens in drinking water comes from the use of chlorine, bromine or their derivatives such as chloramines and chlorine dioxide as a means of water disinfection. Water sterilization byproducts include trihalomethanes, haloacetic acids, haloacetaldehydes, haloacetonitriles, haloamines, nitrosamines, and halobenzoquinones [85]. While most of these have not been extensively studied, the neurotoxicity of several disinfection byproducts, including dibromacetic acid found in water stocks, has been described [86].

An association between extended exposure to trichloroethylene (TCE) and Parkinson's disease has been reported: the exposure of animals to TCE can lead to striatal degeneration and the onset of parkinsonian characteristics. This has been attributed to mitochondrial dysfunction perhaps due to the formation of chloral from TCE and the subsequent synthesis of 1-trichloromethyl-1,2,3,4-tetrahydro-beta-carboline (TaClo) a specific dopaminergic neurotoxin [87].

Acrylamide

Polyacrylamide is a grouting agent and is also an effective flocculent in water clarification. It is useful in lining pipes, wells, and canals in order to reduce loss of water by leakage. Polyacrylamide is also present in herbicide blends in order to increase viscosity. The degradation of polyacrylamide to the monomeric acrylamide can present a toxic hazard to water supplies. There have been cases of poisoning following consumption of water from recently grouted pipes or wells. These have led to severe neurotoxic effects, including ataxia, hallucinations and memory disturbances [88].

There are several reasons why acrylamide contamination is generally not a major issue. Acrylamide is biodegraded in water, with a half-life of around two days, and it is not bioaccumulated. Also, the acrylamide intake from commonly eaten food is several orders of magnitude greater than that to be expected from water [89].

Bisphenol A

Bisphenol A (BPA) is an important chemical used in large amounts in the production of various plastics including polycarbonates and epoxy resins. In addition to being used in manufacture of bottles and linings of cans containing food products, epoxy resins are also used to line water supply pipes. By acting on estrogen receptors, bisphenol A is a developmental toxin and a carcinogen. It also causes neuronal apoptosis together with metabolic and behavioral changes by mechanisms not involving the estrogen receptor [90]. The neurotoxicity of BPA is controversial since another study (supported by the polycarbonate industry) failed to find evidence of neurotoxicity [91]. While the BPA content of bottled drinking water is a major concern, this chemical has also been found in tap water in several countries, including the US, at levels, that can be greater than that of bottled water [92].

Other Organic Contaminants of Anthropogenic Origin

Innocuous materials such as synthetic musks (for example, 1,3,4,6,7,8-hexahydro-4,6,6,7,8,8-hexamethylcyclopenta[γ]-2-benzopyran) are widely used in the perfume industry. They have been reported to contaminate water supplies in China. Since these are very stable compounds, with very low biodegradability, they are capable of bioaccumulation and thus their potential hazard needs to be carefully evaluated [93,94]. Other agents which find widespread use in personal care products, include surfactants such as 4-nonylphenol. These are also found in urban stream water [94].

4-nonylphenol is known to be neurotoxic and to activate retinoid receptors [95].

Mixtures of Contaminants

While a variety of agents may individually be within the safety levels set by regulatory agencies, mixtures of contaminants may produce a significant overall health hazard. In view of all potential interrelationships between toxicants, the health consequences of such multi-component constituents are hard to predict. For example, tap water has been linked to neural tube defects in mice embryos, and since no single contaminant could be linked to the observation, the authors suggest that the effects may be caused by the combinations of low-dose contaminants [96]. Concurrent exposure to a range of toxic heavy metals, including lead, cadmium, arsenic and methylmercury are of notable concern in view of their persistent effects on the brain. The exact toxicological mechanisms invoked by exposure to such mixtures are still unclear, however they influence many common metabolic pathways that are related to cognitive dysfunction [97]. A further consequence of the presence of several agents in water impacting on related processes is that additive, synergistic, or antagonistic interactions between water contaminants are likely. The complexity of this issue is illustrated by a study of four toxic metals in an isolated cell system. With increasing metal concentration, effects were additive, then synergistic, and finally antagonistic [98]. It is significant that at the lower concentrations studied, which are most likely to reflect the real-life situation, effects were additive or synergistic. Similar additive or synergistic interactions have been found in exposures of intact animals to levels of lead, cadmium, and arsenic that reflect Lowest Observed Effect Levels (LOELs) of each component [99]. There is evidence from neurodevelopmental studies in Bangladesh that arsenic, lead, and manganese can potentiate one another's toxicity. In human populations, lead and manganese appear to interact in a non-additive synergistic manner [100]. Synergistic neurotoxic interactions between metals at levels paralleling those found in some Indian ground waters have also been reported in a more defined animal model system [101].

The question thus arises as to whether the regulatory standards for individual compounds are in fact adequate for the real-world situation where the water content of a single constituent cannot be considered in isolation. This is obviously a complex and difficult issue to address but is relevant to water supplies which are rarely affected by a solitary contaminant.

MECHANISMS OF NEUROTOXICITY

While each of the agents potentially contaminating water supplies discussed above has a characteristic profile of neurotoxicity, there are some effects that are common to many chemicals. This is not surprising, as these events frequently constitute signs of an unhealthy cell with sub-optimal metabolic activity. One such process is an enhancement of inflammatory pathways consequent to activation of innate immune responses. Another frequent occurrence is the presence of excess levels of pro-oxidant activity. Both are common mechanisms associated with neurodegenerative diseases. The pathways by which quality of drinking water may be compromised and thus impact on brain health are summarized in Figure 1.

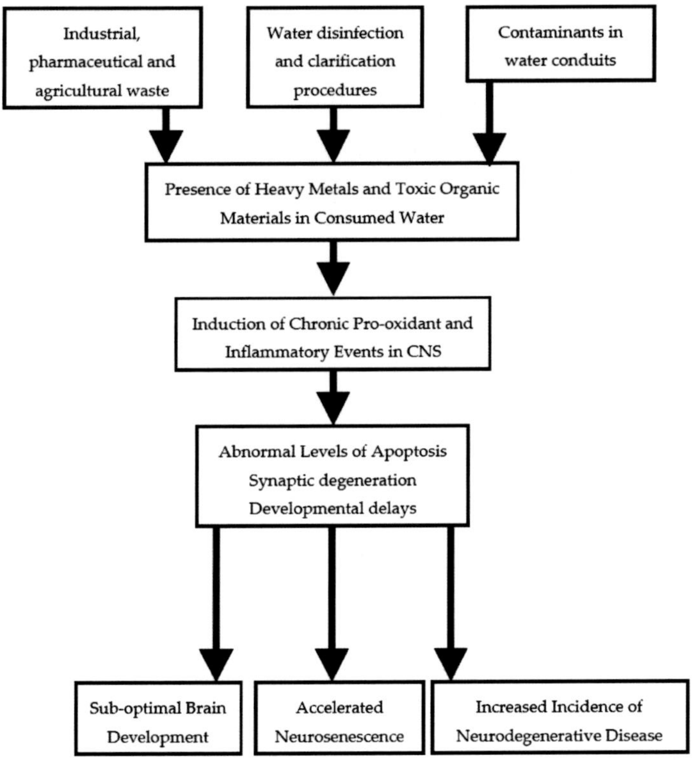

Figure 1: Sources of neurotoxic agents in drinking water and their potential consequences. CNS: central nervous system.

Neuroinflammation

The immune system is composed of a system of cellular and molecular mediators that orchestrate resistance against various insults, including protection against chemicals, which may harm the body. Innate immunity is a first line of defense that uses the inflammatory response to immediately recruit cells to guard against potentially damaging substances. Key participants in inflammation are cytokines and chemokines, which work together in a protective manner [102]. Since the blood brain barrier provides protection for the brain, it was previously thought that the CNS is immunologically privileged. However, it has been shown that this barrier does not confer immune privilege on the brain and that there is active interaction of the CNS with peripheral immune cells [103]. Furthermore, the brain is capable of innate immune activation following environmental insults [15]. Although acute immune responses in the brain that eventually resolve are protective, extended immune activation can occur following a transient toxic insult and this can lead to neurodegeneration [104].

In AD, chronic neuroinflammation is enhanced with disease progression and is considered to be one of the major mechanisms of disease pathology [105,106]. The most important inflammatory cytokines are tumor necrosis factor alpha (TNF-α) and interleukins 1 & 6 (IL-1, IL-6), and levels of these are elevated in AD brains [107]. Inflammatory cytokines can disrupt the integrity of the blood brain barrier resulting in access of environmental toxicants into the brain parenchyma [108].

Oxidative Stress

Reactive oxygen species have a physiological role, and the respiratory burst is a mechanism by which immune cells destroy pathogens [102]. However, abnormal or persistent levels of such oxidant species can be damaging. Contaminants such as redox active metals, once accumulated, may cause neurotoxicity by promotion of free radical formation.

Effective mitochondrial functioning is necessary for maintaining CNS health. This organelle provides energy to actively transport nutrients into the brain and export environmental neurotoxins that may have inadvertently entered the brain. Drinking water contaminants that disrupt mitochondrial function or enhance production of reactive oxygen species may in time exhaust defensive antioxidant processes. The ensuing surge in oxidative stress can lead to abnormal brain function. Indeed, mitochondrial dysfunction and oxidative stress are commonly observed in patients with neurodegenerative disorders such as Alzheimer's disease [109,110,111,112].

CONCLUSIONS

There have been major improvements regarding the quality of residential water supplies. However, there is still considerable variance in different communities, in part due to differing standards. In addition to the World Health Organization and federal government, each state in the US has its individual standards. A unified set of guidelines for drinking water, based on scientific evidence and economic feasibility, would effectively protect global health.

One approach to mitigating the adverse consequences of contaminants in drinking water is to institute green chemistry practices. For instance, the green pharmacy movement is a program aimed at designing drugs that are more biodegradable or improving wastewater treatment by processes such as ozonation, which can further degrade pharmaceutical agents [24,113,114]. Another proposed approach is to enhance planning and technology. An example of this approach would be source separation of wastewater so that "greywater" that is derived from showers or washing machines is collected separately from "blackwater" that is derived from the toilet. The separated wastewater can then be treated more efficiently [115]. Such global developments in technology would limit contaminations and ensure that there will be sufficient clean water to sustain a rapidly developing world population.

REFERENCES

1. Vengosh, A.; Jackson, R.B.; Warner, N.; Darrah, T.H.; Kondash, A. A critical review of the risks to water resources from unconventional shale gas development and hydraulic fracturing in the United States. *Environ. Sci. Technol.* 2014, *48*, 8334–8348.
2. Pérez, H.L.; Osterman-Golkar, S. A sensitive gas chromatographic-tandem mass spectrometric method for detection of alkylating agents in water: Application to acrylamide in drinking water, coffee and snuff. *Analyst* 2003, *128*, 1033–1036.
3. Zhang, L.; Mo, Z.; Qin, J.; Li, Q.; Wei, Y.; Ma, S.; Xiong, Y.; Liang, G.; Qing, L.; Chen, Z.; et al. Change of water sources reduces health risks from heavy metals via ingestion of water, soil, and rice in a riverine area, South China. *Sci. Total Environ.* 2015, *530–531*, 163–170.
4. Islam, M.A.; Romić, D.; Akber, M.A.; Romić, M. Trace metals accumulation in soil irrigated with polluted water and assessment of human health risk from vegetable consumption in Bangladesh. *Environ. Geochem. Health* 2017.
5. Brender, J.D.; Weyer, P.J. Agricultural compounds in water and birth defects. *Curr. Environ. Health Rep.* 2016, *3*, 144–152.
6. Croen, L.A.; Todoroff, K.; Shaw, G.M. Maternal exposure to nitrate from drinking water and diet and risk for neural tube defects. *Am. J. Epidemol.* 2001, *153*, 325–331.
7. Costa, L.G.; Giordano, G.; Guizzetti, M.; Vitalone, A. Neurotoxicity of pesticides: A brief review. *Front. Biosci.* 2008, *13*, 1240–1249.
8. Li, Z.; Jennings, A. Worldwide regulations of standard values of pesticides for human health risk control: A review. *Int. J. Environ. Res. Public Health* 2017, *14*, 826.
9. Li, Z.; Jennings, A.A. Implied maximum dose analysis of standard values of 25 pesticides based on major human exposure pathways. *AIMS Public Health* 2017, *4*, 383–396.
10. Liu, S.; Gunawan, C.; Barraud, N.; Rice, S.A.; Harry, E.J.; Amal, R. Understanding, monitoring, and controlling biofilm growth in drinking water distribution systems. *Environ. Sci. Technol.* 2016, *50*, 8954–8976.
11. Du, Y.; Wu, Q.Y.; Lu, Y.; Hu, H.Y.; Yang, Y.; Liu, R.; Liu, F. Increase of cytotoxicity during wastewater chlorination: Impact factors and surrogates. *J. Hazard. Mater.* 2017, *324*, 681–690.

12. Lytle, D.A.; Liggett, J. Impact of water quality on chlorine demand of corroding copper. *Water Res.* 2016, *92*, 11–21.
13. Gopal, K.; Srivastava, S.B.; Shukla, S.; Bersillon, J.L. Contaminants in drinking water and its mitigation using suitable adsorbents: An overview. *J. Environ. Biol.* 2004, *25*, 469–475.
14. Rondeau, V.; Jacqmin-Gadda, H.; Commenges, D.; Helmer, C.; Dartigues, J.F. Aluminum and silica in drinking water and the risk of Alzheimer's disease or cognitive decline: Findings from 15-year follow-up of the PAQUID cohort. *Am. J. Epidemiol.* 2009, *169*, 489–496.
15. Campbell, A.; Becaria, A.; Lahiri, D.K.; Sharman, K.; Bondy, S.C. Chronic exposure to aluminum in drinking water increases inflammatory parameters selectively in the brain. *J. Neurosci. Res.* 2004, *75*, 565–572.
16. Bondy, S.C. Low levels of aluminum can lead to behavioral and morphological changes associated with Alzheimer's disease and age-related neurodegeneration. *Neurotoxicology* 2016, *52*, 222–229.
17. Santibáñez, M.; Bolumar, F.; García, A.M. Occupational risk factors in Alzheimer's disease: A review assessing the quality of published epidemiological studies. *Occup. Environ. Med.* 2007, *64*, 723–732.
18. Villanueva, C.M.; Gracia-Lavedan, E.; Julvez, J.; Santa-Marina, L.; Lertxundi, N.; Ibarluzea, J.; Llop, S.; Ballester, F.; Fernández-Somoano, A.; Tardón, A.; et al. Drinking water disinfection by-products during pregnancy and child neuropsychological development in the INMA Spanish cohort study. *Environ. Int.* 2018, *110*, 113–122.
19. Liu, J.; Gao, D.; Chen, Y.; Jing, J.; Hu, Q.; Chen, Y. Lead exposure at each stage of pregnancy and neurobehavioral development of neonates. *Neurotoxicology* 2014, *44*, 1–7.
20. Sparks, D.L.; Schreurs, B. Trace amounts of copper in water induce beta-amyloid plaques and learning deficits in a rabbit model of Alzheimer's disease. *Proc. Natl. Acad. Sci. USA* 2003, *100*, 11065–11069.
21. Becaria, A.; Lahiri, D.K.; Bondy, S.C.; Chen, D.; Hamadeh, A.; Li, H.; Taylor, R.; Campbell, A. Aluminum and copper in drinking water enhance inflammatory or oxidative events specifically in the brain. *J. Neuroimmunol.* 2006, *176*, 16–23.
22. Kitazawa, M.; Cheng, D.; Laferla, F.M. Chronic copper exposure exacerbates both amyloid and tau pathology and selectively dysregulates cdk5 in a mouse model of AD. *J. Neurochem.* 2009, *108*, 1550–1560.

23. Kitazawa, M.; Hsu, H.W.; Medeiros, R. Copper exposure perturbs brain inflammatory responses and impairs clearance of amyloid-beta. *Toxicol. Sci.* 2016, *152*, 194–204.
24. Blair, B.D. Potential upstream strategies for the mitigation of pharmaceuticals in the aquatic environment: A brief review. *Curr. Environ. Health Rep.* 2016, *3*, 153–160.
25. Brezina, E.; Prasse, C.; Meyer, J.; Muckter, H.; Ternes, T.A. Investigation and risk evaluation of the occurrence of carbamazepine, oxcarbazepine, their human metabolites and transformation products in the urban water cycle. *Environ. Pollut.* 2017, *225*, 261–269.
26. Schaider, L.A.; Rudel, R.A.; Ackerman, J.M.; Dunagan, S.C.; Brody, J.G. Pharmaceuticals, perfluorosurfactants, and other organic wastewater compounds in public drinking water wells in a shallow sand and gravel aquifer. *Sci. Total Environ.* 2014, *468–469*, 384–393.
27. Brodin, T.; Fisk, J.; Josnsson, M.; Klaminder, J. Dilute concentrations of a psychiatric drug alter behavior of fish from natural populations. *Science* 2013, *339*, 814–815.
28. McKenzie, E.R.; Money, J.E.; Green, P.G.; Young, T.M. Metals associated with stormwater-relevant brake and tire samples. *Sci. Total Environ.* 2009, *407*, 5855–5860.
29. Choudhury, D.; Gupta, S. Impact of waste dump on surface water quality and aquatic insect diversity of Deepor Beel (Ramsar site), Assam, North-east India. *Environ. Monit. Assess.* 2017, *189*, 540.
30. Finkelstein, Y.; Markowitz, M.E.; Rosen, J.F. Low-level lead-induced neurotoxicity in children: An update on central nervous system effects. *Brain Res. Rev.* 1998, *27*, 168–176.
31. Zahran, S.; McElmurry, S.P.; Sadler, R.C. Four phases of the Flint Water Crisis: Evidence from blood lead levels in children. *Environ. Res.* 2017, *157*, 160–172.
32. Edwards, M. Our Sampling of 252 Homes Demonstrates a High Lead in Water Risk: Flint Should Be Failing to Meet the EPA Lead and Copper Rule. 2015. Available online: http://flintwaterstudy.org/2015/09/our-sampling-of-252-homes-demonstrates-a-high-lead-in-water-risk-flint-should-be-failing-to-meet-the-epa-lead-and-copper-rule (accessed on 23 June 2017).
33. Hanna-Attisha, M.; LaChance, J.; Sadler, R.C.; Schnepp, A.C. Elevated blood lead levels in children associated with the flint drinking water

crisis: A spatial analysis of risk and public health response. *Am. J. Public Health* 2016, *106*, 283–290.

34. Barboza, T.; Poston, B. Brain-Damaging Lead Levels near Battery Plant Found as High as 100 Times above Health Limits. 2016. Available online: http://www.latimes.com/local/california/la-me-ln-dangerous-lead-levels-20160714-snap-story.html (accessed on 23 June 2017).
35. Department of Toxic Substances Control (DTSC). An Analysis of Children's Blood Lead Levels in the Area around the Exide Site. 2016. Available online: http://dtsc.ca.gov/HazardousWaste/Projects/upload/An-Analysis-of-Children-s-Blood-Lead-Levels-in-the-Area-Around-the-Exide-Site.pdf (accessed on 23 June 2017).
36. Mazumdar, M.; Xia, W.; Hofmann, O.; Gregas, M.; Ho Sui, S.; Hide, W.; Yang, T.; Needleman, H.L.; Bellinger, D.C. Prenatal lead levels, plasma amyloid β levels, and gene expression in young adulthood. *Environ. Health Perspect.* 2012, *120*, 702–707.
37. Lahiri, D.K.; Maloney, B. The "LEARn" (Latent Early-life Associated Regulation) model integrates environmental risk factors and the developmental basis of Alzheimer's disease, and proposes remedial steps. *Exp. Gerontol.* 2010, *45*, 291–296.
38. Basha, M.R.; Wei, W.; Bakheet, S.A.; Benitez, N.; Siddiqi, H.K.; Ge, Y.W.; Lahiri, D.K.; Zawia, N.H. The fetal basis of amyloidogenesis: Exposure to lead and latent overexpression of amyloid precursor protein and beta-amyloid in the aging brain. *J. Neurosci.* 2005, *25*, 823–829.
39. Wu, J.; Basha, M.R.; Brock, B.; Cox, D.P.; Cardozo-Pelaez, F.; McPherson, C.A.; Harry, J.; Rice, D.C.; Maloney, B.; Chen, D.; et al. Alzheimer's disease (AD)-like pathology in aged monkeys after infantile exposure to environmental metal lead (Pb): Evidence for a developmental origin and environmental link for AD. *J. Neurosci.* 2008, *28*, 3–9.
40. World Health Organization. *Guidelines for Drinking-Water Quality, Second Edition. Volume 2: Health Criteria and Other Supporting Information*; WHO: Geneva, Switzerland, 1996.
41. Perl, D.P.; Brody, A.R. Alzheimer's disease: X-ray spectrometric evidence of aluminum accumulation in neurofibrillary tangle-bearing neurons. *Science* 1980, *208*, 297–299.

42. Xu, N.; Majidi, V.; Markesbery, W.R.; Ehmann, W.D. Brain aluminum in Alzheimer's disease using an improved GFAAS method. *Neurotoxicology* 1992, *13*, 735–743.
43. Mirza, A.; King, A.; Troakes, C.; Exley, C. Aluminium in brain tissue in familial Alzheimer's disease. *J. Trace Elem. Med. Biol.* 2017, *40*, 30–36.
44. Bhattacharjee, S.; Zhao, Y.; Hill, J.M.; Culicchia, F.; Kruck, T.P.; Percy, M.E.; Pogue, A.I.; Walton, J.R.; Lukiw, W.J. Selective accumulation of aluminum in cerebral arteries in Alzheimer's disease (AD). *J. Inorg. Biochem.* 2013, *126*, 35–77.
45. Guerriero, F.; Sgarlata, C.; Francis, M.; Maurizi, N.; Faragli, A.; Perna, S.; Rondanelli, M.; Rollone, M.; Ricevuti, G. Neuroinflammation, immune system and Alzheimer disease: Searching for the missing link. *Aging Clin. Exp. Res.* 2017.
46. McLachlan, D.R.; Bergeron, C.; Smith, J.E.; Boomer, D.; Rifat, S.L. Risk for neuropathologically confirmed Alzheimer's disease and residual aluminum in municipal drinking water employing weighted residential histories. *Neurology* 1996, *46*, 401–405.
47. Meyer-Baron, M.; Schäper, M.; Knapp, G.; van Thriel, C. Occupational aluminum exposure: Evidence in support of its neurobehavioral impact. *Neurotoxicology* 2007, *28*, 1068–1078.
48. Gillette-Guyonnet, S.; Andrieu, S.; Nourhashemi, F.; de La Guéronnière, V.; Grandjean, H.; Vellas, B. Cognitive impairment and composition of drinking water in women: Findings of the EPIDOS Study. *Am. J. Clin. Nutr.* 2005, *81*, 897–902.
49. Foglio, E.; Buffoli, B.; Exley, C.; Rezzani, R.; Rodella, L.F. Regular consumption of a silicic acid-rich water prevents aluminium-induced alterations of nitrergic neurons in mouse brain: Histochemical and immunohistochemical studies. *Histol. Histopathol.* 2012, *27*, 1055–1066.
50. Bondy, S.C.; Campbell, A. Aluminum and neurodegenerative diseases. In *Advances in Neurotoxicology*; Aschner, A., Costa, L.G., Eds.; Elsevier: Atlanta, GA, USA, 2017; Volume 1, pp. 131–156.
51. Campbell, A. The role of aluminum and copper on neuroinflammation and Alzheimer's disease. *J. Alzheimers Dis.* 2006, *10*, 165–172.
52. Hébert, C.D.; Elwell, M.R.; Travlos, G.S.; Fitz, C.J.; Bucher, J.R. Subchronic toxicity of cupric sulfate administered in drinking water

and feed to rats and mice. *Fundam. Appl. Toxicol.* 1993, *21*, 461–475.

53. World Health Organization. Copper in Drinking-Water: Background Document for Development of WHO Guidelines for Drinking Water Quality. 2004. Available online: http://www.who.int/water_sanitation_health/dwq/chemicals/copper.pdf (accessed on 23 June 2017).
54. Environmental Protection Agency (EPA) Lead and Copper Rule. Available online: https://www.epa.gov/dwreginfo/lead-and-copper-rule (accessed on 23 June 2017).
55. Spitalny, K.C.; Brondum, J.; Vogt, R.L.; Sargent, H.E.; Kappel, S. Drinking-water-induced copper intoxication in a Vermont family. *Pediatrics* 1984, *74*, 1103–1106.
56. Araya, M.; Olivares, M.; Pizarro, F.; González, M.; Speisky, H.; Uauy, R. Gastrointestinal symptoms and blood indicators of copper load in apparently healthy adults undergoing controlled copper exposure. *Am. J. Clin. Nutr.* 2003, *77*, 646–650.
57. Squitti, R.; Pasqualetti, P.; Dal Forno, G.; Moffa, F.; Cassetta, E.; Lupoi, D.; Vernieri, F.; Rossi, L.; Baldassini, M.; Rossini, P.M. Excess of serum copper not related to ceruloplasmin in Alzheimer disease. *Neurology* 2005, *64*, 1040–1046.
58. Hozumi, I.; Hasegawa, T.; Honda, A.; Ozawa, K.; Hayashi, Y.; Hashimoto, K.; Yamada, M.; Koumura, A.; Sakurai, T.; Kimura, A.; et al. Patterns of levels of biological metals in CSF differ among neurodegenerative diseases. *J. Neurol. Sci.* 2011, *303*, 95–99.
59. Gorell, J.M.; Johnson, C.C.; Rybicki, B.A.; Peterson, E.L.; Kortsha, G.X.; Brown, G.G.; Richardson, R.J. Occupational exposure to manganese, copper, lead, iron, mercury and zinc and the risk of Parkinson's disease. *Neurotoxicology* 1999, *20*, 239–247.
60. Salustri, C.; Barbati, G.; Ghidoni, R.; Quintiliani, L.; Ciappina, S.; Binetti, G.; Squitti, R. Is cognitive function linked to serum free copper levels? A cohort study in a normal population. *Clin. Neurophysiol.* 2010, *121*, 502–507.
61. Brewer, G.J. Copper toxicity in Alzheimer's disease: Cognitive loss from ingestion of inorganic copper. *J. Trace Elem. Med. Biol.* 2012, *26*, 89–92.
62. Barnham, K.J.; McKinstry, W.J.; Multhaup, G.; Galatis, D.; Morton, C.J.; Curtain, C.C.; Williamson, N.A.; White, A.R.; Hinds, M.G.; Norton, R.S.; et al. Structure of the Alzheimer's disease amyloid

precursor protein copper binding domain. *J. Biol. Chem.* 2003, *278*, 17401–17407.

63. Dong, J.; Atwood, C.S.; Anderson, V.E.; Siedlak, S.L.; Smith, M.A.; Perry, G.; Carey, P.R. Metal binding and oxidation of amyloid-beta within isolated senile plaque cores: Raman microscopic evidence. *Biochemistry* 2003, *42*, 2768–2773.
64. Miller, L.M.; Wang, Q.; Telivala, T.P.; Smith, R.J.; Lanzirotti, A.; Miklossy, J. Synchrotron-based infrared and X-ray imaging shows focalized accumulation of Cu and Zn co-localized with beta-amyloid deposits in Alzheimer's disease. *J. Struct. Biol.* 2006, *155*, 30–37.
65. Dikalov, S.I.; Vitek, M.P.; Mason, R.P. Cupric-amyloid beta peptide complex stimulates oxidation of ascorbate and generation of hydroxyl radical. *Free Radic. Med. Biol.* 2004, *36*, 340–347.
66. Guilloreau, L.; Combalbert, S.; Sournia-Saquet, A.; Mazarguil, H.; Faller, P. Redox chemistry of copper-amyloid-beta: The generation of hydroxyl radical in the presence of ascorbate is linked to redox-potentials and aggregation state. *Chem. Biochem.* 2007, *8*, 1317–1325.
67. Lung, S.; Li, H.; Bondy, S.C.; Campbell, A. Low concentrations of copper in drinking water increase AP-1 binding in the brain. *Toxicol. Ind. Health* 2015, *31*, 1178–1184.
68. Zheng, Z.; White, C.; Lee, J.; Peterson, T.S.; Bush, A.I.; Sun, G.Y.; Weisman, A.; Petris, M.J. Altered microglial copper homeostasis in a mouse model of Alzheimer's disease. *J. Neurochem.* 2010, *114*, 1630–1638.
69. Singh, I.; Sagare, A.P.; Coma, M.; Perlmutter, D.; Gelein, R.; Bell, R.D.; Deane, R.J.; Zhong, E.; Parisi, M.; Ciszewski, J.; et al. Low levels of copper disrupt brain amyloid-β homeostasis by altering its production and clearance. *Proc. Natl. Acad. Sci. USA* 2013, *110*, 14771–14776.
70. Frisbie, S.H.; Ortega, R.; Maynard, D.M.; Sarkar, B. The concentrations of arsenic and other toxic elements in Bangladesh's drinking water. *Environ. Health Perspect.* 2002, *110*, 1147–1153.
71. Mazumder, D.N.; Ghosh, A.; Majumdar, K.K.; Ghosh, N.; Saha, C.; Mazumder, R.N. Arsenic contamination of ground water and its health impact on population of district of Nadia, West Bengal, India. *Indian J. Community Med.* 2010, *35*, 331–338.
72. Tauheed, J.; Sanchez-Guerra, M.; Lee, J.J.; Paul, L.; Ibne Hasan, M.O.S.; Quamruzzaman, Q.; Selhub, J.; Wright, R.O.; Christiani,

D.C.; Coull, B.A.; et al. Associations between post translational histone modifications, myelomeningocele risk, environmental arsenic exposure, and folate deficiency among participants in a case control study in Bangladesh. *Epigenetics* 2017, *12*, 484–491.

73. Tsai, S.Y.; Chou, H.Y.; The, H.W.; Chen, C.M.; Chen, C.J. The effects of chronic arsenic exposure from drinking water on the neurobehavioral development in adolescence. *Neurotoxicology* 2003, *24*, 747–753.
74. Wasserman, G.A.; Liu, X.; Parvez, F.; Ahsan, H.; Factor-Litvak, P.; Kline, J.; van Geen, A.; Slavkovich, V.; Loiacono, N.J.; Levy, D.; et al. Water arsenic exposure and intellectual function in 6-year-old children in Araihazar, Bangladesh. *Environ. Health Perspect.* 2007, *115*, 285–289.
75. Parvez, F.; Wasserman, G.A.; Factor-Litvak, P.; Liu, X.; Slavkovich, V.; Siddique, A.B.; Sultana, R.; Sultana, R.; Islam, T.; Levy, D.; et al. Arsenic exposure and motor function among children in Bangladesh. *Environ. Health Perspect.* 2011, *119*, 1665–1670.
76. Cholanians, A.B.; Phan, A.V.; Ditzel, E.J.; Camenisch, T.D.; Lau, S.S.; Monks, T.J. From the Cover: Arsenic Induces Accumulation of α-Synuclein: Implications for Synucleinopathies and Neurodegeneration. *Toxicol. Sci.* 2016, *153*, 271–281.
77. Dani, S.U. Arsenic for the fool: An exponential connection. *Sci. Total Environ.* 2010, *408*, 1842–1846.
78. Escudero-Lourdes, C. Toxicity mechanisms of arsenic that are shared with neurodegenerative diseases and cognitive impairment: Role of oxidative stress and inflammatory responses. *Neurotoxicology* 2016, *53*, 223–235.
79. US Geological Service. Arsenic in Groundwater of the United States. Available online: http://water.usgs.gov/nawqa/trace/arsenic/index.html (accessed on 17 November 2011).
80. Agnihotri, S.K.; Agrawal, U.; Ghosh, I. Brain most susceptible to cadmium induced oxidative stress in mice. *J. Trace Elem. Med. Biol.* 2015, *30*, 184–193.
81. Stolakis, V.; Tsakiris, S.; Kalafatakis, K.; Zarros, A.; Skandali, N.; Gkanti, V.; Kyriakaki, A.; Liapi, C. Developmental neurotoxicity of cadmium on enzyme activities of crucial offspring rat brain regions. *Biometals* 2013, *26*, 1013–1021.

82. Mandour, R.A.; Azab, Y.A. The prospective toxic effects of some heavy metals overload in surface drinking water of Dakahlia Governorate, Egypt. *Int. J. Occup. Environ. Med.* 2011, *2*, 245–253.
83. Fakhri, F.; Jafarzadeh, S.; Moradi, B.; Zandsalimi, Y.; Langarizadeh, G.; Amirhajeloo, L.R.; Mirzaei, M. The non-carcinogenic risk of cadmium in bottled water in different age groups humans: Bandar Abbas City, Iran. *Mater. Sociomed.* 2015, *27*, 52–55.
84. Needham, L.L.; Barr, D.B.; Caudill, S.P.; Pirkle, J.L.; Turner, W.E.; Osterloh, J.; Jones, R.L.; Sampson, E.J. ConIcentrations of environmental chemicals associated with neurodevelopmental effects in U.S. population. *Neurotoxicology* 2005, *26*, 531–545.
85. Manasfi, T.; Coulomb, B.; Boudenne, J.L. Occurrence, origin, and toxicity of disinfection byproducts in chlorinated swimming pools: An overview. *Int. J. Hyg. Environ. Health* 2017, *220*, 591–603.
86. Moser, V.C.; Phillips, P.M.; Levine, A.B.; McDaniel, K.L.; Sills, R.C.; Jortner, B.S.; Butt, M.T. Neurotoxicity produced by dibromoacetic acid in drinking water of rats. *Toxicol. Sci.* 2004, *79*, 112–122.
87. Zaheer, F.; Slevin, J.T. Trichloroethylene and Parkinson disease. *Neurol. Clin.* 2011, *29*, 657–665.
88. Igisu, H.; Goto, I.; Kawamura, Y.; Kato, M.; Izumi, K. Acrylamide encephaloneuropathy due to well water pollution. *J. Neurol. Neurosurg. Psychiatry* 1975, *38*, 581–584.
89. Gökmen, V. *Acrylamide in Food: Analysis, Content and Potential Health Effects*; Academic Press: Cambridge, MA, USA, 2015.
90. Lee, Y.M.; Seong, M.J.; Lee, J.W.; Lee, Y.K.; Kim, T.M.; Nam, S.Y.; Kim, D.J.; Yun, Y.W.; Kim, T.S.; Han, S.Y.; et al. Estrogen receptor independent neurotoxic mechanism of bisphenol A, an environmental estrogen. *J. Vet. Sci.* 2007, *8*, 27–38.
91. Stump, D.G.; Beck, M.J.; Radovsky, A.; Garman, R.H.; Freshwater, L.L.; Sheets, L.P.; Marty, M.S.; Waechter, J.M., Jr.; Dimond, S.S.; Van Miller, J.P.; et al. Developmental neurotoxicity study of dietary bisphenol A in Sprague-Dawley rats. *Toxicol. Sci.* 2010, *115*, 167–182.
92. Karalius, V.P.; Harbison, J.E.; Plange-Rhule, J.; van Breemen, R.B.; Li, G.; Huang, K.; Durazo-Arvizu, R.A.; Mora, N.; Dugas, L.R.; Vail, L.; et al. Bisphenol A (BPA) found in humans and water in three geographic regions with distinctly different levels of economic development. *Environ. Health Insights* 2014, *8*, 1–3.

93. Zhang, X.; Xu, Q.; Man, S.; Zeng, X.; Yu, Y.; Pang, Y.; Sheng, G.; Fu, J. Tissue concentrations, bioaccumulation, and biomagnification of synthetic musks in freshwater fish from Taihu Lake, China. *Environ. Sci. Pollut. Res. Int.* 2013, *20*, 311–322.
94. Peng, F.J.; Pan, C.G.; Zhang, M.; Zhang, N.S.; Windfeld, R.; Salvito, D.; Selck, H.; Van den Brink, P.J.; Ying, G.G. Occurrence and ecological risk assessment of emerging organic chemicals in urban rivers: Guangzhou as a case study in China. *Sci. Total Environ.* 2107, *589*, 46–55.
95. Litwa, E.; Rzemieniec, J.; Wnuk, A.; Lason, W.; Krzeptowski, W.; Kajta, M. Apoptotic and neurotoxic actions of 4-para-nonylphenol are accompanied by activation of retinoid X receptor and impairment of classical estrogen receptor signaling. *J. Steroid Biochem. Mol. Biol.* 2014, *144*, 334–347.
96. Mallela, M.K.; Were, S.R.; Hrubec, T.C. Neural tube defects in mice exposed to tap water. *Environ. Toxicol.* 2011, *26*, 633–640.
97. Karri, V.; Schuhmacher, M.; Kumar, V. Heavy metals (Pb, Cd, As and MeHg) as risk factors for cognitive dysfunction: A general review of metal mixture mechanism in brain. *Environ. Toxicol. Pharmacol.* 2016, *48*, 203–213.
98. Bae, D.S.; Gennings, C.; Carter, W.H., Jr.; Yang, R.S.; Campain, J.A. Toxicological interactions among arsenic, cadmium, chromium, and lead in human keratinocytes. *Toxicol. Sci.* 2001, *63*, 132–142.
99. Whittaker, M.H.; Wang, G.; Chen, X.Q.; Lipsky, M.; Smith, D.; Gwiazda, R.; Fowler, B.A. Exposure to Pb, Cd, and As mixtures potentiates the production of oxidative stress precursors: 30-day, 90-day, and 180-day drinking water studies in rats. *Toxicol. Appl. Pharmacol.* 2011, *254*, 154–166.
100. Valeri, L.; Mazumdar, M.M.; Bobb, J.F.; Henn, B.C.; Rodrigues, E.; Sharif, O.I.A.; Kile, M.L.; Quamruzzaman, Q.; Afroz, S.; Golam, M.; et al. The joint effect of prenatal exposure to metal mixtures on neurodevelopmental outcomes at 20–40 months of age: Evidence from rural Bangladesh. *Environ. Health Perspect.* 2017, *125*, 067015.
101. Ashok, A.; Rai, N.K.; Tripathi, S.; Bandyopadhyay, S. Exposure to As-, Cd-, and Pb-mixture induces Aβ, amyloidogenic APP processing and cognitive impairments via oxidative stress-dependent neuroinflammation in young rats. *Toxicol. Sci.* 2015, *143*, 64–80.

102. Parham, P. *The Immune System*, 4th ed.; Garland Sciences: New York, NY, USA, 2014.
103. Carson, M.J.; Doose, J.M.; Melchior, B.; Schmid, C.D.; Ploix, C.C. CNS immune privilege: Hiding in plain sight. *Immunol. Rev.* 2006, *213*, 48–65.
104. Qin, L.; Wu, X.; Block, M.L.; Liu, Y.; Breese, G.R.; Hong, J.S.; Knapp, D.J.; Crews, F.T. Systemic LPS causes chronic neuroinflammation and progressive neurodegeneration. *Glia* 2007, *55*, 453–462.
105. Zhang, F.; Jiang, L. Neuroinflammation in Alzheimer's disease. *Neuropsychiatr. Dis. Treat.* 2015, *11*, 243–256.
106. Heppner, F.L.; Ransohoff, R.M.; Becher, B. Immune attack: The role of inflammation in Alzheimer disease. *Nat. Rev. Neurosci.* 2015, *16*, 358–372.
107. McGeer, E.G.; McGeer, P.L. Innate immunity in Alzheimer's disease: A model for local inflammatory reactions. *Mol. Interv.* 2001, *1*, 22–29.
108. Steinman, L. Inflammatory cytokines at the summits of pathological signal cascades in brain diseases. *Sci. Signal.* 2013, *6*, pe3.
109. Lustbader, J.W.; Cirilli, M.; Lin, C.; Xu, H.W.; Takuma, K.; Wang, N.; Caspersen, C.; Chen, X.; Pollak, S.; Chaney, M.; et al. ABAD directly links Abeta to mitochondrial toxicity in Alzheimer's disease. *Science* 2004, *304*, 448–452.
110. Caspersen, C.; Wang, N.; Yao, J.; Sosunov, A.; Chen, X.; Lustbader, J.W.; Xu, H.W.; Stern, D.; McKhann, G.; Yan, S.D. Mitochondrial Abeta: A potential focal point for neuronal metabolic dysfunction in Alzheimer's disease. *FASEB J.* 2005, *19*, 2040–2041.
111. Gibson, G.E.; Karuppagounder, S.S.; Shi, Q. Oxidant-induced changes in mitochondria and calcium dynamics in the pathophysiology of Alzheimer's disease. *Ann. N. Y. Acad. Sci.* 2008, *1147*, 221–232.
112. Coskun, P.E.; Wyrembak, J.; Derbereva, O.; Melkonian, G.; Doran, E.; Lott, I.T.; Head, E.; Cotman, C.W; Wallace, D.C. Systemic mitochondrial dysfunction and the etiology of Alzheimer's disease and down syndrome dementia. *J. Alzheimers Dis.* 2010, *20*, S293–S310.
113. Lubik, N. Opening the "green pharmacy". *Environ. Sci. Technol.* 2008, *42*, 8620–8621.

114. Rastogi, T.; Leder, C.; Kümmerer, K. Re-Designing of Existing Pharmaceuticals for Environmental Biodegradability: A Tiered Approach with β-Blocker Propranolol as an Example. *Environ. Sci. Technol.* 2015, *49*, 11756–11763.

115. Larsen, T.A.; Hoffman, S.; Luthi, C.; Truffer, B.; Maurer, M. Emerging solutions to the water challenges of an urbanizing world. *Science* 2016, *352*, 928–933.

Chapter 7

Water Microbiology. Bacterial Pathogens and Water

João P. S. Cabral

Center for Interdisciplinary Marine and Environmental Research (C. I. I. M. A. R.), Faculty of Sciences, Oporto University, Rua do Campo Alegre, 4169-007 Oporto, Portugal.

ABSTRACT

Water is essential to life, but many people do not have access to clean and safe drinking water and many die of waterborne bacterial infections. In this review a general characterization of the most important bacterial diseases transmitted through water—cholera, typhoid fever and bacillary dysentery—is presented, focusing on the biology and ecology of the causal agents and on the diseases' characteristics and their life cycles in the environment. The importance of pathogenic *Escherichia coli* strains and emerging pathogens in

Citation: Water Microbiology. Bacterial Pathogens and Water. João P. S. Cabral. Int. J. Environ. Res. Public Health 2010, 7, 3657-3703; doi:10.3390/ijerph7103657

drinking water-transmitted diseases is also briefly discussed. Microbiological water analysis is mainly based on the concept of fecal indicator bacteria. The main bacteria present in human and animal feces (focusing on their behavior in their hosts and in the environment) and the most important fecal indicator bacteria are presented and discussed (focusing on the advantages and limitations of their use as markers). Important sources of bacterial fecal pollution of environmental waters are also briefly indicated. In the last topic it is discussed which indicators of fecal pollution should be used in current drinking water microbiological analysis. It was concluded that safe drinking water for all is one of the major challenges of the 21st century and that microbiological control of drinking water should be the norm everywhere. Routine basic microbiological analysis of drinking water should be carried out by assaying the presence of *Escherichia coli* by culture methods. Whenever financial resources are available, fecal coliform determinations should be complemented with the quantification of enterococci. More studies are needed in order to check if ammonia is reliable for a preliminary screening for emergency fecal pollution outbreaks. Financial resources should be devoted to a better understanding of the ecology and behavior of human and animal fecal bacteria in environmental waters.

Keywords: drinking water; cholera; typhoid fever; bacillary dysentery; fecal indicator bacteria; coliforms; ammonia

DRINKING WATER AS A VEHICLE OF DISEASES

Water is essential to life. An adequate, safe and accessible supply must be available to all. Improving access to safe drinking-water can result in significant benefits to health. Every effort should be made to achieve a drinking water quality as safe as possible [1].

Many people struggle to obtain access to safe water. A clean and treated water supply to each house may be the norm in Europe and North America, but in developing countries, access to both clean water and sanitation are not the rule, and waterborne infections are common. Two and a half billion people have no access to improved sanitation, and more than 1.5 million children die each year from diarrheal diseases [2]. According to the WHO, the mortality of water associated diseases exceeds 5 million people per year. From these, more that 50% are microbial intestinal infections, with cholera standing out in the first place.

In general terms, the greatest microbial risks are associated with ingestion of water that is contaminated with human or animal feces. Wastewater discharges in fresh waters and costal seawaters are the major source of fecal microorganisms, including pathogens [1–4].

Acute microbial diarrheal diseases are a major public health problem in developing countries. People affected by diarrheal diseases are those with the lowest financial resources and poorest hygienic facilities. Children under five, primarily in Asian and African countries, are the most affected by microbial diseases transmitted through water [5].

Microbial waterborne diseases also affect developed countries. In the USA, it has been estimated that each year 560,000 people suffer from severe waterborne diseases, and 7.1 million suffer from a mild to moderate infections, resulting in estimated 12,000 deaths a year [6]. The most important bacterial diseases transmitted through water are listed in Table 1.

Table 1: The main bacterial diseases transmitted through drinking water

Disease	**Causal bacterial agent**
Cholera	*Vibrio cholerae*, serovarieties O1 and O139
Gastroenteritis caused by vibrios	Mainly *Vibrio parahaemolyticus*
Typhoid fever and other serious salmonellosis	*Salmonella enterica* subsp. *enterica* serovar Paratyphi
	Salmonella enterica subsp. *enterica* serovar Typhi
	Salmonella enterica subsp. *enterica* serovar Typhimurium
Bacillary dysentery or shigellosis	*Shigella dysenteriae*
	Shigella flexneri
	Shigella boydii
	Shigella sonnei
Acute diarrheas and gastroenteritis	*Escherichia coli*, particularly serotypes such as O148, O157 and O124

CHOLERA

The Genus Vibrio

Vibrio are small, curved-shaped Gram-negative rods, with a single polar flagellum. Vibrios are facultative anaerobes capable of both fermentative and respiratory metabolism. Sodium stimulates growth of all species and is an absolute requirement for most. Most species are oxidase-positive and reduce nitrate to nitrite. Cells of certain species (*V. cholerae*, *V. parahaemolyticus* and *V. vulnificus*) have pili (fimbriae), structures composed of protein TcpA. TcpA formation is co-regulated with cholera toxin expression and is a key determinant of *in vivo* colonization (see below) [7,8].

Several *Vibrio* species can infect humans (Table 2). *V. cholerae* is, by far, the most important of these species. *V. alginolyticus* has been isolated from several types of soft tissue infections.

Table 2: Main species of *Vibrio* and their occurrence in human clinical specimensa

Main species	**Occurrence in human clinical specimens**	
	Intestinal	**Extra-intestinal**
Vibrio alginolyticus	+	++
Vibrio cholerae O1 and O139	+++++	+
Vibrio cholerae non O1 or O139	++	++
Aliivibrio fischeri (*Vibrio fischeri*)	–	–
Vibrio fluvialis	++	–
Vibrio furnissii	++	–
Vibrio harveyi	–	+
Grimontia hollisae (*Vibrio hollisae*)	++	–
Vibrio mimicus	++	+
Vibrio natriegens	–	–
Vibrio parahaemolyticus	++++	+
Vibrio vulnificus	+	+++

[a]Adapted from [7,8]. Nomenclature according to [9]. The symbols give the relative frequency of each organism in human clinical specimens, and apply to the whole World, rather than to a particular country.

V. fluvialis, *Grimontia hollisae* (*V. hollisae*), and *V. mimicus* can cause diarrhea or infections of the gastrointestinal tract. *V. furnissii* has been isolated from a few individuals with diarrhea, but there is no evidence that it can actually cause this pathology. *V. parahaemolyticus* is a well-documented causal agent of acute food-borne gastroenteritis, particularly in Japan and South East Asia. Cases are associated with the consumption of raw or undercooked shellfish such as oysters, shrimp, crabs, and lobster.

V. vulnificus is an important cause of (often fatal) septicemia and wound infections. Other vibrios, namely *Allivibrio fischeri* (*Vibrio fischeri*) and *V. natriegens*, have no relation with humans [7,8].

Vibrios are primarily aquatic bacteria. Species distribution depends on sodium concentration and water temperature. Vibrios are very common in marine and estuarine environments, living free or on the surfaces and in the intestinal contents of marine animals. Species with a low sodium requirement are also found in freshwater habitats [7,8].

The Species Vibrio Cholerae

Vibrio cholerae cells can grow at 40 °C with pH 9–10. The growth is stimulated by the presence of sodium chloride. *Vibrio cholerae* is a very diverse bacterial species (Table 3). It is divided in ca. 200 serovarieties, characterized by the structure of the lipopolysaccharide (LPS) (O antigens). Only serovarieties O1 and O139 are involved in "true" cholera. Some other serovarieties can cause gastroenteritis, but not cholera. The distinction between Classical and El Tor biotypes is based on biochemical and virological characteristics [1,7,8,10,11].

Table 3: Subdivision of *Vibrio cholerae* below the species level[a]

Serovariety	Serotype	Biotype
O1	Inaba	Classical
		El Tor
	Ogawa	Classical
		El Tor
	Hikojima	
O139		
others		

[a]Adapted from [8].

Cholera

Characterization of the disease

The incubation period for cholera is ca. 1–3 days. The disease is characterized by an acute and very intense diarrhea that can exceed one liter per hour. Cholera patients feel thirsty, have muscular pains and general weakness, and show signs of oliguria, hypovolemia, hemoconcentration, followed by anuria. Potassium in blood drops to very low levels. Patients feel lethargic. Finally, circulatory collapse and dehydration with cyanosis occurs [**7**].

The severity of the disease depends on several factors: (1) personal immunity: this may be conferred by both previous infections and by vaccines; (2) inoculum: the disease only occurs after ingestion of a minimum amount of cells, ca. 10^8 [1,7,8,10,11]; (3) The gastric barrier: *V. cholera* cells likes basic media and therefore the stomach, normally very acidic, is an adverse medium for bacterial survival. Patients consuming anti-acidic medications are more susceptible to infection than healthy people; (4) blood group: for still unknown reasons, people with O-group blood are more susceptible than others [1,7,8,10,11].

In the absence of treatment, the mortality of cholera-patients is ca. 50%. It is mandatory to replace not only lost water but also lost salts, mainly potassium. In light dehydrations, water and salts can be orally-administered, but in severe conditions, rapid and intravenous-administration is obligatory. The most efficient antibiotic is currently doxicyclin. If no antibiotic is available for treatment, the administration of water with salts and sugar can, in many cases, save the patient and help in the recovery [1,7,8,10,11].

There are two main determinants of infection: (1) the adhesion of the bacterial cells to the intestinal mucous membrane. This depends on the presence of pili and adesins at the cell's surface; (2) the production of cholera toxin [1,7,8,10,11].

Cholera toxin

Cholera toxin is an exotoxin with a very precise action on target cells. The toxin attaches to a specific receptor (ganglioside Gl) on the cell membrane of intestinal cells and activates the enzyme adenylate cyclase. This results in a non-stop degradation of internal ATP, with release of cAMP and inorganic phosphate. The rise in the internal concentration of cAMP causes an efflux of water, sodium, potassium, chloride and carbonate ions from the cells of the mucous membrane, and this is the main cause of diarrhea [**7**].

Cholera pandemics and the emergence of El Tor biotype and O139 serovariety. New facts about cholera epidemiology

Cholera has been a well known disease since the 19th century. In the 19th and 20th centuries, seven major pandemics are recognized. The first six pandemics occurred during the following periods: 1st: 1816–1826, 2nd: 1829–1851, 3rd: 1852–1860, 4th: 1863–1875, 5th: 1881–1896, 6th: 1899–1923. These pandemics all started in Asia, passed through Europe and then reached South America. The Classical biotype was involved. The seventh pandemic, still in course, started in 1961 in the Celebes Isles, in Asia. In the 1960s, the disease spread through Asia, in the 1970s reached the Middle East and Africa, and in 1991 streaked violently across South America. Now El Tor has replaced the Classical biotype. El Tor biotype had been detected before, in 1905, but only in the development of the seventh pandemic did this biotype replace the Classical one and become dominant [1,7,8,10,11].

In 1992, a new serovariety (O139), which was coined the Bengal serovariety, was detected for the first time in Bangladesh. This new serovariety quickly spread to India and to southeastern Asia, displacing O1. Although serovariety O1 El Tor has reappeared in 1994 and 1995, the Bengal serovariety still remains the dominant one. The illness caused by serovarieties O139 and O1 are indistinguishable [8,12,13].

In 1991, the seventh pandemic entered South America through the coastal area of Peru. On 23 January, in Chancay, north Peru, *Vibrio cholerae* O1 El Tor was isolated from patients with cholera symptoms, confirming the disease. In this region, between 24 January and 9 February, 1,859 people were hospitalized and 66 died. From Peru, the disease spread rapidly to other countries in South America. Two routes have been proposed for the entrance of the bacterium in Peru: (1) ballast water from a boat coming from Asia; (2) the El Niño current may have transported zooplankton harboring *V. cholerae* cells. Shellfish and fish nourishing on this zooplankton became contaminated and the bacterium was transmitted to humans who ate these marine foods [14–17].

The misfortune of people who died in the first months of this disastrous South American cholera epidemic appeared to have unleashed scientists to study the disease harder and, indeed, important epidemiological studies were carried out during this outbreak. These studies confirmed that contaminated uncooked food and beverages can also be a vehicle for transmission of cholera [18].

Genes for toxin and pili protein production

The genes responsible for toxin production are harbored in the CTXΦ segment (7–9.7 kb) of the chromosome (only in toxigenic strains). The CTXΦ segment carries at least six genes. In addition to the gene encoding cholera toxin production, this segment (virulence cassette) include an accessory cholera toxin (*ace*), a zonula occludens toxin (*zot*), core encoded pilin (*cep*), and an open reading frame of unknown function. During the replication of the chromosome, the CTXΦ fragment can form an autonomous copy and this can constitute an independent plasmid. The plasmid can give rise to virus-like particles—CTXΦ bacteriophages, which can infect non-toxigenic strains. The CTXΦ segment incorporates into the chromosome of the infected cells which became toxigenic. This process was demonstrated *in vitro* in cell suspensions and *in vivo* in the gut of the rat [8,13,19,20].

Epidemic and pandemic strains of *V. cholerae* contain another chromosomal segment designated as VPI. VPI is 39.5 kb in size and contains two ToxR-regulated genes: a regulator of virulence genes (ToxT) and a gene cluster containing colonization factors, including the toxin co-regulated pili (TCP). The tcp gene encodes for the 20.5-kDa TcpA pili protein. This VPI segment appears to be transferable from *V. cholerae* O1 to non-O1 strains. *V. cholerae* O139 strains, like O1, carry the structural genes encoded by the CTX operon and TCP. *V. cholerae* strains non-O1 or O139 normally lack cholera toxin genes and have never been found to carry TCP [8].

Ecology of the bacterium and the cycle of the disease

V. cholerae non-O1 or O139 strains are common in the environment, especially in estuaries. They have been isolated from many estuarine animals such as birds, frogs, fishes and shellfish, and survive and multiply on the surface of phytoplankton and zooplankton cells [8,21].

V. cholerae O1 and O139 strains are isolable from the environment only in epidemic areas. They survive in the cultivable state in water and aquatic and marine organisms for a considerable period of time [8,12,22–24]. When *V. cholerae* cells face adverse environmental conditions, they reduce cell size, became coccoid and enter a dormant stage inside exopolysaccharide biofilms. Cells display a certain metabolism, but are not able to growth and multiply on the surface of agarized media and give rise to colonies. Cells in this viable but non-culturable state retain viability as well as the potential for pathogenicity for significant periods of time [25–27].

Viable but non-culturable cells can leave their dormant stage and multiply again, resulting in an explosion of their concentration in the environment. Since the presence of non-toxigenic strains is common in aquatic milieu, especially in estuaries, if a horizontal transfer of cholera exotoxin producing genes occurs between toxigenic and non-toxigenic strains, the number of toxigenic cells in the environment can rise rapidly and pronouncedly. The episodic nature and the sudden appearance of violent cholera outbreaks, followed by a rapid slowing down, are probably related with these phenomena.

SALMONELLOSIS

The Genus Salmonella. Pathogenicity of Main Serovars

The genus *Salmonella* was designated by Lignières in 1900 [28,29]. Antigenic analysis began when Castellani described, in 1902, a method for absorbing antisera. The first antigenic scheme for *Salmonella* was published by White in 1926, and subsequently developed extensively by Kauffmann, in two classical works published in 1966 and 1978 [28,29]. The Kauffmann-White antigenic scheme contained, by 1988, about 2,250 different serovars [28,29].

The genus *Salmonella*, a member of the family *Enterobacteriaceae*, include Gram-negative motile straight rods. Cells are oxidase-negative and catalase-positive, produce gas from D-glucose and utilize citrate as a sole carbon source. Salmonellae have several endotoxins: antigens O, H and Vi [28,29].

The concept "one serovar-one species", in use for many years, is no longer acceptable. The taxonomy and nomenclature of the genus *Salmonella* has been subject of debate since Le Minor and Popoff proposed changes in a paper published in 1987. The issue was settled by a decision of the International Committee on the Systematics of Prokaryotes and published in 2005. The current taxonomy of the genus is presented in Table 4. According to the rules of bacterial nomenclature, the names of the serovars are not italicized and the first letter must be a capital [28–30].

Table 4: Current taxonomy and nomenclature of the genus *Salmonella*. Habitat and pathogenicity of main serovars[a]

Species	**Sub-species**	**Main serovars (from a total of ca. 1,443)**	**Habitat and pathogenicity**
Salmonella enterica	*Salmonella enterica* subsp. *enterica*	Abortusovis	Pathogenic to sheeps.
		Choleraesuis	Pathogenic to humans and animals.
		Enteritidis	Ubiquitous and frequently the cause of infections in humans and animals. Very frequent agent of gastroenteritis in humans.
		Gallinarum	Isolated chiefly from chickens and other birds. Causal agent of fowl thyphoid.
		Paratyphi A	Pathogenic only to humans. Causes paratyphoid fever.
		Paratyphi B	Causes paratyphoid fever in humans and very rarely infects animals.
		Paratyphi C	Causes paratyphoid fever in humans.
		Typhi	Pathogenic only to humans, causing typhoid fever.
			Transmitted by water and food contaminated with feces.
		Typhimurium	Ubiquitous and frequently the cause of infections in humans and animals. Very frequently, the causal agent of gastroenteritis in humans.
		Typhisuis	Pathogenic to swines.
	Salmonella enterica subsp. *arizonae*	At least 94 serovars.	
	Salmonella enterica subsp. *diarizonae*	At least 323 serovars.	
	Salmonella enterica subsp. *houtenae*	At least 70 serovars.	Isolated mainly from cold-blooded animals and from the environment. Not pathogenic to humans.

	Salmonella enterica subsp. *indica*	At least 11 serovars.	
	Salmonella enterica subsp. *salamae*	At least 488 serovars.	
Salmonella bongori		At least 20 serovars.	

[a]Adapted from [29]. Nomenclature according to [9].

Table 5: Current taxonomy and nomenclature of the genus *Shigella*. Habitat and pathogenicity of species[a]

Species	**Main serotypes**	**Habitat and pathogenicity**
Shigella dysenteriae	15 serotypes.	Intestinal pathogens of humans and primates, causing bacillary dysentery.
Shigella flexneri	8 serotypes 9 subserotypes	Humans are the primary reservoir. A long-term carrier state occurs in few cases.
Shigella boydii	19 serotypes	*Shigella dysenteriae* serotype 1 causes more severe disease then other serotypes and produces a potent exotoxin (Shiga toxin). Large epidemics in developing countries are commonly caused by serotype 1. Diseases caused by other serotypes may be mild or severe.
Shigella sonnei	1 serotype	*Shigella sonnei* illness is usually milder than that caused by other *Shigella* species.

[a]Adapted from [34]. Nomenclature according to [9].

S. enterica subsp. *enterica* serovar Enteritidis is the most frequently isolated serovar from humans all over the world. However, locally, other serovars can be predominant. In the period 1994–2004, Tunisia was exposed to salmonellosis outbreaks in 1997, 1999, 2002 and 2004. In 1997, salmonellosis outbreak was caused by serovar Mbandaka. In 1999, three salmonellosis outbreaks were reported from hospitals located in three different regions. Each outbreak was associated with a different serotype: Mbandaka, Livingstone and Typhi Vi+. In 2002, a *S. enterica* subsp. *enterica* serovar Livingstone infection occurred in the same hospital that reported an outbreak caused by serovar Typhi Vi+ in 1999, but in a different unit. In that year, the Livingstone serovar jumped to the first position in human infection in Tunisia. In 2004, a second outbreak by serovar Typhi Vi+ was reported. The source of isolation was a fermented juice traditionally extracted from palm-tree [31].

Characterization of the Diseases

Salmonellae pathogenic to humans can cause two types of salmonellosis: (1) typhoid and paratyphoid fever (do not confuse with typhus, a disease caused by a rickettsia); (2) gastroenteritis [28]. Low infective doses (less than 1,000 cells) are sufficient to cause clinical symptoms. Salmonellosis of newborns and infants presents diverse clinical symptoms, from a grave typhoid-like illness with septicemia to a mild or asymptomatic infection. In pediatric wards, the infection is usually transmitted by the hands of staff [29].

Food-borne *Salmonella* gastroenteritis are frequently caused by ubiquitous *Salmonella* serovars such as Typhimurium. About 12 h following ingestion of contaminated food, symptoms (diarrhea, vomiting and fever) appear and last 2–5 days. Spontaneous cure usually occurs. *Salmonella* may be associated with all kinds of food. Prevention of *Salmonella* food-borne infection relies on avoiding contamination (improvement of hygiene), preventing multiplication of *Salmonella* in food (constant storage of food at 4 °C), and use of pasteurization (milk) or sterilization when possible (other foods). Vegetables and fruits may carry *Salmonella* when contaminated with fertilizers of fecal origin, or when washed with polluted water [28].

The incidence of typhoid fever decreases when the level of development of a country increases (*i.e.*, controlled water sewage systems, pasteurization of milk and dairy products). Where these hygienic conditions are missing, the probability of fecal contamination of water and food remains high and so is the incidence of typhoid fever [29].

Ecology of Salmonellae and the Cycle of Salmonellosis

The principal habitat of *Salmonella* is the intestinal tract of humans and animals [28]. Salmonellae are constantly found in environmental samples, because they are excreted by humans, pets, farm animals, and wild life. Municipal sewage, agriculture pollution, and storm water runoff are the main sources of these pathogens in natural waters [1,32]. Salmonellae do not seem to multiply significantly in the natural environment, but they can survive several weeks in water and in soil if conditions of temperature, humidity, and pH are favorable [28].

Salmonellae isolated from environmental sources are predominantly non-Typhi or Paratyphi serovars. In a study carried out in Tunisia during 1994–2004, *S. enterica* subsp. *enterica* serovars Anatum, Enteritidis and Corvallis were the most common serotypes isolated from food. The great majority of the strains were isolated from poultry, red meat, milk and dairy products, vegetables and fruits. From environmental sources, 73% of the isolates were from tap water. Serovars Corvallis, Enteritidis, and Anatum were the commonest [31]. Arvanitidou *et al.* [32] reported a comparative study carried out in Rivers Aliakmon and Axios, in northern Greece, during a 1-year period, from May 2002 to April 2003. A total of 29 *Salmonella* species were recovered from the water samples. Many of the isolated *Salmonella* serovars were of non-human animal origin such as Mbantaka, Virchow, Hadar, Infantis and Senftenberg, commonly isolated from poultry farm.

Unlike cholera, humans infected with salmonellae can carry the bacteria in the gut without signs of disease. Infected humans can harbor the bacteria for considerable periods of time. About 5% of patients clinically cured from typhoid fever remain carriers for months or even years. These people can be chronic holders of the bacterium in the gut, and constitute the main reservoir of the bacteria in the environment [29].

The salmonellosis cycle in the environment can involve shellfish. Salmonellae survive sewage treatments if suitable germicides are not used in sewage processing. If effluent from the sewage plant passes into a coastal area, edible shellfish (mussels, oysters) can become contaminated. Shellfish concentrate bacteria as they filter several liters of water per hour. Ingestion by humans of these seafoods (uncooked or superficially cooked) may cause typhoid fever or other salmonellosis. Evidence of such a cycle has been obtained by the use of strain markers, including phage typing [29].

SHIGELLOSIS OR BACILLARY DYSENTERY

The Genus Shigella

Shigella are Gram-negative, non-sporeforming, non-motile, straight rod-like members of the family *Enterobacteriaceae*. Cells ferment sugars without gas production. Salicin, adonitol and myo-inositol are not fermented. Cells do not utilize citrate, malonate and acetate as sole carbon source and do not produce H_2S. Lysine is not decarboxylated. Cells are oxidase-negative and catalase-positive. Members of the genus have a complex antigenic pattern, and taxonomy is based on their somatic O antigens [1,33,34].

Characterization of the Disease

The incubation period is 1–4 days. The disease usually begins with fever, anorexia, fatigue and malaise. Patients display frequent bloody stools of small volume (sometimes grossly purulent) and abdominal cramps. Twelve to 36 hours later, diarrhea progresses to dysentery, blood, mucus and pus appearing in feces that decreases in volume (no more than 30 mL of fluid per kg per day) [34–36].

Although the molecular basis of shigellosis is complex, the initial step in pathogenesis is penetration of the colonic mucosa. The resulting focus of *Shigella* infection is characterized by degeneration of the epithelium and by an acute inflammatory colitis in the lamina propria. Ultimately, desquamation and ulceration of the mucosa cause leakage of blood, inflammatory elements, and mucus into the intestinal lumen. Under these conditions the absorption of water by the colon is inhibited and the volume of stool is dependent upon the ileocecal flow. As a result, the patient will pass frequent, scanty, dysenteric stools [37,38].

In order for *Shigella* to enter an epithelial cell, the bacterium must first adhere to its target cell. Generally, the bacterium is internalized via an endosome, which it subsequently lyses to gain access to the cytoplasm where multiplication occurs [37,38].

Virulence Factors

S. dysenteriae serotype 1 produces high levels of a cytotoxic Shiga toxin. *S. sonnei* and *S. flexneri* produce much lower amounts of this toxin. Shiga toxin binds to Galotl-4Galp (galabiose) glycolipid receptors and inhibits mammalian protein synthesis by cleaving the N-glycosidic bond at adenine 4324 in 28S rRNA. The toxic mechanism is identical to that of the plant toxin ricin, produced by *Ricinus communis*. Shigella also release a LPS endotoxin (O antigens), that cause an inflammatory response [37,38].

Shigella 180- to 230-kb plasmids encode genes essential for virulence, namely for: production of adhesins involved in the adherence of bacteria onto the surface of target epithelial cells; production of invasion plasmid antigens (Ipa) that have a direct role in the *Shigella* invasion process; transport or processing functions that ensure the correct surface expression of the Ipa proteins; induction of endocytic uptake of bacteria and disruption of endocytic vacuoles; regulation of plasmid-encoded virulence genes [37,38].

Shigella emerged from *E. coli* during evolution. The acquisition and evolution of the pathogenicity island which encodes all of the genes required for cell invasion and phagolysosomal lysis, permitted a major alteration in pathogenesis [37,38].

Risk Factors

Many studies have identified risk factors and protective effects for shigellosis incidence and fatality. Despite gradual improvements in water supply, shigellosis continues to be endemic among the disadvantaged populations living in the tropics, often among displaced populations following natural disasters and political crises. In Guatemala, young children, the elderly, and 15–44-year-old males were found to be most susceptible to *S. dysenteriae* serotype 1. In Sierra Leone, the attack rate was higher among children younger than 5 years of age than in the rest of the population. In rural Bangladesh, shigellosis was most common in children aged 1–2 years and in people 60 years or older. In Dhaka, Bangladesh, it was found that shigellosis mortality was most common in severely malnourished people of all ages, in children under 2 who were not being breastfed, and in all children under 1. In a 3-year study carried out in Matlab, Bangladesh, during 1992 to 1994, it was found that the incidence of *S. dysenteriae* serotype 1 and *S. flexneri* was highest in children under 2 followed by children from 2 to 5. The location of *S. dysenteriae* serotype 1 risk varies in time but *S. flexneri* risk areas were persistent in time. Neighborhoods near bazaars with many non-septic latrines were at highest risk for *S. dysenteriae* serotype 1. *S. flexneri* was most common in flood-controlled areas. It was concluded that *S. dysenteriae* serotype 1 risk was more related to hygiene and sanitation whereas *S. flexneri* was more related to the environment [35].

Shigellosis through the World

The total number of *Shigella* episodes that occur each year throughout the World is estimated to be 164.7 million, including 163.2 million cases in developing countries, 1.1 million of which result in death. Children under 5 account for 61% of all deaths attributable to shigellosis [35,36].

Shigella species are not uniformly distributed in the world. *S. dysenteriae* is usually found in densely populated areas of South America, Africa and Asia. Infections usually result in significant epidemic outbreaks. Serotype 1 has been distinguished by both its virulence and its ability to produce ravaging epidemics. It predominates in India, Malaysia and Guatemala.

Serotype 2 predominates in Yemen and Nigeria. *S. flexneri* is usually found in areas where endemic shigellosis occurs. *S. boydii* occurs sporadically, except in the Indian subcontinent where it was first identified. *S. sonnei* usually occurs in Western developed countries, such as France and USA [35,36].

Important epidemics were reported in the last decades: (1) in 1970 in Central America where 112,000 people were affected and 13,000 died; (2) in 1985, in Texas (USA), 5,000 people became infected after ingestion of contaminated lettuce; (3) in May–June 1994, domestic cases of *S. sonnei* infection were detected in several European countries, including Norway, Sweden, and the United Kingdom. Epidemiological evidence incriminated imported iceberg lettuce as the vehicle of transmission; (4) in 1996, in Paris, with 153 reported patients [33].

Ecology of Shigellae and the Cycle of Shigellosis

Shigella is typically an inhabitant of the intestinal tract of humans and other primates [1,33,34,36,39]. It is typically spread by fecal-contaminated drinking water or food, or by direct contact with an infected person. In water, shigellae can survive for at least six months at room temperature, and this high survival favors transmission through water. Flies have been implicated on the transmission of *Shigella* cells from human feces to foods. The hand is an important vehicle for transmission of shigellosis, since *S. dysenteriae* serotype 1 cells survives for up to one hour on a human's skin and a very small inoculum is required to unchain infection and disease. Indeed, studies on American volunteers experimentally infected with *Shigella* have shown that as few as one hundred *Shigella* cells given orally cause the disease in 25–50% of the cases. Resistance of *Shigella* to gastric juice certainly accounts, although not exclusively, for this high infectivity [36,40]. Asymptomatic and inappropriately-treated patients with shigellosis can harbor the bacteria in the gut and these appear to be the main reservoirs of the bacteria in the environment [41].

Recent reports on the ecology of shigellae have brought new elements for the understanding of the cycle of the disease. In environmental waters of regions with high numbers of shigellosis' cases, it has been found that, although numbers of cultivable cells were low, genetic elements such as plasmids and genetic fragments with bacteriophage origin, could be detected. Many of the genes that code for exotoxin production are precisely found in these genetic elements. These results suggest that the sudden rise of the number of virulent strains in the environment can result from the incorporation, by cells with reduced virulence, of this type of genetic elements present in the waters. If this hypothesis is confirmed, there is a certain similarity between the cholera and the shigellosis cycles in the environment. It remains to be elucidated if shigellae can also exist in environmental waters in a viable but non-culturable state, as vibrios [41].

PATHOGENIC *ESCHERICHIA COLI* STRAINS

E. coli strains isolated from intestinal diseases have been grouped into at least six different main groups, based on epidemiological evidence, phenotypic traits, clinical features of the disease and specific virulence factors. From these, enterotoxigenic (ETEC, namely O148), enterohemorrhagic (EHEC, namely O157) and enteroinvasive serotypes (EIEC, namely O124) are of outstanding importance and can be transmitted through contaminated water [**42**,**43**].

Enterotoxigenic E. coli (ETEC) Strains

Enterotoxigenic *E. coli* (ETEC) serotypes can cause infantile gastroenteritis. The number of reports of their occurrence in developed countries is comparatively small, but it is an extremely important cause of diarrhea in the developing world, where there is no adequate clean water and poor sanitation. In developing countries, these strains are the most commonly isolated bacterial enteropathogen in children below 5 years of age, and account for several hundred million cases of diarrhea and several ten of thousand deaths each year [42–44].

Disease caused by ETEC follows ingestion of contaminated food or water and is characterized by profuse watery diarrhea lasting for several days that often leads to dehydration and malnutrition in young children [42–44]. ETEC also are the most common cause of "travelers' diarrhea" that affects individuals from industrialized countries travelling to developing regions of the World [42–44].

Enterohemorrhagic E. coli (EHEC) Strains

Reported outbreaks had been associated mainly with the consumption of contaminated foods, such as raw or undercooked ground meat products and raw milk. The primary reservoir of this bacterium has been found to be healthy cattle [42,45,46].

E. coli serotype O157:H7 causes abdominal pain, bloody diarrhea, and hemolytic uremic syndrome. This bacterium produces Shiga-like toxins. The incubation period is 3–4 days, and the symptoms occur for 7–10 days. It is estimated that 2–7% of *E. coli* O157:H7 infections result in acute renal failure [42,45,46].

Although *E. coli* O157:H7 is not usually a concern in treated drinking water, outbreaks involving consumption of drinking water contaminated with human sewage or cattle feces have been documented. An increasing number of outbreaks are associated with the consumption of fruits and vegetables (sprouts, lettuce, coleslaw, salad) contaminated with feces from domestic or wild animals at some stage during cultivation or handling. EHEC has also been isolated from bodies of water (ponds, streams), wells and water troughs, and has been found to survive for months in manure and water-trough sediments [45,46].

Person-to-person contact is an important mode of transmission through the oral-fecal route. An asymptomatic carrier state has been reported, where individuals show no clinical signs of disease but are capable of infecting others [45,46].

Enteroinvasive E. coli (EIEC) Strains

Enteroinvasive *E. coli* (EIEC) behave in many respects like shigellae. They are capable of invading and multiplying in the intestinal epithelial cells of the distal large bowel in humans. The illness is characterized by abdominal cramps, diarrhea, vomiting, fever, chills, a generalized malaise, and the appearance of blood and mucus in the stools of infected individuals. [42,43,47].

EIEC strains were isolated, for instance, from 28 subjects in the Jesreel district of Israel during a peak period for dysentery. An investigation in Croatia showed that *E. coli* O124 could frequently be isolated from cases of gastroenteritis, enterocolitis, and dysentery. The dysentery was more common among the older age groups, while the two other types of disease occurred equally in all age groups. A 1985 survey was carried out in Bankok, Thailand in which 410 children with diarrhea and an equal number of control children without diarrhea were examined for the presence of strains of *Shigella*, EIEC, and other pathogens. It was found that 17 of the children with diarrhea and six without yielded EIEC [42,43].

Any food contaminated with human feces from an ill individual, either directly or via contaminated water, could cause disease in others. Outbreaks have been associated with hamburger meat and unpasteurized milk [47].

Emerging Waterborne Bacterial Pathogens

The emerging pathogenic bacteria of concern outlined here have the potential to be spread through drinking water, but they do not correlate with the presence of *E. coli* or with other commonly used drinking water quality indicators, such as coliform bacteria. In most cases, there are no satisfactory microbiological indicators of their presence. More studies are needed in order to understand the real significance and dimension of the diseases caused by water contaminated with these bacteria, and the ecology of these pathogens [45].

Mycobacterium Avium Complex (Mac)

The *Mycobacterium avium* complex (Mac) consists of 28 serovars of two distinct species: *Mycobacterium avium* and *Mycobacterium intracellulare.* The importance of Mac organisms was recognized with the discovery of disseminated infection in immunocompromised people, particularly people with HIV and AIDS. Members of MAC are considered opportunistic human pathogens [45,48].

Mac organisms have been identified in a broad range of environmental sources, including marine waters, rivers, lakes, streams, ponds, springs, soil, piped water supplies, plants, and house dust. Mac organisms have been isolated from natural water and drinking water distribution systems in the USA [45,49,50].

The ubiquitous nature of Mac organisms results from their ability to survive and grow under varied conditions. Mac organisms can proliferate in water at temperatures up to 51 °C and can grow in natural waters over a wide pH range [45]. These mycobacteria are highly resistant to chlorine and the other chemical disinfectants used for the treatment of drinking-water. Standard drinking-water treatments will not eliminate Mac organisms but, if operating satisfactorily, will significantly reduce the numbers that may be present in the source water to a level that represents a negligible risk to the general population. The entryway of these mycobacteria in distribution systems is through leaks. Growth of Mac organisms in biofilms is probably important for their continuous presence in distribution systems. Slow growing mycobacteria can be found at densities greater than 4,000 per cm^2 in the surface biofilm, creating a potentially high level of exposure [48].

The symptoms encountered with Mac infections result from colonization of either the respiratory or the gastrointestinal tract, with possible dissemination to other locations in the body. Exposure to Mac organisms may occur through the consumption of contaminated food, the inhalation of air with contaminated soil particles, or contact with or ingestion, aspiration, or aerosolization of potable water containing the organisms [45].

With respect to water supplies, infection with *M. avium* and *M. intracellulare* has been well documented. Unlike gastrointestinal pathogens, where *E. coli* can be used to indicate their potential presence, no suitable indicators have been identified to signal increasing concentrations of Mac organisms in water systems [45].

Helicobacter Pylori

Helicobacter pylori has been cited as a major etiologic agent for gastritis and has been implicated in the pathogenesis of peptic and duodenal ulcer disease and gastric carcinoma. However, most individuals that are infected by this pathogen remain asymptomatic [45].

Using culture-based methods, *H. pylori* has not been isolated from environmental sources, including water [45,51]. On the contrary, molecular methods have been successful in detecting this pathogen. Fluorescence *in situ* hybridization has been successfully used to detect this pathogen in drinking water distribution systems and other water bodies. Polymerase chain reaction has also been used to detect the presence of *H. pylori* DNA in drinking water, especially associated with biofilms [45,51,52]. In drinking-water biofilms, *H. pylori* cells rapidly lose culturability, entering a viable but non-culturable state. In these biofilms, cells can persist for more than one month, with densities exceeding 10^6 cells per cm^2 [51].

How the organism is transmitted is still not fully understood. However, the fact that it has been recovered from saliva, dental plaques, the stomach, and fecal samples strongly indicates oral-oral or fecal-oral transmission. Water and food appear to be of lesser direct importance, but they can still play a significant role in situations with improper sanitation and hygiene [45].

Aeromonas Hydrophyla

In recent years, *A. hydrophila* has gained public health recognition as an opportunistic pathogen. It has been implicated as a potential agent of gastroenteritis, septicemia, meningitis, and wound infections. It can play a significant role in intestinal disorders in children under five years old, the elderly, and immunosuppressed people. [45,53,54].

Aeromonas hydrophila are Gram-negative, non-sporeforming, rod-shaped, facultative anaerobic bacilli belonging to the family *Aeromonadaceae*. Although *A. hydrophila* is usually the dominant species, other aeromonads, such as *A. caviae* and *A. sobria*, have also been isolated from human feces and from water sources [45,54].

Aeromonas species, including *A. hydrophila*, are ubiquitous in the environment. It is frequently isolated from food, drinking water, and aquatic environments [45,53,54]. In clean rivers and lakes, concentrations of *Aeromonas* spp. are usually around 10^2 colony-forming units (CFU)/mL. Groundwaters generally contain less than 1 CFU/mL. Drinking water immediately leaving the treatment plant has been found to contain between 0 and 10^2 CFU/mL. Drinking water in distribution systems can display higher *Aeromonas* concentrations, due to the growth in biofilms [45,55]. *Aeromonas* spp. have been found to grow between 5 °C and 45 °C [44,54]. *A. hydrophila* is resistant to standard chlorine treatments, probably surviving inside biofilms [56].

The common routes of infection suggested for *Aeromonas* are the ingestion of contaminated water or food or contact of the organism with a break in the skin. Drinking or natural mineral water can be a possible source of contamination for humans. No person-to-person transmission has been reported [45,54].

MICROBIOLOGICAL WATER ANALYSIS

The Rationale of the Use of Fecal Indicator Bacteria

The most important bacterial gastrointestinal diseases transmitted through water are cholera, salmonellosis and shigellosis. These diseases are mainly transmitted through water (and food) contaminated with feces of patients. Drinking water can be contaminated with these pathogenic bacteria, and this is an issue of great concern. However, the presence of pathogenic bacteria in water is sporadic and erratic, levels are low, and the isolation and culture of these bacteria is not straightforward. For these reasons, routine water microbiological analysis does not include the detection of pathogenic bacteria. However, safe water demands that water is free from pathogenic bacteria [57].

The conciliation of the two needs was met by the discovery and testing of indicator bacteria. Water contaminated with pathogenic species also has the normal inhabitants of the human intestine. A good bacterial indicator of fecal pollution should fulfill the following criteria: (1) exist in high numbers in the human intestine and feces; (2) not be pathogenic to humans; (3) easily, reliably and cheaply detectable in environmental waters. Additionally, the following requisites should be met if possible: (4) does not multiply outside the enteric environment; (5) in environmental waters, the indicator should exist in greater numbers than eventual pathogenic bacteria; (6) the indicators should have a similar die-off behavior as the pathogens; (7) if human fecal pollution is to be separated from animal pollution, the indicator should not be very common in the intestine of farm and domestic animals [1,4,6,57,58]. The usefulness of indicator bacteria in predicting the presence of pathogens was well illustrated in many studies, namely by Wilkes *et al.* [59].

The Composition of Human and Animal Feces

Microbiological analysis of the human feces was important in order to structure and validate the use of fecal indicator bacteria in environmental waters. Bacteria present in feces are naturally derived from the microbiota of the human gastrointestinal tract.

Although bacteria are distributed throughout the human gastrointestinal tract, the major concentration of microbes and metabolic activity can be found in the large intestine. The upper bowel (stomach, duodenum, and jejunum) has a sparse microbiota with up to 10^5 CFU/ml of contents. From the ileum on, bacterial concentrations gradually increase reaching in the colon 10^{10} to 10^{11} CFU/g [60].

It has been estimated that at least 500–1,000 different microbial species exist in the human gastrointestinal microbiota, although on a quantitative basis 10–20 genera usually predominate (Table 6). The total number of microbial genes in the human gastrointestinal tract has been estimated as 2–4 million. This represents an enormous metabolic potential which is far greater than that possessed by the human host [60,64].

Table 6: Total viable count in feces of healthy humans (children, adults and elderly)**a**

Microbial group	**Log_{10} CFU/g feces**
Bacteroides	11.3*
Eubacterium	10.7*
Bifidobacterium	10.2*
Ruminococcus	10.2*
Peptostreptococcus	10.1*
Peptococcus	10.0*
Clostridium	9.8*
Lactobacillus	9.6*
Propionobacterium	9.4*
Actinomyces	9.2*
Methanobrevibacter	8.8*
Desulphovibrio	8.4*
Fusobacterium	8.4*
Enterococci	3.5–7.2**
Enterobacteriaceae	5.9–8.0**
Escherichia coli	7.5–7.7**
Citrobacter	3.3**
Klebsiella	2.4**
Yeasts	1.0–2.5**

[a]Adapted from [61–63]** and [64]*.
*Values expressed as dry weight.
**Values expressed as wet weight.

The composition of feces from an individual is stable at genus level, but the species composition can vary markedly from day to day. The relative proportion of intestinal bacterial groups can vary between individuals [60,64].

The microflora of the human gastrointestinal tract is dominated by obligate anaerobes, which are ca. 10^3 more abundant than facultative anaerobes. The main anaerobic genera are *Bacteroides*, *Eubacterium* and Bifidobacteria. These organisms account for ca. 90% of the cultivable human fecal bacteria. *Bacteroides* (mainly *B. thetaiotaomicron* and *B. vulgatus*) are the most abundant organism in the human feces and account for 20–30% of cultivable bacteria. The most abundant facultative anaerobes are Enterococci and *Enterobacteriaceae*. The main *Enterobacteriaceae* genera are *Escherichia*, *Citrobacter*, *Klebsiella*, *Proteus* and *Enterobacter*. *Citrobacter* and *Klebsiella* are present in most individuals although in low numbers. *Proteus* and *Enterobacter* are only present in a minority of humans [64].

A variety of molecular techniques have been used to study the microbial composition of the human gastrointestinal tract. Results yielded by these studies have shown that many microbes detected by molecular techniques are not isolable by conventional culture-based methods. The presence of high proportions of bifidobacteria detected by culture-based methods is not supported by the results of molecular-based studies. However, the results of molecular-based approaches support many of the findings derived from culture-based methods: the dominance of the obligate anaerobes over facultative anaerobes; the presence of high counts of *Bacteroides*, *Clostridium* and *Eubacterium* [64].

Anaerobic bacteria such as *Bacteroides* and *Eubacterium* are not easily cultured by conventional techniques since require incubation chambers with nitrogen atmosphere. *Bifidobacterium* and *Lactobacillus* tolerate some oxygen but are fastidious bacteria growing very slowly in culture media. Therefore, these four genera are not adequate to be used as indicators of fecal pollution (the introduction of molecular techniques may improve the situation). *Citrobacter*, *Klebsiella* and *Enterobacter* are present in low numbers in the human intestine and are widespread in environmental waters, and therefore are also not suitable as indicators of fecal pollution. *Clostridium*, *Streptococcus* and *Escherichia* do not suffer from these drawbacks. Therefore, their suitability as fecal indicators has been tested since several decades.

Fecal Bacteria in Their Hosts and in the Environment

Bacteroides

The traditional genus *Bacteroides* included Gram-negative, non-sporeforming, anaerobic pleiomorphic rods. Many species have been transferred to other genera—*Mitsuokella, Porphyromonas, Prevotella, Ruminobacter*. Bacteroides are the most abundant bacteria in human feces. In animal feces, on the contrary, *Bacteroides* are present at low numbers. Although anaerobic, *Bacteroides* are among the most tolerant to oxygen of all anaerobic human gastrointestinal species. *B. thetaiotaomicron* is one of the most abundant species in the lower regions of the human gastrointestinal tract. *Bacteroides* have a high pathogenic potential and account for approximately two-thirds of all anaerobes isolated from clinical specimens. The most frequently isolated species has been *B. fragilis*. The survival of *Bacteroides* in environmental waters is usually much lower than the survival of coliforms [64,65].

Eubacterium

The traditional genus *Eubacterium* included anaerobic non-sporeforming Gram-positive rods. Some species have been transferred to other genera—*Actinobaculum, Atopobium, Collinsella, Dorea, Eggerthella, Mogibacterium, Pseudoramibacter* and *Slackia*. Cells are not very aerotolerant. Species isolated from the human gastrointestinal tract include: *E. barkeri*, E. *biforme*, *E. contortum*, *E. cylindrioides*, *E. hadrum*, *E. limosum*, *E. moniliforme, E. rectal* and *E. ventricosum* [64].

Bifidobacterium

Bifidobacteria are Gram-positive, non-sporeforming, pleiomorphic rods. Bifidobacteria are anaerobic (some species tolerate oxygen in the presence of carbon dioxide) or facultative anaerobic. The optimum growth temperature is 35–39 °C. The genus *Bifidobacterium* contains ca. 25 species, most of which have been detected in the human gastrointestinal tract [64–66].

Bifidobacteria are present in high numbers in the feces of humans and some animals. Several *Bifidobacterium* species are specific either for humans or for animals. *B. cuniculi* and *B. magnum* have only been found in rabbit fecal samples, *B. gallinarum* and *B. pullorum* only in the intestine of chickens and *B. suis* only in piglet feces. In human feces, the species composition changes with the age of the individual. In the intestine of infants *B. breve* and *B. longum* generally predominate. In the adult, *B. adolescentis*, *B. catenulatum*, *B. pseudocatenulatum* and *B. longum* are the dominant species. In both human and animal feces, bifibobacteria are always much more abundant than coliforms [64–66].

Bifidobacteria have been found in sewage and polluted environmental waters, but appears to be absent from unpolluted or pristine environments such as springs and unpolluted soil. This results from the fact that upon introduction into the environment, bifidobacteria decrease appreciably in numbers, probably due to their stringent growth requirements. Bifidobacteria grow poorly below 30 °C and have rigorous nutrient requirements. Reports on the survival of bifidobacteria in environmental waters indicate that their survival is lower than that of coliforms [64–66].

The presence of bifidobacteria in the environment is therefore considered an indicator of fecal contamination. Since some species are specific for humans and animals, the identification of *Bifidobacterium* species present in the polluted water could, in principle, provide information on the origin of fecal pollution [64–66].

A study carried out in a highly contaminated stream near Bologna, Italy, revealed that *B. adolescentis*, *B. catenulatum*, *B. longum*, *B. pseudocatenulatum* and *B. thermophilum* were the most representative species, whereas *B. angulatum*, *B. animalis* subsp. *animalis (B. animalis)*, *B. breve*, *B. choerinum*, *B. minimum*, *B. pseudolongum* subsp. *globosum* (*B. globosum*) and *B. subtile* occurred only in low numbers [66].

Bifidobacteria are the less studied of all fecal bacteria, due to the technical difficulties in their isolation and cultivation. Other Gram-positive bacteria, such as *Streptococcus* and *Lactobacillus*, which may occur in higher numbers than bifidobacteria, can inhibit their growth. Although selective media has been designed for the isolation of bifidobacteria from environmental waters, the outcome is still unsatisfactory, with appreciable numbers of false positives and low recovery percentages [64–66].

Clostridia

The genus *Clostridium* is one of the largest genera of the prokaryotes containing 168 validly published species. From these, 77 (including *C. perfringens*) are considered to belong to a united group—*Clostridium sensu stricto* [64,67,68].

Clostridia are Gram-positive rods, forming endospores. Most of the clostridial species are motile with peritrichous flagellation. Cells are catalase-negative and do not carry out a dissimilatory sulphate reduction. Clostridia usually produce mixtures of organic acids and alcohols from carbohydrates and proteins. Many species are saccharolytic and proteolytic. Some species fix atmospheric dinitrogen [64,67,68].

The genus *Clostridium* includes psychrophilic, mesophilic, and thermophilic species. The major role of these organisms in nature is in the degradation of organic material to acids, alcohols, CO_2, H_2, and minerals. Frequently, a butyric acid smell is associated with the proliferation of clostridia. The ability to form spores that resist dryness, heat, and aerobic conditions makes the clostridia ubiquitous [64,67,68].

Most species are obligate anaerobic, although tolerance to oxygen occurs. Oxygen sensitivity restricts the habitat of the clostridia to anaerobic areas or areas with low oxygen tensions. Growing and dividing clostridia will, therefore, not be found in air saturated surface layers of lakes and rivers or on the surface of organic material and soil. Clostridial spores, however, are present with high probability in these environments, and will germinate when oxygen is exhausted and when appropriate nutrients are present [64,67,68].

C. perfringens ferment lactose, sucrose and inositol with the production of gas, produce a stormy clot fermentation with milk, reduce nitrate, hydrolyze gelatin and produce lecithinase and acid phosphatase. The species is divided into five types, A to E, on the basis of production of major lethal toxins [68,69].

C. perfringens appears to be a universal component of the human and animal intestine, since has been isolated from the intestinal contents of every animal that has been studied. Humans carry *C. perfringens* as part of the normal endogenous flora. The main site of carriage is the distal gastrointestinal tract. The principal habitats of type A are the soil and the intestines of humans, animals, and birds. Types B, C, D, and E appears to be obligate parasites of animals and occasionally are found in humans [68,69].

Clostridium perfringens is the most frequently isolated *Clostridium* in clinical microbiology laboratories, although it seldom causes serious infections. *C. perfringens* is isolated from infections in humans and the organism most commonly found in gas gangrene in humans. *C. perfringens* is most commonly isolated from infections derived from the colonic flora, namely peritonitis or abdominal abscess [68,69].

This organism is a common cause of food poisoning due to the formation of the enterotoxin in the intestine. *C. perfringens* food poisoning is seldom fatal, being marked by diarrhea and nausea, with no vomiting and no fever [68,69].

Sources yielding *C. perfringens* include soil and marine sediment samples worldwide, clothing, raw milk, cheese, semi-preserved meat products, and venison. Like *E. coli*, *C. perfringens* does not multiply in most water environments and is a highly specific indicator of fecal pollution. Berzirtzoglou *et al.* [70] reported a comparative study on the occurrence of vegetative cells and spores of *Clostridium perfringens* in a polluted station of the lake Pamvotis, in rural North-West Greece. The numbers of *C. perfringens* varied according to the water depth. Sporulated forms were found in all sampling sites with the exception of the surface sampling.

Lactobacillus

Lactobacilli are non-sporeforming Gram-positive long rods. There are more than thirty species in the genus. Most are microaerophillic, although some are obligate anaerobes. Cells are catalase-negative and obtain their energy by the fermentation of sugars, producing a variety of acids, alcohol and carbon dioxide. Lactobacilli have complex nutritional requirements and in agarized media may need the supplementation with aminoacids, peptides, fatty-acid esters, salts, nucleic acid derivatives and vitamins. Lactobacilli very rarely cause infections in humans [64].

Enterococci

Enterococci are Gram-positive, non-sporeforming, catalase-negative ovoid cells. Cells occur singly, in pairs or short chains. Optimal growth for most species is 35–37 °C. Some will grow at 42–45 °C and at 10 °C. Growth requires complex nutrients but is usually abundant on commonly used bacteriological media. Cells are resistant to 40% bile, 0.4% azide, 6.5% sodium chloride, have β-glucosidase and hydrolyze esculin. The enterococci are facultative anaerobic but prefer anaerobic conditions [64,71].

The genus was separated from *Streptococcus* in the 1980s. Enterococci form relatively distinct groups. Members of such groups exhibit similar phenotypic characteristics and species delimitation can be difficult. The *E. faecalis* group contains, among others, *E. faecalis*. The *E. avium* group contains, among others, *E. avium*. The *E. faecium* group contains, among others, *E. faecium*, *E. durans* and *E. hirae*. The *E. gallinarum* group contains, among others, *E. gallinarum* [64,71].

Most species are part of the intestinal flora of mammals, reptiles, birds, and other animals. In the human digestive tract, *E. faecalis* is the prevailing species, although in particular situations, *E. faecium* may predominate. In poultry, *E. cecorum*, *E. durans*, *E. faecalis*, *E. faecium* and *E. hirae* and dominate the intestinal flora [64,71].

Enterococci have been increasingly isolated from a variety of nosocomial and other infections, mainly from the urinary tract and wound infections, bacteremias, and endocarditis [64,71].

Although enterococci are considered only a temporary part of the microflora of plants, in optimal conditions, cells can proliferate on their surfaces. *E. casseliflavus*, *E. faecalis*, *E. faecium*, *E. hirae*, *E. mundtii* and *E. sulfureus* have been isolated from plants. They are generally isolated more often from flowers than from buds or leaves [64,71].

Enterococci are naturally present in many kinds of foods, especially those of animal origin such as milk and milk products, meat and fermented sausages. Enterococci are usually considered secondary contaminants of food, although they often play a positive role in ripening and aroma development of some types of cheeses [64,71]. Although soil is not a natural habitat for enterococci, cells can be found in this habitat due to the transport by rain [64,71].

Environmental waters are not a natural habitat for enterococci and their presence in this milieu is considered the result of fecal pollution. The most common species found in environmental waters are *E. durans, E. faecalis, E. faecium* and *E. hirae*, and less commonly, *E. avium*, *E. cecorum*, *E. columbae* and *E. gallinarum*. However, pristine waters in Finland have been reported to contain *E. casseliflavus* [64,71].

In environmental samples (compost, sewage effluent, harbor sediments, brackish water and swimming pool water), Pinto *et al.* [72] reported the isolation of *E. casseliflavus*, *E. durans*, *E. faecalis*, *E. faecium*, *E. gallinarum* and *E. hirae. E. durans*, *E. faecium* and *E. hirae* were isolated from all sources except from harbor sediments. *E. raffinosus* was only isolated from compost and swimming pool water. *E. faecalis* and *E. faecium* accounted for the vast majority of enterococcal strains.

Escherichia

Escherichia, a member of *Enterobacteriaceae*, are oxidase-negative catalase-positive straight rods that ferment lactose. Cells are positive in the Methyl-Red test, but negative in the Voges-Proskauer assay. Cells do not use citrate, do not produce H_2S or lipase, and do not hydrolyze urea [73]. *E. coli* is a natural and essential part of the bacterial flora in the gut of humans and animals. Most *E. coli* strains are nonpathogenic and reside harmlessly in the colon. However, certain serotypes do play a role in intestinal and extra-intestinal diseases, such as urinary tract infections [43]. In a study of the enteric bacteria present in the feces of Australian mammals, Gordon and FitzGibbon [74] reported that *E. coli* was the commonest species, being isolated from nearly half of the species studied.

Citrobacter

Citrobacter, a member of *Enterobacteriaceae*, are motile straight rods. Cells are oxidase-negative, catalase-positive and positive in the Methyl-Red test. Cells use citrate, are negative in the Voges-Proskauer test and do not decarboxylate lysine [73].

In a study of the enteric bacteria present in the feces of Australian mammals, Gordon and FitzGibbon [74] reported the isolation of *C. amalonaticus, C. freundii* and *C. koseri* (*C. diversus*). *Citrobacter* species can be isolated from different clinical sites. In particular, *C. freundii* is an intestinal inhabitant of humans that may sometimes have—or acquire—the ability to produce an enterotoxin and thus become an intestinal pathogen. *Citrobacter* is reported to occur in environments such as water, sewage, soil and food [75,76].

Klebsiella and Raoultella

Klebsiella and *Raoultella* are *Enterobacteriaceae*, oxidase-negative catalase-positive non-motile straight rods, surrounded by a capsule. Cells decarboxylate lysine, but are ornithine and arginine dihydrolase negative. Cells grow on KCN, do not produce H_2S and ferment most carbohydrates [73].

In humans, *K. pneumoniae* is present as commensal in the nasopharynx and in the intestinal tract. *Klebsiella* spp. can cause human diseases, ranging from asymptomatic colonization of the intestinal, urinary, or respiratory tract to fatal septicemia. *Klebsiella* are mostly considered nosocomial pathogens.

K. pneumoniae and *Enterobacter aerogenes* (*K. mobilis*) are most frequently involved, although *K. oxytoca* and *R. planticola*, and rarely *R. terrigena*, can be found. In the hospital, the principal reservoir of *K. pneumoniae* is the gastrointestinal tract of patients. The principal vectors are the hands of personnel [77,78]. In a study of the enteric bacteria present in the feces of Australian mammals, Gordon and FitzGibbon [74] reported the isolation of *K. pneumoniae* and *K. oxytoca*.

Klebsiellae are ubitiquous in the environment. They have been found in a variety of environmental situations, such as soil, vegetation, or water, and they influence many biochemical and geochemical processes. They have been recovered from aquatic environments receiving industrial wastewaters, plant products, fresh vegetables, food with a high content of sugars and acids, frozen orange juice concentrate, sugarcane wastes, living trees, and plants and plant byproducts. They are commonly associated with wood, sawdust, and waters receiving industrial effluents from pulp and paper mills and textile finishing plants (see below). *Klebsiella* have been isolated from the root surfaces of various plants. *K. pneumoniae*, *K. oxytoca*, and *R. planticola* are all capable of fixing dinitrogen [77,78].

Enterobacter

Enterobacter a member of *Enterobacteriaceae*, are motile straight rods. Cells are positive in the Voges-Proskauer test VP and in Simmons citrate agar. Cells do not decarboxylate lysine, but are ornithine positive. Malonate is usually utilized and gelatin is slowly liquefied. Cells do not produce H_2S, deoxyribonuclease and lipase [73].

In a study of the enteric bacteria present in the feces of Australian mammals, Gordon and FitzGibbon [74] reported the isolation of *Enterobacter cloacae* subsp. *cloacae* (*E. cloacae*), *E. cancerogenus* (*E. taylorae*) and *E. aerogenes* (*Klebsiella mobilis*).

Before the widespread use of antibiotics, *Enterobacter* species were rarely found as pathogens, but these organisms are now increasingly encountered, causing nosocomial infections such as urinary tract infections and bacteremia. In addition, they occasionally cause community-acquired infections [79,80].

In the USA, the Surveillance and Control of Pathogens of Epidemiological Importance project analyzed 24,179 nosocomial bloodstream infections, from 1995–2002. *Enterobacter* species were the second most common gram-negative organism, behind *Pseudomonas aeruginosa*. Both bacteria were reported to each represent 4.7% of bloodstream infections in intensive care units. *Enterobacter* species represented 3.1% of bloodstream infections in non-intensive care units. Of nearly 75,000 gram-negative organisms collected from intensive care units' patients in the USA, between 1993 and 2004, *Enterobacter* species comprised 13.5% of the isolates. Multidrug resistance increased over time, especially in infections caused by *E. cloacae* [81].

In the USA, the National Healthcare Safety Network reported a study on healthcare-associated infections between 2006 and 2007. They found *Enterobacter* species to be the eighth most common cause of healthcare-associated infections (5% of all infections) and the fourth most common gram-negative cause of these infections [82].

Enterobacter cloacae subsp. *cloacae* (*E. cloacae*) occurs in the intestinal tracts of humans and animals, in hospital environments, the skin, in water, sewage, soil, meat. Nitrogen-fixing strains have been isolated from the roots of rice plants. *E. amnigenus* has been mostly isolated from water, but some strains were isolated from clinical specimens from the respiratory tract, wounds and feces. *E. asburiae* strains were isolated from clinical specimens, mostly urine, respiratory tract, feces, wounds, and blood [79,80].

Origin of the Use of Fecal Indicator Bacteria

Historically, the design and use of indicators of fecal pollution comes from the end of the 19th to beginning of the 20th century. In 1880, von Fritsch described *Klebsiella pneumoniae* and *K. pneumoniae* subsp. *rhinoscleromatis* (*Klebsiella rhinoscleromatis*) as micro-organisms characteristically found in human feces [83]. In 1885, Escherich described several microorganisms in the feces of newborn and suckling babies. This included a motile, rod-shaped microorganism that caused milk to clot, which was named "*Bacterium coli commune*". He observed that within a few weeks after birth, this bacterium became the dominant organism in the infant colon [6]. Also in 1885, Percy and Grace Frankland started the first routine bacteriological examination of water in London, using Robert Koch's solid gelatin media to count bacteria [83]. In 1891, Percy and Grace Frankland came up with the concept that organisms characteristic of sewage must be identified to provide evidence of potentially dangerous pollution [83]. In 1892, Schardinger proposed that since "*Bacterium coli*" was a characteristic component of the fecal flora, its presence in water could be taken as an indication of the presence of fecal pollution and therefore of the potential presence of enteric pathogens [6].

Soon after the description of "*Bacterium coli*", other bacteria were isolated from stools and water—*Klebsiella* in 1882 and *Enterobacter* in 1890 [6]. By 1893, the "Wurtz method" of enumerating "*Bacterium coli*", by direct plating water samples on litmus lactose agar, was being used by sanitary bacteriologists. This was based on the concept of acid and gas production (detected by the Durham tube) from lactose as a diagnostic feature [6]. In 1905, MacConkey described his now famous MacConkey's broth, which was diagnostic for lactose-fermenting bacteria tolerant of bile salts. Coliforms were already considered to be a heterogeneous group of organisms, many of which were not of fecal origin. The origins of the critical observation that "*Bacterium coli*" was largely fecal in origin while other coliforms were not, could be claimed by Winslow and Walker in 1907 [83].

Various classification schemes for coliforms have emerged. The earliest were those of MacConkey in 1909, which recognized 128 different coliform types, while Bergey and Deehan in 1908, identified 256. By the early 1920s, differentiation of coliforms had come to a series of correlations that suggested that indole production, gelatin liquefaction, sucrose fermentation and the Voges–Proskauer reaction were among the more important tests for determining fecal contamination. These developments culminated in the IMViC (Indole, Methyl Red, Voges–Proskauer and Citrate) tests for the differentiation of so-called fecal coliforms, soil coliforms and intermediates [83].

Fecal Indicator Bacteria

Coliforms

Total coliforms are Gram-negative, oxidase-negative, non-sporeforming rods, that ferment lactose with gas production at 35–37 °C, after 48h, in a medium with bile salts and detergents [1,4,6,57,84]. When the test of coliforms is carried out with environmental waters, several species of the four *Enterobacteriaceae* genera *Escherichia*, *Klebsiella*, *Enterobacter* and *Citrobacter* give positive results and therefore are coliforms according to this definition. However, the environmental significance of these four genera is very disparate as discussed in the present text. Therefore, total coliform counts are not necessarily a measure of fecal pollution and indeed can have no relation with this cause [1,4,6,84].

Fecal coliforms (or thermotolerant coliforms) are traditionally defined as coliforms that ferment lactose at 44.5 °C in a medium with bile salts

[1,4,57,84]. The range of species detected by the experimental procedure is much lower than that of total coliforms. With environmental polluted waters, only *E. coli*, and *K. oxytoca* and *K. pneumoniae* gave positive results in the test [85].

Traditional tests for total and fecal coliforms are carried out either by the multiple-tube fermentation technique or by filtration through membrane. The multiple-tube fermentation technique is used for medium or highly contaminated waters, and the filtration through membrane for low or very low contaminated waters. Filtration through membrane is a very sensitive technique since can detect one (culturable) cell in 500 or even 1,000 mL of water. However, both methods take several days to complete and do not detect viable but non-culturable bacteria [3,57,86]. These limitations stimulate the discovery of alternative methods, faster and, if possible, less prone to false negative results such as those caused by the viable but non-culturable bacteria.

The detection of β-D-galactosidase activity (at 37 °C) is usually a good marker for total coliforms in environmental waters, since most of these bacteria display this enzymatic activity [1,3,57,87–91]. Most *Escherichia*, *Citrobacter*, *Enterobacter*, *Klebsiella* and *Raoultella* strains have galactosidase. *Hafnia*, *Serratia* and *Yersinia* also possess this enzymatic activity. Most *Proteus*, *Salmonella* and *Edwardsiella* strains do not display β-galactosidase [92–95]. Ca. 10% of the coliform strains isolated from the environment do not have an active formic hydrogenolyase (cleaves formate with the formation of CO_2) and therefore do not produce gas being undetected by the traditional techniques but are detected by the assay of β-galactosidase activity [57,96,97].

β-galactosidase cleaves lactose in glucose and galactose, and can be detected by using colored or fluorescent markers that change color after enzyme action, such as XGAL (5-bromo-4-chloro-3-indol-β-galactopyranoside) and ONPG (O-nitrophenyl-β-D-galactopyranoside) or MUGAL (4-methylumbelliferyl-β-D-galactopyranoside), respectively [57,96,97].

In environmental waters, the presence of *Aeromonas* or *Vibrio cholerae* can be a source of false positives in the β-D-galactosidase assay, since these bacteria have galactosidase, but are not coliforms [93,95,96,98]. Additionally, in particular environments, such as estuaries, β-galactosidase activity can overestimate total coliform count due to UV-stimulated enzymatic activity in certain bacteria such as *E. coli* [86].

The detection of β-D-glucuronidase activity (at 44.5 °C) is, generally, a good marker for fecal coliforms in environmental polluted waters and very specific for *E. coli* [13,85,87–91,97,99–101]. In Gram-negative bacteria, this enzymatic activity if found in most *E. coli* strains and in some *Salmonella* and *Shigella* strains [92–95,97,102]. *Aeromonas*, *Citrobacter*, *Enterobacter*, non-*coli Escherichia*, *Hafnia*, *Klebsiella*, *Proteus*, *Serratia*, *Vibrio*, *Yersinia*, and most *Salmonella* strains do not display β-glucuronidase activity [93–95,101,102].

β-D-glucuronidase activity can be detected by using colored or fluorescent markers that change color after enzyme action, such as XGLUC (5-bromo-4-chloro-3-indoxyl-β-d-glucuronide), IBDG (indoxil-β-glucuronide), and MUGLU (4-methylumbelliferyl-β-d-glucuronide), respectively [57,97,100].

The presence of this enzyme in some strains of *Bacteroides*, *Flavobacterium*, *Staphylococcus*, *Streptococcus*, in anaerobic corynebacteria and *Clostridium*, has also been reported [93,95–97,102]. β-D-glucuronidase activity in fecal bacteria other then *E. coli* (*Bacteroides*, bifidobacteria, clostridia, enterococci and *Lactobacillus*) is very limited [61]. Although all these glucuronidase positive bacteria could lead to false positive detections in the fecal coliform test, experimental results for environmental polluted waters indicate a significant correlation between fecal coliform detection using conventional techniques and the glucuronidase assay, suggesting that false positives are not significant [85,96].

The detection of total coliforms and fecal coliforms by enzymatic methods are much less time consuming than traditional techniques. With fluorescent markers and the use of a spectrofluorimeter the detection of coliforms can be performed in minutes [57,101]. However, in very low contaminated waters, enzymatic methods might not be able to detect coliform cells. Moreover, on-line monitoring of glucuronidase activity is currently too insensitive to replace culture based detection of *E. coli*. Nevertheless, on-line enzymatic methods can be a valuable complementary tool for high temporal resolution monitoring. More research is needed in order to enhance sensitive and lower detection limits of available on-line glucuronidase techniques.

The seminal work of Leclerc *et al.* [103] clarified the diversified roles that coliforms have in the environment and the real meanings of the tests on total coliforms and fecal coliforms. It was shown that *Enterobacteriaceae* encompass three groups of bacteria with very different roles in the environment. Group I harbored only *E. coli*. Since this species usually do not survive for long periods outside this environment (but see topic 10), it was considered a good and reliable indicator of fecal pollution (both animal and human). Group II, the "ubiquitary" group, encompassed several species of *Klebsiella* (*K. pneumoniae* and *K. oxytoca*), *Enterobacter* (*Enterobacter cloacae* subsp. *cloacae*, *E. aerogenes*) and *Citrobacter* (*C. amalonaticus*, *C. koseri* and *C. freundii*). These bacteria live in the animal and human gut, but also in the environment, and are easily isolated from the soil, polluted water and plants. Their presence in polluted waters does not necessarily indicate fecal pollution. Finally Group III was composed of *Raoultella planticola*, *R. terrigena*, *Enterobacter amnigenus* and *Kluyvera intermedia* (*Enterobacter intermedius*), *Serratia fonticola*, and the genera *Budvicia*, *Buttiauxella*, *Leclercia*, *Rahnella*, *Yersinia*, and most species of *Erwinia* and *Pantoea*. These bacteria live in fresh waters, plants and small animals. They grow at 4 °C, but not at 41 °C. They are not indicators of fecal pollution, although can be detected in the total coliform test. Leclerc *et al.* concluded that: (1) in the enterobacteria, *E. coli* is the only true and reliable indicator of fecal pollution in environmental waters; (2) the traditional total coliform test should be abandoned, since it can detect bacteria that have no connection with fecal pollution; (3) the detection of fecal coliforms must be carried out at 44.5 °C, and positive results confirmed by identification to species levels in order to exclude false positives such as *K. pneumoniae*.

Streptococci and Enterococci

Fecal streptococci also belong to the traditional indicators of fecal pollution. Fecal streptococci are Gram-positive, catalase-negative, non-sporeforming cocci that grow at 35 °C in a medium containing bile salts and sodium azide. Cells hydrolyze esculin [1,4,57]. Azide is a strong inhibitor of the respiratory chain. Since streptococci are one of the very few bacteria that have no respiratory chain, the test is very specific for this group, and false positives are rarely found [104,105].

Fecal enterococci (*E. faecalis*, *E. faecium, E. avium* and *E. gallinarum*) are fecal streptococci that grow in the presence of 6.5% NaCl at 45 °C. Selective media use these particular characteristics in order to separate enterococci from the other streptococci [104,105].

Several studies [104,106] have reported on the microbiological composition of human and animal (cattle, chicken, deer, dog, fowl, goose, and swine) feces. *E. faecalis* and *E. faecium* were present in human and animal feces. However, whereas human feces almost have only these two enterococci, in the animals others species co-occur, like *E. avium*, *E. cecorum*, *E. durans*, *E. gallinarum* and *E. hirae*. It was concluded that in urban areas where contamination with dog and chicken feces is not likely, the best marker for human fecal pollution was *E. faecalis*.

The intestinal enterococci group has been used as an index of fecal pollution. In human feces, the numbers of intestinal enterococci are generally about an order of magnitude lower than those of *E. coli* (Table 6). Most species do not grow in environmental waters. In this milieu, fecal enterococci are able to survive longer, are more resistant to drying and chlorination, than *E. coli* [1,84].

However, caution should be taken with interpreting the results obtained by the enterococci procedure in water analysis. Enterococci and other group D-streptococci are present in many foods, especially those of animal origin. The isolation of *E. faecalis* and *E. faecium* was used to indicate fecal contamination of food. However, enterococci are now also considered as normal parts of the food microflora and not only as indicators for poor hygiene [104]. In addition, agricultural soils and crops with added with manure also harbor enterococci [105].

The use of ratios between indicator counts

The ratio of counts of fecal coliforms to fecal streptococci has been proposed as a means to differentiating between contamination from human and animal sources. Ratios greater than 4 have been suggested to indicate a human source whereas ratios less than 0.7 suggest an animal source. This results from the fact that streptococcal concentrations in human feces are generally less than coliforms (Table 6). In contrast, in animal feces fecal streptococci generally outnumber fecal coliforms (Table 7). In urban sewage, fecal streptococci tend to be present in concentrations 10–100 times less than fecal coliforms [65].

Table 7: Bacteria in the feces of farm and domestic warm-blooded animals[a]

Animal	**Log_{10} cells/g wet weight feces**		
	Fecal coliforms	**Fecal strepto-cocci**	***Clostridium perfringens***
Chicken	5.4	6.1	2.3
Duck	7.5	7.7	-
Horse	4.1	6.8	< 0
Pig	6.5	7.9	3.6
Sheep	7.2	7.6	5.3
Turkey	5.5	6.4	-
Cat	6.9	7.4	7.4
Dog	7.1	9.0	8.4

[0za]Adapted from [83].

Geldreich [107] summarized the information available on the fecal coliforms to fecal streptococci ratios in the feces of warmblooded animals, and reported the following values: human feces, 4.3; cattle, sheep, and poultry, from 0.104 to 0.421; and wild animals (including rabbits, field mice, chipmunks, and birds), 0.0008 to 0.043. Fecal coliforms to fecal streptococci ratios for the feces of wild animals appear to be at least 10-fold lower than those of domestic livestock.

Doran and Linn [108] reported a study of the runoff from a cow-calf pasture in eastern Nebraska (USA), monitored during a three-year period. It was concluded that the fecal coliforms to fecal streptococci ratio in pasture runoff was useful in identifying the relative contributions of cattle and wildlife and in evaluating the effects of cattle management and distribution on runoff water quality. Ratios below 0.05 were indicative of wildlife sources and ratios above 0.1 were characteristic of grazing cattle. Fecal coliforms to fecal streptococci ratios of diluted cattle waste in excess of 1 were interpreted as the result of differential aftergrowth and die-off between fecal coliforms and fecal streptococci. Ratios between 0.7 and 4.0 may indicate situations where cattle are localized close to sampling or outflow points.

However, the interpretation of this ratio should be cautious. It has been observed a shift in the ratio with time and distance from the fecal pollution source. This resulted from the fact that both in surface and groundwaters, fecal streptococci are more persistent than fecal coliforms. Therefore increasing the distance from the pollution point and with passing time, the ratio tends to decrease without a change in the nature of the pollution source [65]. For these reasons, this ratio has been considered by some authors as too unreliable to be useful in characterizing pollution sources [65,84].

The ratio of fecal enterococci to fecal streptococci differs among vertebrate species. Humans have a predominance of enterococci, whereas animals contain appreciable amounts of streptococci. However, since enterococci are also present in animals and are more persistent in the environment than other fecal streptococci, the identification of the enterococci and streptococci species present in polluted waters, and the concomitant calculation of this ratio is generally considered unreliable as an indicator of the source of fecal pollution [65].

Limitations of coliform and enterococcus counts as indicator of fecal pollution

An extreme case of uselessness of the determination of total and fecal coliforms and enterococci in the assessment of fecal pollution has been demonstrated by several authors studying the microbiology of pulp and paper mill effluents.

Caplenas and Kanarek [109] reported a study of pulp and paper mills located in Wisconsin (USA). Fresh water supplies, re-cycled water within mills, treated effluent wastewater and waters receiving effluent wastes downstream, were assessed for the presence of fecal coliforms and *Klebsiella*. Wastewaters prior to treatment contained fecal coliforms and *Klebsiella*. Up to 84% of the fecal coliforms (detected by the standard test procedure) were indeed *Klebsiella*. In treated effluent wastewaters this value reached 90%. Treatment of the wastewater lowered the concentration of "true" fecal bacterial contamination, but since *Klebsiella* grew rapidly in the wastewaters, fecal coliform counts were high, although no true fecal contamination was involved. The source of *Klebsiella* was traced to the early pulping stages in the mills. *Klebsiella* maintains a wood, bark or soil reservoir. It was concluded that: (1) *Klebsiella* are ubiquitous in the pulp and paper mill industry processing stages; (2) the standard procedure for fecal coliform estimation is useless to assess the microbiological quality of the effluents of these industries; (3) The assay of *E. coli* should replace the fecal coliform detection procedure.

Gauthier *et al.* [110] and Gauthier and Archibald [58] reported two studies of seven pulp and paper mills in Ontario and Quebec, Canada. Total and fecal coliforms and enterococci were detected in nearly all the biotreatment, biosolids (sludges), and in-mill water system samples. In the mill samples, the majority of the fecal coliforms (detected by the standard test procedure) were *K. pneumoniae*, *Raoultella terrigena* and *Raoultella planticola*, with *E. coli* in minority. *E. faecalis* and *E. faecium* were detected in relatively large numbers in most samples from all of the seven mills examined. Other coliforms such as *Enterobacter* spp. and *Citrobacter freundii* were occasionally recovered from total and fecal coliform tubes. Biofilms established in the piping, tanks, and machinery where thermal and pH conditions permit were the most likely source of these bacteria. Analyses using two independent *Salmonella* detection/enumeration methods showed no detectable *Salmonella* cells in the sludges and final effluents of the five mills tested. It was concluded that for these particular systems, the determination of total and fecal coliforms and enterococci is useless and have no relationship with real fecal pollution. These studies also demonstrated the importance of checking the identities of bacteria causing the positive results in the tests. Both *Escherichia* and *Klebsiella* can give positive results in the fecal coliform test, but their ecological meaning is opposite.

Another important case of failure of the use of coliforms to detect fecal pollution was the 1993 *Cryptosporidium* outbreak in Milwaukee (USA).

Cryptosporidium parvum, a protozoan parasite that causes gastrointestinal illness, is transmitted by ingestion of oocysts excreted in human or animal feces. Typical modes of transmission include person to person, animal to person, by exposure to contaminated food or water [111,112].

From 1990 to 2000, at least 10 cryptosporidiosis outbreaks associated with contaminated drinking water were reported in the USA. In 1993, an estimated 403,000 residents of the greater Milwaukee area (Wisconsin, population, ca. 1.61 million) became ill when an ineffective filtration process led to the inadequate removal of *Cryptosporidium* oocysts in one of two municipal water-treatment plants [111,112].

It was the largest waterborne disease outbreak in documented USA history. Over the span of approximately two weeks, people became ill with stomach cramps, fever, diarrhea and dehydration caused by the pathogen. More than half the people who received residential drinking water from the southern water-treatment plant became ill, which was twice the rate of illness among people whose residential drinking water came mainly from the northern water-treatment plant. Over 54 deaths were attributed to this outbreak, mostly among the elderly and immunocompromised people, such as AIDS patients [111,112].

Standard microbiological water analysis was ineffective in detecting this parasite. Indeed, throughout the period from February to April, samples of treated water from both plants were negative for coliforms. The origin of the contamination was determined as water from Lake Michigan. No specific source of the *Cryptosporidium* was ever identified but runoff from abnormally heavy spring rains most likely carried the parasite to the lake [111,112].

Clostridium perfringens

Sulphite-reducing clostridia, namely *Clostridium perfringens*, are spore-forming Gram-positive, non-motile, anaerobic, sulfite-reducing rods. *C. perfringens* is present in higher numbers in the feces of some animals, such as dogs, than in the feces of humans and less often in the feces of many other warm-blooded animals. The numbers excreted in feces are normally substantially lower than those of *E. coli.*

Clostridium spores are exceptionally resistant to unfavorable conditions in water environments, including UV irradiation, temperature and pH extremes, and disinfection processes, such as chlorination. Although clostridia probably do not growth in surface waters, the high resistance of their spores makes their presence ubiquitous in environmental waters [1,4,84,113].

The presence of chlorine in water rapidly inactivates indicator bacteria such as *E. coli* and coliforms, but it leaves the most resistant pathogens almost unaffected for several hours. This creates a false sense of security by providing negative coliform and negative *E. coli* results to authorities responsible for water testing. *Giardia* cysts, *Crystosporidium* oocysts, and human enteric viruses all have higher resistance to disinfectants and constitute a major public health risk if distribution system integrity is breached. *C. perfringens* spores are less affected by the residual concentrations of chlorine. Testing for the spores of this bacterium can probably provide an added margin of safety in the evaluation of treatment [114].

Correlations between parameters used to assess fecal pollution

In environmental waters, several studies have reported significant correlations between indicators of fecal pollution and between indicators and pathogenic gastrointestinal bacteria.

Charriere *et al.* [115] reported a study of deep aquifer waters (raw waters and piped chlorinated waters) in Normandy, France. In heavily contaminated raw waters and in slightly contaminated treated waters, fecal coliforms and enterococci were correlated.

Martins *et al.* [116] reported a study of 60 public outdoor swimming pools in Sao Paulo city, Brazil. Total coliforms, fecal coliforms and fecal streptococci levels increased with number of bathers and water temperature, and decreased with chlorine levels. All these indicators were significantly correlated with each other.

Ferguson *et al.* [117] reported a study carried out in Georges River, in the Sydney region, Australia. In the water column, concentrations of fecal coliforms, fecal streptococci, *C. perfringens* spores were all positively correlated with each other. Isolation of *Salmonella* spp. were most frequent during rainfall and sewage overflow events. In the water column, 55% of samples contained *Salmonella* when fecal coliform densities exceeded 2,000 CFU/100 mL.

Medema *et al.* [118] reported a study of seven different fresh water sites normally used for triathlon competitions. Sites were small rivers, channels, lakes and harbors, and were influenced by sewage effluents and agricultural run-off. When data from all triathlons were pooled, geometric mean densities of fecal coliforms and *E. coli*, of *E. coli* and fecal enterococci and of fecal coliforms and fecal enterococci, were significantly correlated.

Polo *et al.* [119] reported a study of water samples obtained from 213 beaches, eight rivers and 14 freshwaters in north-eastern Spain. In freshwaters and heavily contaminated seawaters, *Salmonella* and fecal coliforms were correlated, while in less contaminated seawaters, the highest correlation was with between *Salmonella* and fecal streptococci.

Byamukama *et al.* [99] reported a study of the microbiology of Nakivubo channel, Uganda. This channel receives raw sewage from slums, industrial effluents, and discharges from a sewage treatment plant and from a complex of slaughterhouses. Water from eight sampling sites was assessed for the presence of total and fecal coliforms, *E. coli* and sulphite-reducing clostridia. All microbiological parameters were significantly correlated.

Noble *et al.* [120] reported a comparative determination of total coliforms, fecal coliforms and enterococci in 108 sites along the southern California coastline, USA. Results by traditional and enzymatic methods and from all three parameters were correlated.

Harwood *et al.* [121] reported a study on indicator and pathogenic microorganisms carried out in six wastewater reclamation facilities in the USA. Data from disinfected effluent (reclaimed water) samples were analyzed separately (by facility) and as a pooled data set (all facilities). Significant correlations between indicator organism concentrations were observed in the pooled data sets, namely for total and fecal coliforms.

Cabral and Marques [85] reported a microbiological study of a polluted river (Febros) in the Great Oporto area, northwest Portugal. Total and fecal coliforms, fecal streptococci and enterococci were all significantly correlated with each other.

Touron *et al.* [122] reported a study carried out in the Seine estuary, France. Water was sampled at nine stations (along an upstream/downstream transect of 156 km), during nine years, for fecal coliforms, *E. coli*, enterococci and *Clostridium perfringens* spores. At the upstream part of the estuary (at Poses), *Salmonella* and fecal coliforms, and *E. coli* and enterococci counts were correlated. At the mouth of the estuary (at Honfleur), significant correlation was found for *Salmonella* and enterococci counts. No significant correlation between concentrations of any combination of indicator organism and pathogen was observed.

Wilkes *et al.* [59] reported a comparative study on the presence and concentration of several pathogenic and indicator bacteria in the surface water of a Canadian river. Surface water was collected within the South Nation River basin in eastern Ontario, from the river proper, and from several lower stream order tributaries. Using data aggregated during the entire multi-year study, significant correlations were found among all indicator bacteria - total and fecal coliforms, *E. coli*, *Enterococcus*, and *C. perfringens*.

However, in others studies no correlation was found between the different fecal indicator bacteria. Garrido-Pérez *et al.* [123] reported a study of the bathing seawater quality in 18 Spanish beaches near the Strait of Gibraltar. Sample locations were selected as a single point located in the area of highest bather density of each beach. No significant correlation was found between fecal coliforms and *Clostridium perfringens* counts in the bathing seawater.

FECAL INDICATOR CHEMICAL COMPOUNDS

Several chemical substances have been used as markers of fecal pollution in environmental waters. Caffeine is present in several beverages and in many pharmaceutical products. It is excreted in the urine of individuals who have ingested the substance. The main source of caffeine in domestic wastewaters is excretion following consumption of coffee, tea, soft drinks, or medication. Levels of caffeine in domestic wastewater have been measured to be between 20 and 300 μg/L. Levels in receiving waters are much lower due to significant dilution. Due to its high solubility, low octanol-water partition coefficient, insignificant volatility and clear anthropogenic origin, the presence of caffeine in environmental waters can be a good marker for human fecal pollution [124–129].

However, relationships between fecal indicator bacteria and caffeine are variable. Wu *et al.* [129] reported a study carried out in the Rochor Canal and Marina Bay, Singapore. In Rochor Canal, the highest concentration of caffeine (1.35 ng/mL) was found at downstream, and the lowest (0.68 and 0.37 ng/mL) were determined at middle and upstream points. At Marina Bay, the concentration of caffeine was in the range of 0.41–0.96 ng/mL. Fecal coliform concentrations were very high, exceeding 5,000 CFU/100 mL. Caffeine and fecal coliform concentrations were significantly correlated in Rochor Canal samples, but no significant correlation was observed in Martina Bay water samples.

The use of caffeine as marker of fecal pollution has additional important limitations. Caffeine is often present in the urban environment from numerous plant species debris as well as from human "dumping" of coffee wastes. In addition, the current analytical methods used are relatively complex and expensive [124].

Coprostanol (5β-cholestan-3β-ol) is a fecal stanol that is formed by indigenous bacteria present in the gut of humans and higher animals, during catabolism of cholesterol. It is the main stanol present in human feces (24 to 89% of total steroids) and in domestic wastewater. Based on these facts, it has been proposed as a chemical indicator of human fecal pollution. Feces from pigs and cats also contain coprostanol, but at much lower levels. Additional fecal stanols, such as 24-ethylcoprostanol, were found to be predominant in herbivores, such as cows, horses, and sheep, suggesting potential use of this chemical as an indicator of fecal pollution from these sources. Reported half-lives of coprostanol in aerobic conditions are generally lower than 10 days at 20 °C. Thus, the presence of coprostanol in an aerobic environment can be considered an indication of recent fecal input to the waters. [124,128,130].

Isobe *et al.* [130] reported a study carried out in the Mekong Delta (Vietnam) and in the Tokyo metropolitan area. During the wet season in the Mekong Delta, higher bacterial densities were observed in rivers, probably due to the higher bacterial inputs from soil particles with runoff. In Tokyo, higher bacterial densities were usually observed during summer, followed by those in the typhoon aftermath and winter. Significant correlations between the concentrations of *E. coli* and coprostanol (log scale) were found in all surveys. It was concluded that the determination of coprostanol can improve standard microbiological assays of fecal pollution.

Reports from several regions throughout the world indicate variable quantitative relationships between fecal coliform densities and coprostanol concentrations. In the Derwent Estuary and Sydney region (Australia), coprostanol concentration of 400 ng/L corresponded to 1,000 CFU of fecal coliforms/100 mL. In the Mekong Delta, during the wet season, this coliform density corresponded to 30 ng coprostanol/L, and in the dry seasons, 100 ng/L. In Tokyo metropolitan area, these values were 30 ng coprostanol/L, in summer, and 100 ng coprostanol/L, in a typhoon aftermath. These differences were interpreted as a result of differences in water temperature and soil particle concentration [130].

Fecal sterol analysis, although expensive and complex, has resolved problems of source attribution in urban and rural environments not possible with use of traditional fecal indicator bacteria [124]. These chemical indicators are especially useful in environments which allow survival and growth of fecal bacteria. For instance, in tropical regions, characterized by high temperatures and frequent rainstorms that facilitate erosion of soils, fecal bacteria can proliferate in environmental waters reaching densities that are not representative of real sewage inputs in the environment [130].

SOURCES OF FECAL BACTERIAL POLLUTION OF ENVIRONMENTAL WATERS

Sources of Surface and Groundwater Contamination

Determinations carried out in the sewage systems of urbanized areas have confirmed the presence of high numbers of intestinal bacteria. Treatment of sewage reduces the concentration of these bacteria by 1–2 logs, but effluent still contains high levels of intestinal bacteria (Table 8). Effluents from sewage treatment plants can be a source of contamination of surface waters with fecal bacteria.

Table 8: Typical concentrations of selected bacteria in raw and treated domestic wastewater[a]

Bacterial group	Raw sewage (cells/ml)	Treated effluent (cells/ml)
Salmonella	$10^{-1} – 10^{1}$	$10^{-1} – 10^{1}$
Total coliforms	$10^{4} – 10^{6}$	$10^{3} – 10^{5}$
Fecal coliforms	$10^{3} – 10^{5}$	$10^{2} – 10^{4}$
Enterococci	$10^{3} – 10^{4}$	$10^{1} – 10^{3}$
Clostridium perfringens	$10^{2} – 10^{3}$	$10^{1} – 10^{2}$

[a]Adapted from Medema *et al.* [131].

Septic tanks, cesspools, latrines and other on-site systems are widely used for wastewater storage and treatment. The water percolating from these facilities contains bacteria that may contaminate groundwater supplies.

Many farmers use cellars, tanks or landfills to store manure. Water leaching from these storage sites may also contaminate groundwater, especially during periods of rainfall. The application of animal manure to agricultural lands as fertilizer is common practice throughout the world. Bacteria present in the manure may leach into the groundwater.

The potential for bacteria present in human and animal wastes to contaminate water in nearby wells needs special attention [131]. An important source of contamination of surface and ground waters is runoff water from agricultural and pasture lands, and urban areas.

Fecal bacteria enter surface water by direct deposit of feces and by overland runoff. The movement of animal wastes into surface waters can be a major factor contributing to the pollution of available water in many regions. Over one-third of the land area of USA is used for grazing livestock and receives 50% of all livestock wastes.

In a study reported by Doran and Linn [108], runoff from a cow-calf pasture in eastern Nebraska was monitored during a three-year period. Rainfall runoff from the grazed area contained 5 to 10 times more fecal coliforms than runoff from the fenced, ungrazed area. However, fecal streptococci counts were higher in runoff from the ungrazed area and reflected the contributions from wildlife.

Urban and suburban areas are dominated by impervious cover. During storms, rainwater flows across these impervious surfaces, mobilizing contaminants. The pollutants carried in runoff originate from a variety of urban and suburban nonpoint sources. Contaminants commonly found in stormwater runoff include fecal and pathogenic bacteria. Stormwater transports pollutants to water bodies such as lakes and streams [132].

Enterococci and *E. coli* can be found in high numbers in most storm drains and creeks. In Southern California (USA), Ferguson *et al.* [133] found high levels of enterococci (*Enterococcus faecalis*, *Enterococcus faecium*, *Enterococcus hirae*, *Enterococcus casseliflavus* and *Enterococcus mundtii*) in intertidal sediments in a seasonal river, and near a storm drain outlet.

Survival in Surface Water

Most intestinal bacteria that contaminate environmental waters are not able to survive and multiply in this environment. Survival rates vary widely among fecal bacteria introduced in environmental waters. Pathogenic enteric bacteria and *E. coli* display low survival rates (Table 9).

Table 9: Reduction times for fecal bacteria in surface waters[a]

Bacterial group	**Time for 50% reduction in concentration (days)**
Total coliforms	0.9
E. coli	1.5–3
Enterococci	0.9–4
Clostridium perfringens	60 – > 300
Salmonella	0.1–0.67
Shigella	1

[a]Adapted from Medema *et al.* [131].

Table 10: Disappearance rates of fecal bacteria in groundwaters[a]

Bacterial group	**Disappearance rate (per day)**
E. coli	0.063–0.36
Fecal streptococci	0.03–0.24
Clostridium bifermentans spores	0.00
Salmonella enterica subsp. *enterica* serovar Typhimurium	0.13–0.22

[a]Adapted from Medema *et al.* [131].

The ability of fecal bacteria to survive in environmental waters generally increases as the temperature decreases. Others factors that influence survival include dissolved organic carbon concentration, sunlight intensity and the ability to enter the viable but non-culturable state [131].

In a comparative study on the survival of 10 different coliform species (*E. coli*, *Citrobacter freundii*, *Citrobacter youngae*, *Klebsiella pneumoniae*, *K. oxytoca*, *Enterobacter amnigenus*, *Enterobacter cloacae* subsp. *cloacae*, and *Pantoea agglomerans* (*Enterobacter agglomerans*)) inoculated in sterilized river water with different concentrations of dissolved organic carbon, Boualam *et al.* [134] found that only *C. freundii*, *K. pneumoniae* and *E. cloacae* subsp. *cloacae* remained cultivable after 96 hours of incubation. In a posterior study, using the same bacteria and medium, Boualam *et al.* [135] found that after 28 days, only *C. freundii* and *E. cloacae* subsp. *cloacae* survived.

Baudišová [136] reported a comparative study on the survival of total coliforms, fecal coliforms and *E. coli*, in sterile and non-sterile river water. In sterile water, all bacteria survived for many months. However, in non-sterile conditions (closer to true environmental conditions), the elimination rate of all bacteria was considerably faster. Total coliforms survived the longest and *E. coli* the shortest.

Survival in Groundwater

Survival of bacteria in groundwater is influenced by several factors, namely the survival in soil, since in order to reach the groundwater bacteria have to percolate through the soil. Generally, survival in soil (and concomitantly in groundwater) is enhanced by low temperatures, high soil humidity, neutral or alkaline soil pH and the presence of organic carbon [131].

WHICH INDICATORS OF FECAL POLLUTION SHOULD BE USED?

Several fecal indicator bacteria in environmental waters are in current use. From these stand out fecal coliforms, *E. coli* and enterococci [1,6,137]. In environmental waters, most fecal coliform strains are *E. coli*.

In particular situations, the presence of *E. coli* is definitively not associated with fecal pollution. These situations were firstly detected in some African countries, namely Nigeria, Ivory Coast and New Guinea (although not in others, such as Uganda) [42,99].

Recent studies carried out in temperate zones indicated that *E. coli* can persist in secondary, nonhost habitats, outside the hot tropical areas, and become naturalized in these habitats. Byappanahalli *et al.* [138] reported that *E. coli* could be isolated from coastal temperate forest soils in Indiana (USA). The aquatic alga *Cladophora glomerata* (L.) from several Lake Michigan beaches was shown to harbor high densities of *E. coli* [139].

Ishii *et al.* [140] reported a study on the survival of *E. coli* in temperate riverine soils of northern Minnesota (USA). Viable *E. coli* populations were repeatedly isolated from northern temperate soils in three Lake Superior watersheds. Seasonal variation in the population density of soilborne *E. coli* was observed; the greatest cell densities were found in the summer to fall, and the lowest numbers, occurred during the winter to spring months. Horizontal, fluorophore-enhanced repetitive extragenic palindromic PCR (HFERP) DNA fingerprint analyses indicated that identical soilborne *E. coli* genotypes, overwintered in frozen soil and were present over time, and that these strains were different from *E. coli* strains obtained from wildlife commonly found in the studied habitats or river water. Soilborne *E. coli* strains had HFERP DNA fingerprints that were unique to specific soils and locations. In laboratory studies, naturalized *E. coli* strains had the ability to grow and replicate to high cell densities, in nonsterile soils when incubated at 30 or 37 °C and survived longer than 1 month when soil temperatures were lower than 25 °C. It was concluded that these *E. coli* strains became naturalized, autochthonous members of the soil microbial community.

In a latter paper, Ksoll *et al.* [141] studied epilithic periphyton communities at three sites on the Minnesota shoreline of Lake Superior (USA). Fecal coliform densities increased up to 4 orders of magnitude in early summer, and decreased during autumn. HFERP DNA fingerprint analyses indicated that waterfowl (geese, terns, and gulls) were the major primary source of periphyton *E. coli* strains that could be identified. Periphyton and sewage effluent were also major potential sources. Several periphyton *E. coli* isolates were genotypically identical, repeatedly isolated over time. Inoculated *E. coli* rapidly colonized natural periphyton in laboratory microcosms and persisted for several weeks, and some cells were released to the overlying water. It was concluded that *E. coli* had became a naturalized member of the bacterial periphyton communities.

The presence, persistence, and possible naturalization of *E. coli* in these habitats can confound the use of fecal coliforms as a reliable indicator of recent fecal contamination of environmental waters. Future studies should consider other nonhost habitats as potential sources of fecal coliform bacteria in aquatic environments.

Considering these limitations, it appears to be advisable, in order to check the microbiological quality of drinking water, to complement the determination of *Escherichia coli* with the assay of enterococci. This rationale that has been followed, for many years, in the making of the drinking-water legislation in the European Union.

However, for many developing countries, where limited financial resources are the norm and reality, the routine determination of these two parameters can be difficult to implement. In these circumstances, it appears common sense that is better to determine a (good) parameter, such as *Escherichia coli*, than have no analysis done.

In this context, USA legislation emerges as a pragmatic approach to the problem. According to the American legislation, total coliforms are the routine parameter to be determined. Only when these determinations are repeatedly positive, it is mandatory to assess fecal coliforms [142,143]. Although total coliforms are not necessarily fecal bacteria, the rationale behind this system is correct, since: (1) a positive test in fecal coliforms (which is our target) is necessarily positive in the total coliform procedure; (2) the inverse is not necessarily true; (3) total coliforms are easily and cheaply assayed in waters.

As an alternative to the determination of both *E. coli* and enterococci, the assay of ammonia in environmental waters can be useful and complement the determination of fecal coliforms.

Ammonia is one of the key molecules in the nitrogen cycle. The presence of ammonia in surface waters can be due to direct contamination by agricultural fertilizers, and/or to microbial degradation of proteins, nucleic acids and urea, implying therefore the presence of a considerable concentration of organic matter in the water. Ammonia is rapidly oxidized in the environment and is typically found in natural waters at concentrations less than 0.1 mg/L. Concentrations significantly above this indicate gross contamination by fresh sanitary waste, where ammonia levels are typically very high (tens or hundreds of mg/L) [84].

Espigares *et al.* [144] reported a comparative study of chemical and microbiological indicators (total and fecal coliforms, fecal streptococci and sulphite-reducing clostridia) in a stretch of the Guadalquivir River (Spain) and its affluents. Total coliforms were correlated with fecal coliforms, but were not correlated with fecal streptococci and clostridia. Fecal coliforms were correlated with the other indicators. Fecal streptococci and sulphite-reducing clostridia were correlated with the other indicators except for total coliforms. All these microbiological indicators were correlated with dissolved oxygen (negatively), dissolved organic carbon and ammonia (positively). Cabral and Marques [85] in a study of a polluted river (Febros) in the Great Oporto area (northwest Portugal) found that ammonia was significantly correlated with all the microbiological assayed parameters—total and fecal coliforms, fecal streptococci and enterococci. These correlations are most probably due to the carry-over of organic matter in wastewaters, and to a high microbial ammonification activity [85].

Simple and rapid in-field tests and automated and continuous systems are available for the assay of ammonia in environmental waters. More studies are needed in order to confirm the use of ammonia as a reliable parameter in a preliminary screening for emergency fecal pollution outbreaks.

CONCLUSIONS

- Safe drinking water for all is one of the major challenges of the 21st century.
- Microbiological control of drinking water should be the norm everywhere.
- Routine basic microbiological analysis of drinking water should be carried out by assaying the presence of *Escherichia coli* by the culture methods. On-line monitoring of glucuronidase activity is currently too insensitive to replace culture based detection of *E. coli* but is a valuable complementary tool for high temporal resolution monitoring. Whenever financial resources are available, coliform determinations should be complemented with the quantification of enterococci.
- More studies are needed in order to check if ammonia is reliable for a preliminary screening for emergency fecal pollution outbreaks.
- Financial resources should be devoted to a better understanding of the ecology and behavior of human and animal fecal bacteria in environmental waters.

REFERENCES

1. WHO (World Health Organization). *Guidelines for Drinking-water Quality, Incorporating 1st and 2nd Addenda, Volume 1, Recommendations*, 3rd ed; WHO: Geneva, Switzerland, 2008.
2. Fenwick, A. Waterborne Diseases—Could they be Consigned to History? *Science* 2006, *313*, 1077–1081.
3. George, I; Crop, P; Servais, P. Use of β-D-Galactosidase and β-D-Glucuronidase Activities for Quantitative Detection of Total and Faecal Coliforms in Wastewater. *Can. J. Microbiol* 2001, *47*, 670–675.
4. Grabow, WOK. Waterborne Diseases: Update on Water Quality Assessment and Control. *Water SA* 1996, *22*, 193–202.
5. Seas, C; Alarcon, M; Aragon, JC; Beneit, S; Quiñonez, M; Guerra, H; Gotuzzo, E. Surveillance of Bacterial Pathogens Associated with Acute Diarrhea in Lima, Peru. *Int. J. Infect. Dis* 2000, *4*, 96–99.
6. Medema, GJ; Payment, P; Dufour, A; Robertson, W; Waite, M; Hunter, P; Kirby, R; Anderson, Y. Safe drinking water: an ongoing challenge. In *Assessing Microbial Safety of Drinking Water Improving Approaches and Method*; WHO & OECD, IWA Publishing: London, UK, 2003; pp. 11–45.
7. Farmer, JJ; Hickam-Brenner, FW. The Genus *Vibrio* and *Phtotobacterium*. In *The Prokaryotes: An Evolving Electronic Resource for the Microbiological Community*electronic release 314,, 3th ed; Dworkin, M, Falkow, S, Rosenberg, E, Eds.; Springer-Verlag: New-York, NY, USA, 2003.
8. Farmer, JJ; Janda, JM; Brenner, FW; Cameron, DN; Birkhead, KM. Genus *Vibrio*. In *Bergey's Manual of Systematic Bacteriology*, 2nd ed; Brenner, DJ, Krieg, NR, Staley, JT, Eds.; Springer: New York, NY, USA, 2005; Volume 2, Part B, pp. 494–546.
9. DSMZ (Deutsche Sammlung von Mikroorganismen und Zellkulturen GmbH). *Bacterial Nomenclature Up-to-Date (Approved Lists; Validation Lists)*; DSMZ: Braunschweig, Germany, 2010; List 05/2010.
10. Sack, DA; Sack, RB; Nair, GB; Siddique, AK. Cholera. *Lancet* 2004, *363*, 223–233.
11. Todar, K. *Vibrio cholerae* and asiatic cholera. *Todar's Online Textbook of Bacteriology*. 2009. Available online: http://www.textbookofbacteriology.net/cholera.html (assessed on 4 May 2010).

12. Ali, M; Emch, M; Yunus, M; Sack, RB. Are the Environmental Niches of *Vibrio cholerae* 0139 Different from those of *Vibrio cholerae* 01 El Tor? *Int. J. Infect. Dis* 2001, *5*, 214–219.
13. Ramamurthy, T; Yamasaki, S; Takeda, Y; Nair, GB. *Vibrio cholerae* O139 Bengal: Odyssey of a Fortuitous Variant. *Microbes Infect* 2003, *5*, 329–344.
14. Arbora, S; Crum, S. *Medical Geography and Cholera in Perú*; The Geographer's Craft Project, Department of Geography, The University of Colorado at Boulder: Boulder, CO, USA, 1996. Available online: http://www.colorado.edu/geography/gcraft/warmup/cholera/cholera_f.html (assessed on 4 May 2010).
15. Glass, RI; Claeson, M; Blake, PA; Waldmann, RJ; Pierce, NF. Cholera in Africa: Lessons on Transmission and Control for Latin America. *Lancet* 1991, *338*, 791–795.
16. Glass, RI; Libel, M; Brandling-Bennet, AD. Epidemic Cholera in the Americas. *Science* 1992, *256*, 1524–1525.
17. Tamplin, ML; Parodi, CC. Environmental Spread of *Vibrio cholerae* in Peru. *Lancet* 1991, *338*, 1216–1217.
18. Steinberg, EB; Greene, KD; Bopp, CA; Cameron, DN; Wells, JG; Mintz, ED. Cholera in the United States, 1995–2000: Trends at the End of the Twentieth Century. *J. Infect. Dis* 2001, *184*, 799–802.
19. Meibom, KL; Blokesch, M; Dolganov, NA; Wu, C-Y; Schoolnik, GK. Chitin Induces Natural Competence in *Vibrio cholerae*. *Science* 2005, *310*, 1824–1827.
20. Waldor, MK; Mekalanos, JJ. Lysogenic Conversion by a Filamentous Phage Encoding Cholera Toxin. *Science* 1996, *272*, 1910–1914.
21. Hervio-Heath, D; Colwell, RR; Derrien, A; Robert-Pillot, A; Fournier, JM; Pommepuy, M. Occurrence of Pathogenic Vibrios in Coastal Areas of France. *J. Appl. Microbiol* 2002, *92*, 1123–1135.
22. Alam, M; Hasan, NA; Sadique, A; Bhuiyan, NA; Ahmed, KU; Nusrin, S; Nair, GB; Siddique, AK; Sack, RB; Sack, DA; Huq, A; Colwell, RR. Seasonal Cholera Caused by *Vibrio cholerae* Serogroups O1 and O139 in the Coastal Aquatic Environment of Bangladesh. *Appl. Environ. Microbiol* 2006, *72*, 4096–4104.
23. Bhanumathi, R; Sabeena, F; Isac, SR; Shukla, BN; Singh, DV. Molecular Characterization of *Vibrio cholerae* O139 Bengal Isolated from Water and the Aquatic Plant *Eichhornia crassipes* in the River

Ganga, Varanasi, India. *Appl. Environ. Microbiol* 2003, *69*, 2389–2394.

24. Munro, PM; Colwell, RR. Fate of *Vibrio cholerae* O1 in Seawater Microcosms. *Water Res* 1996, *30*, 47–50.
25. Alam, M; Sultana, M; Nair, GB; Sack, RB; Sack, DA; Siddique, AK; Ali, A; Huq, A; Colwell, RR. Toxigenic *Vibrio cholerae* in the Aquatic Environment of Mathbaria, Bangladesh. *Appl. Environ. Microbiol* 2006, *72*, 2849–2855.
26. Chaiyanan, S; Chaiyanan, S; Huq, A; Maugel, T; Colwell, RR. Viability of the Nonculturable *Vibrio cholerae* O1 and O139 System. *Appl. Microbiol* 2001, *24*, 331–341.
27. Molloy, S. Biofilms in the Time of Cholera. *Nature Rev. Microbiol* 2007, *5*, 906–907.
28. Le Minor. The genus *Salmonella*. In *The Prokaryotes: An Evolving Electronic Resource for the Microbiological Community*electronic release 314,, 3th ed; Dworkin, M, Falkow, S, Rosenberg, E, Eds.; Springer-Verlag: New York, NY, USA, 2003.
29. Popoff, MY; Le Minor, LE. Genus *Salmonella*. In *Bergey's Manual of Systematic Bacteriology*, 2nd ed; Brenner, DJ, Krieg, NR, Staley, JT, Eds.; Springer: New York, NY, USA, 2005; Volume 2, Part B, pp. 764–799.
30. Tindall, BJ; Grimont, PAD; Garrity, GM; Euzéby, JP. Nomenclature and Taxonomy of the Genus *Salmonella*. *Int. J. Syst. Evolut. Microbiol* 2005, *55*, 521–524.
31. Ben Aissa, R; Al-Gallas, N; Troudi, H; Belhadj, N; Belhadj, A. Trends in *Salmonella enterica* Serotypes Isolated from Human, Food, Animal, and Environment in Tunisia, 1994–2004. *J. Infect* 2007, *55*, 324–339.
32. Arvanitidou, M; Kanellou, K; Vagiona, DG. Diversity of *Salmonella* spp. and Fungi in Northern Greek Rivers and their Correlation to Faecal Pollution Indicators. *Environ. Res* 2005, *99*, 278–284.
33. Kapperud, G; Rørvik, LM; Hasseltvedt, V; Høiby, EA; Iversen, BG; Staveland, K; Johnsen, G; Leitão, J; Herikstad, H; Andersson, Y; Langeland, G; Gondrosen, B; Lassen, J. Outbreak of *Shigella sonnei* Infection Traced to Imported Iceberg Lettuce. *J. Clin. Microbiol* 1995, *33*, 609–614.
34. Strockbine, NA; Maurelli, AT. Genus *Shigella*. In *Bergey's Manual of Systematic Bacteriology*, 2nd ed; Brenner, DJ, Krieg, NR, Staley,

JT, Eds.; Springer: New York, NY, USA, 2005; Volume 2, Part B, pp. 811–823.

35. Emch, M; Ali, M; Yunus, M. Risk Areas and Neighborhood-Level Risk Factors for *Shigella dysenteriae* 1 and *Shigella flexneri. Health Place* 2008, *14*, 96–105.

36. Germani, Y; Sansonetti, PJ. The Genus *Shigella*. In *The Prokaryotes: An Evolving Electronic Resource for the Microbiological Community*electronic release 314,, 3th ed; Dworkin, M, Falkow, S, Rosenberg, E, Eds.; Springer-Verlag: New York, NY, US, 2003.

37. Hale, TL. Genetic Basis of Virulence in *Shigella* Species. *Microbiol. Rev* 1991, *55*, 206–224.

38. Todar, K. *Shigella* and shigellosis. *Todar's Online Textbook of Bacteriology*. 2009. Available online: http://www.textbookofbacteriology.net/Shigella.html (assessed on 1 September 2010).

39. Tetteh, GL; Beuchat, LR. Survival, Growth, and Inactivation of Acid-Stressed *Shigella flexneri* as Affected by pH and Temperature. *Int. J. Food Microbiol* 2003, *87*, 131–138.

40. Chompook, P; Todd, J; Wheeler, JG; von Seidlein, L; Clemens, J; Chaicumpa, W. Risk Factors for Shigellosis in Thailand. *Int. J. Infect. Dis* 2006, *10*, 425–433.

41. Faruque, SM; Khan, R; Kamruzzman, M; Yamasaki, S; Ahmad, QS; Azim, T; Nair, GB; Takeda, Y; Sack, DA. Isolation of *Shigella dysenteriae* type 1 and *S. flexneri* Strains from Surface Waters in Bangladesh: Comparative Molecular Analysis of Environmental *Shigella* Isolates *versus* Clinical Strains. *Appl. Environ. Microbiol* 2002, *68*, 3908–3913.

42. Bettelheim, KA. The genus *Escherichia*. In *The Prokaryotes: An Evolving Electronic Resource for the Microbiological Community*electronic release 314,, 3th ed; Dworkin, M, Falkow, S, Rosenberg, E, Eds.; Springer-Verlag: New York, NY, USA, 2003.

43. Scheutz, F; Strockbine, NA. Genus *Escherichia*. In *Bergey's Manual of Systematic Bacteriology*, 2nd ed; Brenner, DJ, Krieg, NR, Staley, JT, Eds.; Springer: New York, NY, USA, 2005; Volume 2, Part B, pp. 607–623.

44. World Health Organization. Enterotoxigenic *Escherichia coli* (ETEC). *Diarrhoeal Diseases*. Available online: http://www.who.int/vaccine_

research/diseases/diarrhoeal/en/index4.html (assessed on 4 September 2010).

45. Health Canada. *Guidelines for Canadian Drinking Water Quality: Guideline Technical Document. Bacterial Waterborne Pathogens. Current and Emerging Organisms of Concern*; Health Canada: Ottawa, ON, Canada, 2006.
46. World Health Organization. Enterohaemorrhagic *Escherichia coli* (EHEC). Fact sheet N°125; Available online: http://www.who.int/mediacentre/factsheets/fs125/en/ (assessed on 4 September 2010).
47. Enteroinvasive *Escherichia coli* (EIEC); US Department of Health & Human Services, U.S. Food and Drug Administration. Available online: http://www.fda.gov/Food/FoodSafety/FoodborneIllness/FoodborneIllnessFoodbornePathogensNaturalToxins/BadBugBook/ucm071298.htm (assessed on 4 September 2010).
48. WHO (World Health Organization). *Pathogenic Mycobacteria in Water: A Guide to Public Health Consequences, Monitoring and Management*; Pedley, S, Bartram, J, Rees, G, Dufour, A, Cotruvo, J, Eds.; IWA Publishing: London, UK, 2004.
49. Holtzman, A; Aronson, T; Froman, S; Berlin, OGW; Dominguez, P; Kunkel, KA; Overturf, G; Stelma, G; Smith, C; Yakrus, M. The Isolation and Identification of *Mycobacterium avium* Complex (MAC) Recovered from Los Angeles Potable Water, a Possible Source of Infection in AIDS Patients. *Int. J. Environ. Health Res* 1994, *4*, 63–72.
50. Von Reyn, CF; Maslow, JN; Barber, TW; Falkinham, JO; Arbeit, RD. Persistent Colonisation of Potable Water, a Source of *Mycobacterium avium* Infection in AIDS. *Lancet* 1994, *343*, 1137–1141.
51. Gião, MS; Azevedo, NF; Wilks, SA; Vieira, MJ; Keevil, CW. Persistence of *Helicobacter pylori* in Heterotrophic Drinking Water Biofilms. *Appl. Environ. Microbiol* 2008, *74*, 5898–5904.
52. Hulten, K; Han, SW; Enroth, H; Klein, PD; Opekun, AR; Gilman, RH; Evans, DG; Engstrand, L; Graham, DY; El-Zaatari, FA. *Helicobacter pylori* in the Drinking Water in Peru. *Gastroenterology* 1996, *110*, 1031–1035.
53. Handfield, M; Simard, P; Couillard, M; Letarte, R. *Aeromonas hydrophila* Isolated from Food and Drinking Water: Hemagglutination, Hemolysis, and Cytotoxicity for a Human Intestinal Cell Line (HT-29). *Appl. Environ. Microbiol* 1996, *62*, 3459–3461.

54. Daskalov, H. The Importance of *Aeromonas hydrophila* in Food Safety. *Food Control* 2006, *17*, 474–483.
55. Chauret, C; Volk, C; Creason, R; Jarosh, J; Robinson, J; Warnes, C. Detection of *Aeromonas hydrophila* in a Drinking-Water Distribution System: A Field and Pilot Study. *Can. J. Microbiol* 2001, *47*, 782–786.
56. Fernández, MC; Beatriz, N; Giampaolo, SB; Ibañez, M; Guagliardo, V; Esnaola, MM; Conca, L; Valdivia, P; Stagnaro, SM; Chiale, C; Frade, H. *Aeromonas Hydrophila* and its Relation with Drinking Water Indicators of Microbiological Quality in Argentine 2000. *Genetica* 2000, *108*, 35–40.
57. George, I; Servais, P. *Sources et Dynamique des Coliformes dans le Basin de la Sein*; Rapport de Synthèse; Programme PIREN-Seine 1998–2001, Sources et dynamique des coliformes dand le bassin de la Seine; C. N. R. S.: Paris, France, 2002.
58. Gauthier, F; Archibald, F. The Ecology of "Faecal Indicator" Bacteria Commonly Found in Pulp and Paper Mill Water Systems. *Water Res* 2001, *35*, 2207–2218.
59. Wilkes, G; Edge, T; Gannon, V; Jokinen, C; Lyautey, E; Medeiros, D; Neumann, N; Ruecker, N; Topp, E; Lapena, DR. Seasonal Relationships Among Indicator Bacteria, Pathogenic Bacteria, *Cryptosporidium* Oocysts, *Giardia* Cysts, and Hydrological Indices for Surface Waters Within an Agricultural Landscape. *Water Res* 2009, *43*, 2209–2223.
60. Saarela, M; Lähteenmäki, L; Crittenden, R; Salminen, S; Mattila-Sandholm, T. Gut Bacteria and Health Foods—The European Perspective. *Int. J. Food Microbiol* 2002, *78*, 99–117.
61. Hawksworth, G; Drasar, BS; Hill, MJ. Intestinal Bacteria and the Hydrolysis of Glycoside Bonds. *J. Med. Microbiol* 1971, *4*, 451–459.
62. Hopkins, MJ; Sharp, R; Macfarlane, GT. Variation in Human Intestinal Microbiota with Age. *Digest Liver Dis* 2002, *34*(Suppl.2), S12–S18.
63. Leclerc, H; Moriamez, JC. Étude Quantitative de la Flore Fécale de l'Adulte et du Nourrison Alimenté Artificiellement. *Path. Bio* 1980, *28*, 217–226.
64. Wilson, M. *Microbial Inhabitants of Humans Their Ecology and Role in Health and Disease*; Cambridge University Press: Cambridge, UK, 2005.

65. Sinton, LW; Finlay, RK; Hannah, DJ. Distinguishing Human from Faecal Contamination in Water: A Review. *New Zealand J. Marine Freshwater Res* 1998, *32*, 323–348.

66. Biavati, B; Mattarelli, P. The family *Bifidobacteriaceae*. In *The Prokaryotes: An Evolving Electronic Resource for the Microbiological Community*electronic release 314,, 3th ed; Dworkin, M, Falkow, S, Rosenberg, E, Eds.; Springer-Verlag: New York, NY, USA, 2003.

67. Hippe, H; Andreesen, JR; Gottschalk, G. The genus *Clostridium*—Nonmedical. In *The Prokaryotes: An Evolving Electronic Resource for the Microbiological Community*electronic release 314,, 3th ed; Dworkin, M, Falkow, S, Rosenberg, E, Eds.; Springer-Verlag: New York, NY, USA, 2003.

68. Rainey, FA; Hollen, BJ; Small, A. Genus *Clostridium*. In *Bergey's Manual of Systematic Bacteriology*, 2nd ed; De Vos, P, Garrity, GM, Jones, D, Krieg, NR, Ludwig, W, Rainey, FA, Schleifer, K-H, Whitman, WB, Eds.; Springer: New York, NY, USA, 2009; Volume 3, pp. 738–828.

69. Smith, LDS. The genus *Clostridium*—Medical. In *The Prokaryotes: An Evolving Electronic Resource for the Microbiological Community*electronic release 314,, 3th ed; Dworkin, M, Falkow, S, Rosenberg, E, Eds.; Springer-Verlag: New York, NY, USA, 2003.

70. Bezirtzoglou, E; Panagiou, A; Savvaidis, I; Maipa, V. Distribution of *Clostridium perfringens* in Polluted Lake Environments. *Anaerobe* 1997, *3*, 169–172.

71. Švec, P; Devriese, LA. Genus *Enterococcus*. In *Bergey's Manual of Systematic Bacteriology*, 2nd ed; DE Vos, P, Garrity, GM, Jones, D, Krieg, NR, Ludwig, W, Rainey, FA, Schleifer, K-H, Whitman, WB, Eds.; Springer: New York, NY, USA, 2009; Volume 3, pp. 594–607.

72. Pinto, B; Pierotti, R; Canale, G; Reali, D. Characterization of 'Faecal Streptococci' as Indicators of Faecal Pollution and Distribution in the Environment. *Lett. Appl. Microbiol* 1999, *29*, 258–263.

73. Bergey's Manual of Determinative Bacteriology, 9th ed; Holt, JG; *et al.* (Eds.) Williams & Wilkins: Baltimore, MD, USA, 1994; pp. 175–190.

74. Gordon, DM; FitzGibbon, F. The Distribution of Enteric Bacteria from Australian Mammals: Host and Geographical Effects. *Microbiology* 1999, *145*, 2663–2671.

75. Frederiksen, W; Søgaard, P. The genus *Citrobacter*. In *The Prokaryotes: An Evolving Electronic Resource for the Microbiological Community*electronic release 314,, 3th ed; Dworkin, M, Falkow, S, Rosenberg, E, Eds.; Springer-Verlag: New York, NY, USA, 2003.
76. Frederiksen, W. Genus *Citrobacter*. In *Bergey's Manual of Systematic Bacteriology*, 2nd ed; Brenner, DJ, Krieg, NR, Staley, JT, Eds.; Springer: New York, NY, USA, 2005; Volume 2, Part B, pp. 651–656.
77. Grimont, F; Grimont, PAD; Richard, C. The Genus *Klebsiella*. In *The Prokaryotes: An Evolving Electronic Resource for the Microbiological Community*electronic release 314,, 3th ed; Dworkin, M, Falkow, S, Rosenberg, E, Eds.; Springer-Verlag: New York, NY, USA, 2003.
78. Grimont, PAD; Grimont, F. Genus *Klebsiella*. In *Bergey's Manual of Systematic Bacteriology*, 2nd ed; Brenner, DJ, Krieg, NR, Staley, JT, Eds.; Springer: New York, NY, USA, 2005; Volume 2, Part B, pp. 685–693.
79. Grimont, F; Grimont, PAD. The Genus *Enterobacter*. In *The Prokaryotes: An Evolving Electronic Resource for the Microbiological Community*electronic release 314,, 3th ed; Dworkin, M, Falkow, S, Rosenberg, E, Eds.; Springer-Verlag: New York, NY, USA, 2003.
80. Grimont, PAD; Grimont, F. Genus *Enterobacter*. In *Bergey's Manual of Systematic Bacteriology*, 2nd ed; Brenner, DJ, Krieg, NR, Staley, JT, Eds.; Springer: New York, NY, USA, 2005; Volume 2, Part B, pp. 661–669.
81. Lockhart, SR; Abramson, MA; Beekmann, SE; Gallagher, G; Riedel, S; Diekema, DJ; Quinn, JP; Doern, GV. Antimicrobial Resistance Among Gram-negative Bacilli Causing Infections in Intensive Care Unit Patients in the United States between 1993 and 2004. *J. Clin. Microbiol* 2007, *45*, 3352–3359.
82. Hidron, AI; Edwards, JR; Patel, J; Horan, TC; Sievert, DM; Pollock, DA. NHSN Annual Update: Antimicrobial Resistant Pathogens Associated with Healthcare-Associated Infections: Annual Summary of Data Reported to the National Healthcare Safety Network at the Centers for Disease Control and Prevention, 2006–2007. *Infect. Control Hosp. Epidemiol* 2008, *29*, 996–1011.
83. Ashbolt, NJ; Grabow, OK; Snozzi, M. Indicators of microbial water quality. In *Water Quality: Guidelines, Standards and Health*; Fewtrell, L, Bartram, J, Eds.; World Health Organization (WHO), IWA Publishing: London, UK, 2001; pp. 289–316.

84. Payment, P; Waite, M; Dufour, A. Introducing parameters for the assessment of drinking water quality. In *Assessing Microbial Safety of Drinking Water Improving Approaches and Method*; WHO & OECD, IWA Publishing: London, UK, 2003; pp. 47–77.
85. Cabral, JP; Marques, C. Faecal Coliform Bacteria in Febros river (Northwest Portugal): Temporal Variation, Correlation with Water Parameters, and Species Identification. *Environ. Monit. Assess* 2006, *118*, 21–36.
86. Villarino, A; Toribio, AL; Brena, BM; Grimont, PAD; Bouvet, OMM. On the Relationship Between the Physiological State of Bacteria and Rapid Enzymatic Assays of Faecal Coliforms in the Environment. *Biotechnol. Lett* 2003, *25*, 1329–1334.
87. George, I; Petit, M; Servais, P. Use of Enzymatic Methods for Rapid Enumeration of Coliforms in Freshwaters. *Lett. Appl. Microbiol* 2000, *88*, 404–413.
88. George, I; Petit, M; Theate, C; Servais, P. Distribution of Coliforms in the Seine River and Estuary (France) Studied by Rapid Enzymatic Methods and Plate Counts. *Estuaries* 2001, *24*, 994–1002.
89. Nelis, H; van Poucke, S. Enzymatic Detection of Coliforms and *Escherichia coli* Within 4 Hours. *Water, Air, and Soil Pollution* 2000, *123*, 43–52.
90. Rompré, A; Servais, P; Baudart, J; de-Roubin, M-R; Laurent, P. Detection and Enumeration of Coliforms in Drinking Water: Current Methods and Emerging Approaches. *J. Microbiol. Method* 2002, *49*, 31–54.
91. Van Poucke, SO; Nelis, HJ. Rapid Detection of Fluorescent and Chemiluminescent Total Coliforms and *Escherichia coli* on Membrane Filters. *J. Microbiol. Method* 2000, *42*, 233–244.
92. Kämpfer, P; Rauhoff, O; Dott, W. Glycosidase Profiles of Members of the Family *Enterobacteriaceae. J. Clin. Microbiol* 1991, *29*, 2877–2879.
93. Kilian, M; Bülow, P. Rapid Diagnoses of *Enterobacteriaceae.* I. Detection of Bacterial Glycosidases. *Acta Pathol. Microbiol. Scandinavica* 1976, *84*, 245–251.
94. Muytjens, HL; van der Ros-van de Repe, J; nan Druten, HAM. Enzymatic Profiles of *Enterobacter sakazakii* and Related Species with Special Reference to the α-Glucosidase Reaction and Reproducibility of the Test System. *J. Clin. Microbiol* 1984, *20*, 684–686.

95. Tryland, I; Fiksdal, L. Enzyme Characteristics of β-D-Galactosidase- and β-D-Glucuronidase-Positive Bacteria and their Interference in Rapid Methods for Detection of Waterborne Coliforms and *Escherichia coli. Appl. Environ. Microbiol* 1998, *64*, 1018–1023.
96. Köster, W; Egli, T; Ashbolt, N; Botzenhart, K; Burlion, N; Endo, T; Grimont, P; Guillot, E; Mabilat, C; Newport, L; Niemi, M; Payment, P; Prescott, A; Renaud, P; Rust, A. Analytical methods for microbiological water quality testing. In *Assessing Microbial Safety of Drinking Water Improving Approaches and Method*; WHO & OECD, IWA Publishing: London, UK, 2003; pp. 237–292.
97. Manafi, M; Kneifel, W; Bascomb, S. Fluorogenic and Chromogenic Substrates Used in Bacterial Diagnostics. *Microbiol. Rev* 1991, *55*, 335–348.
98. Geissler, K; Manafi, M; Amorós, I; Alonso, JL. Quantitative Determination of Total Coliforms and *Escherichia coli* in Marine Waters with Chromogenic and Fluorogenic Media. *J. Appl. Microbiol* 2000, *88*, 280–285.
99. Byamukama, D; Kansiime, F; Mach, RL; Farnleitner, AH. Determination of *Escherichia coli* Contamination with Chromocult Coliform Agar Showed a High Level of Discrimination Efficiency for Differing Faecal Pollution Levels in Tropical Waters of Kampala, Uganda. *Appl. Environ. Microbiol* 2000, *66*, 864–868.
100. Eccles, JP; Searle, R; Holt, D; Dennis, PJ. A Comparison of Methods Used to Enumerate *Escherichia coli* in Conventionally Treated Sewage Sludge. *J. App. Microbiol* 2004, *96*, 375–383.
101. Farnleitner, AH; Hocke, L; Beiwl, C; Kavka, GG; Zechmeister, T; Kirschner, AKT; Mach, RL. Rapid Enzymatic Detection of *Escherichia coli* Contamination in Polluted River Water. *Lett. Appl. Microbiol* 2001, *33*, 246–250.
102. Edberg, SC; Kontnick, CM. Comparison of β-Glucuronidase-Based Substrate Systems for Identification of *Escherichia coli. J. Clin. Microbiol* 1986, *24*, 368–371.
103. Leclerc, H; Mossel, DAA; Edberg, SC; Struijk, CB. Advances in the Bacteriology of the Coliform Group: their Suitability as Markers of Microbial Water Safety. *Ann. Rev. Microbiol* 2001, *55*, 201–234.
104. Klein, G. Taxonomy, Ecology and Antibiotic Resistance of Enterococci from Food and the Gastro-Intestinal Tract. *Int. J. Food Microbiol* 2003, *88*, 123–131.

105. Kühn, I; Iversen, A; Burman, LG; Olsson-Liljequist, B; Franklin, A; Finn, M; Aarestrup, F; Seyfarth, AM; Blanch, AR; Vilanova, X; Taylor, H; Caplin, J; Moreno, MA; Dominguez, L; Herrero, IA; Möllby, R. Comparison of Enterococcal Populations in Animals, Humans, and the Environment—An European Study. *Int. J. Food Microbiol* 2003, *88*, 133–145.

106. Wheeler, AL; Hartel, PG; Godfrey, DG; Hill, JL; Segars, WI. Potentital of *Enterococcus faecalis* as a Human Fecal Indicator for Microbial Source Tracking. *J. Environ. Qual* 2002, *31*, 1286–1293.

107. Geldreich, EE. Fecal Coliform and Fecal Streptococcus Density Relationships in Waste Discharges and Receiving Waters. *Crit. Rev. Environ. Control* 1976, *6*, 349–369.

108. Doran, JW; Linn, DM. Bacteriological Quality of Runoff Water from Pasteureland. *Appl. Environ. Microbiol* 1979, *37*, 985–991.

109. Caplenas, NJ; Kanarek, MS. Thermotolerant Non-Fecal Source of *Klebsiella pneumoniae*: Validity of the Fecal Coliform Test in Recreational Waters. *Am. J. Public Health* 1984, *74*, 1273–1275.

110. Gauthier, F; Neufeld, JD; Driscoll, BT; Archibald, FS. Coliform Bacteria and Nitrogen Fixation in Pulp and Paper Mill Effluent Treatment Systems. *Appl. Environ. Microbiol* 2000, *66*, 5155–5160.

111. Corso, PS; Kramer, MH; Blair, KA; Addiss, DG; Davis, JP; Haddix, AC. Cost of Illness in the 1993 Waterborne Cryptosporidium Outbreak, Milwaukee, Wisconsin. *Emerg. Inf. Dis* 2003, *9*, 426–431.

112. MacKenzie, WR; Hoxie, NJ; Proctor, ME; Gradus, MS; Blair, KA; Peterson, DE; Kazmierczak, JJ; Addiss, DG; Fox, KR; Rose, JB; Davis, JP. A Massive Outbreak in Milwaukee of *Cryptosporidium* Infection Transmitted Through the Public Water Supply. *N. Engl. J. Med* 1994, *331*, 161–167.

113. Vilanova, X; Manero, A; Cerdà-Cuéllar, M; Blanch, AR. The Composition and Persistence of Faecal Coliforms and Enterococcal Populations in Sewage Treatment Plants. *J. Appl. Microbiol* 2004, *96*, 279–288.

114. Araujo, M; Sueiro, RA; Gómez, MJ; Garrido, MJ. Enumeration of *Clostridium perfringens* Spores in Groundwater Samples: Comparison of Six Culture Media. *J. Microbiol. Methods* 2004, *57*, 175–180.

115. Charriere, G; Mossel, DAA; Beaudeau, P; Leclerc, H. Assessment of the Marker Value of Various Components of the *Coli-Aerogenes* Group

of *Enterobacteriaceae* and of a Selection of *Enterococcus* spp. for the Official Monitoring of Drinking Water Supplies. *J. Appl. Bacteriol* 1994, *76*, 336–344.

116. Martins, MT; Sato, MIZ; Alves, MN; Stoppe, NC; Prado, VM; Sanchez, PS. Assessment of Microbiological Quality for Swimming Pools in South America. *Water Res* 1995, *29*, 2417–2420.
117. Ferguson, CM; Coote, BG; Ashbolt, NJ; Stevenson, IM. Relationships Between Indicators, Pathogens and Water Quality in an Estuary System. *Water Res* 1996, *30*, 2045–2054.
118. Medema, GJ; van Asperen, IA; Havelaar, AH. Assessment of the Exposure of Swimmers to Microbiological Contaminants in Fresh Waters. *Water Sci. Technol* 1997, *35*, 157–163.
119. Polo, F; Figueras, MJ; Inza, I; Sala, J; Fleisher, JM; Guarro, J. Relationship Between Presence of *Salmonella* and Indicators of Faecal Pollution in Aquatic Habitats. *FEMS Microbiol. Lett* 1998, *160*, 253–256.
120. Noble, RT; Leecaster, MK; McGee, CD; Weisberg, SB; Ritter, K. Comparison of Bacterial Indicator Analysis Methods in Stormwater-Affected Coastal Waters. *Water Res* 2004, *38*, 1183–1188.
121. Harwood, VJ; Levine, AD; Scott, TM; Chivukula, V; Lukasik, J; Farrah, SR; Rose, JB. Validity of the Indicator Organism Paradigm for Pathogen Reduction in Reclaimed Water and Public Health Protection. *Appl. Environ. Microbiol* 2005, *71*, 3163–3170.
122. Touron, A; Berthe, T; Gargala, G; Fournier, M; Ratajczak, M; Servais, P; Petit, F. Assessment of Faecal Contamination and the Relationship Between Pathogens and Faecal Bacterial Indicators in an Estuarine Environment (Seine, France). *Mar. Pollut. Bull* 2007, *54*, 1441–1450.
123. Garrido-Pérez, MC; Anfuso, E; Acevedo, A; Perales-Vargas-Machuca, JA. Microbial Indicators of Faecal Contamination in Waters and Sediments of Beach Bathing Zones. *Int. J. Hyg. Environ. Health* 2008, *211*, 510–517.
124. Ashbolt, N; Fujioka, R; Glymph, T; McGee, C; Schaub, S; Sobsey, M; Toranzos, G. Pathogen indicators, and indicators of fecal contamination. In *Report of the Experts Scientific Workshop on Critical Research Needs for the Development of New or Revised Recreational Water Quality*; EPA 823-R-07–006; U.S. Environmental Protection Agency, Office of Water, Office of Research and Development: Warrenton, VA, 2007; Chapter 2,; pp. 35–56.

125. Ferreira, AP. Caffeine as an Environmental Indicator for Assessing Urban Aquatic Ecosystems. *Cad. Saúde Pública* 2005, *21*, 1884–1892.

126. Gardinali, PR; Zhao, X. Trace Determination of Caffeine in Surface Water Samples by Liquid Chromatography-Atmospheric Pressure Chemical Ionization–Sass spectrometry (LC-APCI-MS). *Environ. Int* 2002, *28*, 521–528.

127. Peeler, KA; Opsahl, SP; Chanton, JP. Tracking Anthropogenic Inputs Using Caffeine, Indicator Bacteria, and Nutrients in Rural Freshwater and Urban Marine Systems. *Environ. Sci. Technol* 2006, *40*, 7616–7622.

128. Scott, TM; JB; Rose, TM; Jenkins, S; Farrah, R; Lukasik, J. Microbial Source Tracking: Current Methodology and Future Directions. *Appl. Environ. Microbiol* 2002, *68*, 5796–5803.

129. Wu, J; Yue, J; Hu, R; Yang, Z; Zhang, L. Use of Caffeine and Human Pharmaceutical Compounds to Identify Sewage Contamination. *World Acad. Sci. Eng. Technol* 2008, *44*, 438–442.

130. Isobe, KO; Tarao, M; Chiem, NH; Minh, LY; Takada, H. Effect of Environmental Factors on the Relationship between Concentrations of Coprostanol and Fecal Indicator Bacteria in Tropical (Mekong Delta) and Temperate (Tokyo) Freshwaters. *Appl. Environ. Microbiol* 2004, *70*, 814–821.

131. Medema, GJ; Shaw, S; Waite, M; Snozzi, M; Morreau, A; Grabow, W. Catchment characteristics and source water quality. In *Assessing Microbial Safety of Drinking Water Improving Approaches and Method*; WHO & OECD, IWA Publishing: London, UK, 2003; pp. 111–158.

132. EPA. *Source Water Protection Practices Bulletin. Managing Stormwater Runoff to Prevent Contamination of Drinking Water*; Office of Water (4606); United States Environmental Protection Agency (EPA): Washington, DC, USA, 2009.

133. Ferguson, DM; Moore, DF; Getrich, MA; Zhowandai, MH. Enumeration and Speciation of Enterococci Found in Marine and Intertidal Sediments and Coastal Water in Southern California. *J. Appl. Microbiol* 2005, *99*, 598–608.

134. Boualam, M; Mathieu, L; Fass, S; Cavard, J; Gatel, D. Relationship between Coliform Culturability and Organic Matter in Low Nutritive Waters. *Water Res* 2002, *36*, 2618–2626.

135. Boualam, M; Fass, S; Saby, S; Lahoussine, V; Cavard, J; Gatel, D; Mathieu, L. Organic Matter Quality and Survival of Coliforms in Low-Nutrient Waters. *J. AWWA* 2003, *95*, 119–126.

136. Baudišová, D. Evaluation of *Escherichia coli* as the Main Indicator of Faecal Pollution. *Water Sci. Technol* 1997, *35*, 333–336.

137. Edberg, SC; Rice, EW; Karlin, RJ; Allen, MJ. *Escherichia coli*: The Best Biological Drinking Water Indicator for Public Health Protection. *J. Appl. Microbiol* 2000, *88*, 106S–116S.

138. Byappanahalli, MN; Whitman, RL; Shively, DA; Sadowsky, MJ; Ishii, S. Population Structure, Persistence, and Seasonality of Autochthonous *Escherichia coli* in Temperate, Coastal Forest Soil from a Great Lakes Watershed. *Environ. Microbiol* 2006, *8*, 504–513.

139. Ishii, S; Yan, T; Shively, DA; Byappanahalli, MN; Whitman, RL; Sadowsky, MJ. *Cladophora* (Chlorophyta) spp. Harbor Human Bacterial Pathogens in Nearshore Water of Lake Michigan. *Appl. Environ. Microbiol* 2006, *72*, 4545–4553.

140. Ishii, S; Ksoll, WB; Hicks, RE; Sadowsky, MJ. Presence and Growth of Naturalized *Escherichia coli* in Temperate Soils from Lake Superior Watersheds. *Appl. Environ. Microbiol* 2006, *72*, 612–621.

141. Ksoll, WB; Ishii, S; Sadowsky, MJ; Hicks, RE. Presence and Sources of Fecal Coliform Bacteria in Epilithic Periphyton Communities of Lake Superior. *Appl. Environ. Microbiol* 2007, *73*, 3771–3778.

142. Hecq, P; Hulsmann, A; Hauchman, FS; McLain, JL; Schmitz, F. Drinking water regulations. In *Analytical Methods for Drinking Water*; Quevauviller, P, Thompson, KC, Eds.; John Wiley: London, UK, 2006; Chapter 1.

143. Total Coliform Rule: A Handbook for Small Noncommunity Water Systems Serving less than 3,300 Persons; EPA 816-B-06–001; Office of Water (4606M), EPA: Washington, DC, USA, 2006.

144. Espigares, M; Coca, C; Fernández-Crehuet, M; Moreno, O; Gálvez, R. Chemical and Microbiologic Indicators of Faecal Contamination in the Guadalquivir (Spain). *Eur. Water Pollut. Control* 1996, *6*, 7–13.

Chapter 8

Mechanisms of Fluid Movement into, through and out of the Brain: Evaluation of the Evidence

Stephen B Hladky and Margery A Barrand

Department of Pharmacology, University of Cambridge, Cambridge CB2 1PD, UK

THE CONTEXT

Introduction

The fluids that surround the brain and bathe the cells within it play several important roles. They provide the most suitable ionic and nutrient microenvironment to allow neurons and associated cells to function correctly: they deliver all the substances the cells require and remove unwanted material and they provide a cushion against physical damage.

Citation: Mechanisms of fluid movement into, through and out of the brain: evaluation of the evidence. Stephen B Hladky* and Margery A Barrand. Hladky and Barrand Fluids and Barriers of the CNS 2014, 11:26 http://www.fluidsbarrierscns.com/content/11/1/26.

Hence regulation of composition, volume and turnover of these fluids is vital. To allow this regulation there are barriers that prevent free exchange of material between brain and blood, mechanisms that secrete fluid of controlled composition into the brain, and mechanisms that reabsorb, eventually to blood, the extracellular fluids whatever their composition. Structures associated with this regulation include those generating the fluid, i.e. the choroid plexuses and the brain capillaries that make up the blood-CSF and blood-brain barriers, respectively, and those able to remove the fluid, e.g. the arachnoid villi and the perineural spaces of nerves penetrating the cribriform plate. This review considers the relative contribution of these structures in secreting and removing the fluids and evaluates the evidence for the various ideas of how fluid secreted into the brain may move through the brain parenchyma and thence exit to the periphery.

The traditional view is that the choroid plexuses secrete the major portion of the cerebrospinal fluid (CSF) surrounding the brain while the blood-brain barrier has been thought to make a much smaller contribution to fluid production, generating interstitial fluid (ISF) that drains into CSF. Comparisons between these interfaces with regard to their rates of secretion are considered in sections 3 and 4. Ideas regarding the relative importance of these structures in regulating brain fluid may need to be reassessed and will be discussed further in a later publication. Many of the topics considered in this review have been discussed on previous occasions. The reader interested in accounts of earlier work with references should consult a number of excellent reviews: [1–3] for anatomy and generally [4–14]. Recently, Damkier *et al.*[15] have presented a comprehensive account of the mechanisms of secretion by the choroid plexuses, Pollay [16] has discussed the routes for absorption of CSF and Brinker *et al.*[17] have surveyed topics of current interest about the circulation of CSF. O'Donnell [18] has reviewed the mechanisms at the blood-brain barrier with attention to factors that can regulate or modulate secretion and has considered the changes that occur in focal oedema. Strazielle and Ghersi-Egea [19] have discussed the structure and function of the blood-brain interfaces and Engelhardt and Sorokin [20] and Luissant *et al.*[21] have reviewed the molecular basis and properties of the tight junctions important for barrier function.

The Anatomy and Terminology

The contents of the skull (Figure 1) consist of the following: the brain parenchyma comprising cells surrounded by ISF together with connective tissue; the vasculature; the meninges overlying the brain; and the CSF that fills the ventricles within the brain and the spaces on its outer surfaces. The ISF, blood plasma and CSF collectively make up the extracellular fluid of the brain and spinal cord. Working inwards from the skull (Figure 1c), there is a thick supporting layer of tissue (the dura), another connective tissue layer (the arachnoid membrane), the pia mater and the brain parenchyma with its glial outer surface. The arachnoid membrane and pia mater are separated by the subarachnoid space that is bridged by chordae and larger trabeculae (Figure 1d). The space between the pia mater and the parenchyma, called the subpial space or space of His, may be a virtual space, i.e. with zero thickness [1, 22]. The parenchyma is made up of cells, e.g. neurons and glia, the extracellular matrix and ISF that pervades the matrix and fills the spaces between the cells. Arteries and veins run through the subarachnoid spaces with branches penetrating into the parenchyma [23]. Veins leaving the parenchyma drain into the various venous sinuses and thence to the jugular veins.

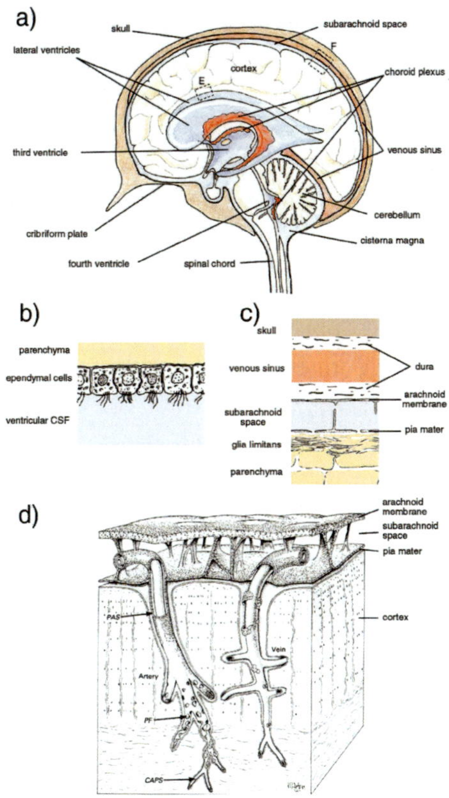

Figure 1: Structures of the brain considered in this review. a) Mid-sagittal section from nose to the back of the head incorporating images of the lateral

ventricles that lie to each side of the section. CSF filled spaces are shown in blue, blood filled spaces in pale red. The choroid plexuses are shown in darker red. b) Enlarged view of surface of a lateral ventricle at (E). c) Enlarged view of the cortical surface at (F). The glia limitans is a mat of glial processes. d) Drawing showing the relationships between the arachnoid membrane, the pia mater, the subarachnoid space, and the vasculature supplying the cortex. As described by Zhang *et al.* for the artery "the sheath has been cut away to show that the periarterial spaces (PAS) of the intracerebral and extracerebral arteries are in continuity. The layer of pial cells becomes perforated (PF) and incomplete as smooth muscle cells are lost from the smaller branches of the artery. The pial sheath finally disappears as the perivascular spaces are obliterated around capillaries (CAPS). Perivascular spaces around the vein (right of picture) are confluent with the subpial space and only small numbers of pial cells are associated with the vessel wall" [23]. The cortical boundary along these vessels is formed by glial foot-processes. The graphic elements in a) and b) and in c) with minor extensions are taken with permission from Figure one of Strazielle *et al.*[19] and relabelled. d) is reproduced with permission from Figure ten in [23] and partially relabelled. R.O. Weller (personal communication) has emphasized that the spaces shown between the arterial wall and the pial sheath and between the sheath and the glial end-feet were virtual spaces in their electron micrographs. The periarterial spaces are also portrayed in Figure 6.

When considering fluid balance, the brain can be thought of as a greatly distorted tube, sealed at one end. Starting from the blind end, the spaces inside this tube include (Figure 1a), a pair of lateral ventricles, one in each cortical hemisphere, connected to a midline third ventricle, which in turn is connected via the cerebral aqueduct (or aqueduct of Sylvius) to a midline fourth ventricle that leads to both the central spinal canal and, via foramina of Magendie and Luschka, to the subarachnoid spaces outside the tube. All of these spaces including the central canal and the subarachnoid space of the spinal cord are filled with CSF. Because the skull is a rigid box, the sum of the volumes occupied by the parenchyma with its ISF and associated connective tissue, the vasculature, the meninges and the CSF must be constant, a concept known as the Monro-Kellie dogma [24] which has implications for CSF movements during the cardiac cycle, see below. The spinal cord has the same components but less rigid constraints on its total volume [6].

Functions of the Fluids

The ISF provides the environment within which the brain cells survive and function. Water and solutes enter and leave the cells from and to this fluid. As such its composition and volume must be controlled to allow reproducible and reliable cell activity. There must be mechanisms for rapid delivery of both oxygen and nutrients, especially glucose and possibly lactate but also amino acids and other substances. There must also be mechanisms for rapid removal of the principal waste products, especially carbon dioxide and possibly lactate. These requirements are met by having microvessels within the parenchyma sufficiently close to all cells that diffusion distances for these solutes are small. Regulation of Na^+, K^+, Cl^-, Ca^{2+}, Mg^{2+}, HCO_3^- and other ion concentrations in the ISF is important for correct nervous activity and regulation of the sum total of the solute concentrations, the osmolality of ISF, is crucial for regulation of cell volume.

The CSF serves a number of functions which include providing a) partial buoyancy for the brain, cushioning it within its rigid box, b) a means to compensate for the changes in blood volume within the skull during the cardiac cycle and c) a route for removal of waste products from the brain, particularly those of high molecular weight. CSF is also important in removing protein and other debris during recovery from injury or microhemorrhage. It may also serve as a route for dispersal of certain nutrients and hormones and other molecular signals which may be particularly important in brain development [25].

Composition of Fluids

CSF and ISF are very similar in composition. K^+, Ca^{2+}, Na^+ and Cl^- concentrations in ISF have been determined directly using ion selective microelectrodes [26, 27]. They have also been inferred from the modest changes in composition of CSF measured as it flows over the surface of the brain exchanging components with ISF [6, 9, 28, 29]. The composition of ISF and of CSF does, however, differ significantly from that of blood plasma (Table 1). This is because many hydrophilic substances are effectively prevented from entering the brain from the blood and many substances, including neurotransmitters, are greatly hindered in exiting from the brain (e.g. [6, 30]). The barriers preventing these movements include the blood-brain barrier separating blood and ISF within the parenchyma and the blood-CSF barriers comprising the choroid plexuses and the arachnoid and dura that surround the brain (Figure 1).

Table 1: Comparison of composition of CSF and blood plasma

Component	Ref.	CSF	Blood plasma
		Concentration/mg (100 ml)$^{-1}$	
Protein	a	16 to 38	6300 to 8500
Sugar	a	45 to 80	80 to 120
Amino acids	a	1.5 to 3	4.5 to 9
Creatinine	a	0.5 to 2.2	0.7 to 2
Uric acid	a	0.4 to 2.8	2.9 to 6.9
Urea	a	5 to 39	22 to 42
Cholesterol	a	trace	100 to 150
Lactic acid	a	8 to 25	10 to 32
Phosphate (inorganic)	b	3.4	4.7
		Concentration/mmol kg^{-1}	
Na+	b	147	150
K+	b	2.86	4.63
Ca2+	b	1.14	2.35
Mg2+	b	1.1	0.8
Cl^-	b	113	99
HCO_3^-	b	23.3	26.8
pH	b	7.3	7.4

Values are taken from the compilations in a) Millen and Woollam (1962) [2] based on data from Merritt and Fremont-Smith (1937) [31] and b) Davson and Segal (1996) [9].

Anatomical Basis of Barriers and Routes of Fluid Transfer

The blood-brain barrier (Figure 2a) is formed primarily by the layer of endothelial cells lining the microvessels within the brain parenchyma [32, 33]. These cells are joined to each other via tight junctions that effectively block free passage of molecules and ions between cells across the endothelial layer, i.e. the paracellular route. Substances required to enter or leave the brain across the barrier must of necessity be transferred through the cells. They must either pass through the lipid portions of the plasma membranes, if they are not too hydrophilic, or be transferred via specific transporters (e.g. GLUT1 for glucose) located on each side of the cells.

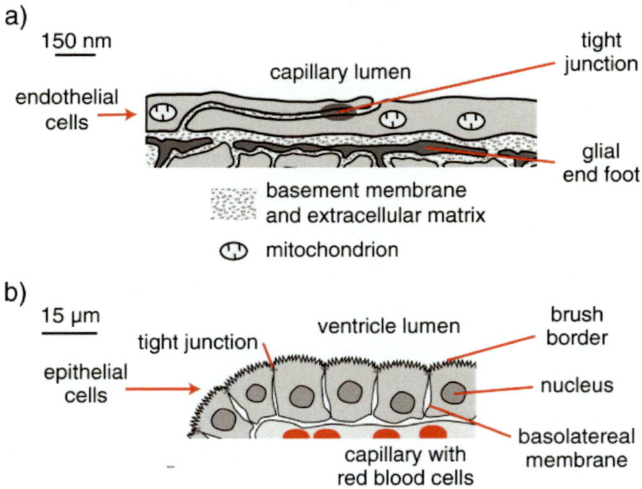

Figure 2: Key features of a) blood-brain barrier and b) choroid plexus. a) Sketch based on electron micrographs [34] of brain microvessel walls showing endothelial cells linked by tight junctions. These cells form a complete lining of the microvessel lumen and are in turn surrounded by glial end-feet. Occasional pericytes embedded in the basement membrane that separates the endothelial cells and the end-feet are not shown. The tight junctions limit the transport of solutes between the endothelial cells but the basement membrane, the gaps between the glial end-feet and the extracellular spaces within the parenchyma allow passage of molecules as big as proteins. Mitochondria occupy up to 10 % of the endothelial cell volume [35] and provide a substantial source of energy for transport. b) Diagram of one side of a choroid plexus villus as seen in light micrographs (e.g. [2]). Note the 100-fold difference in scales. Each choroid plexus has many such villi, each consisting of a layer of epithelial cells linked by tight junctions and surrounding a core of connective tissue containing capillaries. The endothelial cell layer lining these capillaries is not sealed by tight junctions. The epithelial cells have a prominent apical brush border facing the ventricle and a band of tight junctions separating the apical and basolateral domains of the cell membrane. There is folding of the basolateral membranes, which like the individual microvilli of the brush border are not visible in light micrographs (for electron micrographs see Figures two & twelve in [15]). The microvilli and basolateral folds increase the surface area of the membrane domains and hence the number of transporters that can be located on the two sides of the cells. Choroid plexus epithelium has the structure and properties of a leaky epithelium capable of transporting large quantities of isosmotic fluid.

The CSF-containing spaces are bounded by: the arachnoid membrane on the outside surface of the subarachnoid spaces, the pia mater on the inner surface of those spaces, the ependyma lining the ventricles and the choroid plexuses. The arachnoid membrane consists of one or more layers of cells connected by tight junctions [36] and these layers are presumed to be the basis for the impermeability of the arachnoid membrane (see [9] for further discussion). The choroid plexuses (Figure 2b) consist of a layer of epithelial cells facing ventricular CSF and overlying permeable capillaries of the fenestrated type. The barrier separating CSF from blood is provided by the epithelial layer rather than the underlying capillary endothelium [37]. The cells of the choroidal epithelium are sealed together by tight junctions that limit paracellular movement of most materials between CSF and peripheral circulation [20].

Small solutes appear to move freely across the boundaries between ISF and CSF. The inner surfaces of the parenchyma facing the ventricles (Figure 1b) are lined with ependymal cells, which do not appear to possess tight junctions that would provide a barrier such that even substances as large as inulin can pass between them [12, 32, 38–40]. Nevertheless loss of inulin or albumin from ventricular CSF into ISF is slow (at least in the absence of oedema), simply because diffusion of large molecules is slow and there is almost no convection. The outer surfaces of the brain parenchyma, defined by the glia limitans and the pia mater (Figure 1c), also do not provide a barrier to the movement of water and small solutes (section 4.1.1.1). Albumin injected into the parenchyma is slowly cleared from the ISF: some of it reaching CSF and some transported back to blood via other routes (section 4.1). The interfaces and routes of transfer important in production, circulation and removal of ISF and CSF are indicated schematically in Figure 3. An overriding principle is that what goes in must come out. An important way to address the rates of formation of CSF and ISF is to measure their rates of removal and this has been the subject of much investigation.

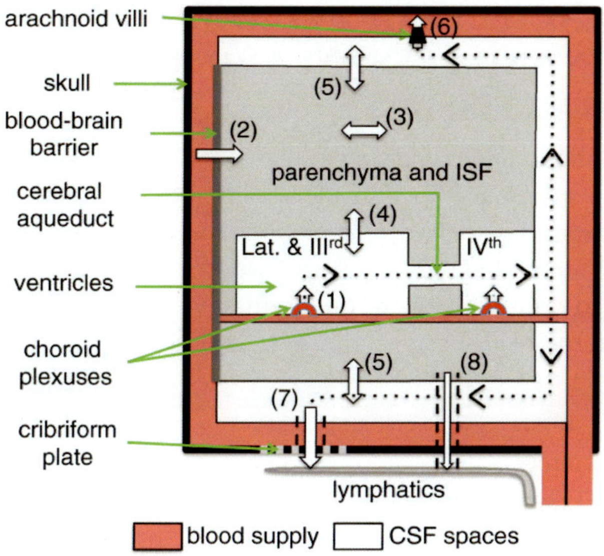

Figure 3: Schematic plan of the whole brain indicating the fluid movements considered in this review. Fluid is secreted into the ventricles across the choroid plexuses (1) and into the brain parenchyma across the blood-brain barrier (2). Fluid components can move through the parenchyma (3) and there are exchanges of water and solutes (4) and (5) between the interstitial fluid (ISF) of the parenchyma and cerebrospinal fluid (CSF) contained in the ventricles and in the subarachnoid spaces respectively. There is net fluid outflow across the arachnoid villi (6) leading to the dural venous sinuses (including but not restricted to the superior sagittal sinus) and along cranial nerves, most notably the olfactory nerve leading to the cribriform plate (7) and thence to the nasal mucosa. There may also be outflow of fluid in the walls of arteries or veins (8) leading to lymph nodes in the neck. The traditional view of the directions of net CSF flow is indicated by the dotted lines with arrowheads.

There is abundant evidence that much, probably most, of the CSF is formed by secretion by the choroid plexuses that are located in the brain ventricles, but the importance of ISF from the brain parenchyma as a source of CSF remains controversial. The principal sites at which CSF leaves the CNS are currently thought to include [16] the arachnoid villi linking cranial subarachnoid spaces to low pressure venous sinuses, the cribriform plate through which pass the olfactory nerves, and arachnoid villi located at the origins of the spinal nerves (Figure 1a).

Go to:

BASIC PRINCIPLES OF FLUID MOVEMENTS IN THE BRAIN AND LESSONS FROM STUDIES ON PERIPHERAL TISSUES

In gaining a better understanding of how ISF and CSF are formed and removed, it is necessary to consider a number of basic principles: not least, the distinctions between *diffusion* and *bulk flow/convection* and between *filtration* and *secretion*. Some aspects of these issues particularly as they apply to drug delivery to the brain have been discussed recently by Wolak and Thorne [41].

Diffusion

Diffusion is the observable movement of substances resulting from their random thermal motions. For a neutral solute in a stationary fluid, diffusion occurs down a concentration gradient. For ions, diffusion occurs down the resultant of the concentration and electrical gradients. For solutes crossing a lipid membrane, diffusion is best thought of as occurring down the gradient of electrochemical potential, defined in thermodynamics to include effects of differences in concentration, pressure and electrical potential. Diffusion occurs because all molecules are constantly in motion colliding and exchanging energy with their neighbours. For a neutral solute in a fluid, these random movements can generate net movement of the solute simply because there will be more solute molecules moving away from regions with high concentrations than there will be solute molecules moving away from regions with low concentrations. In a dilute solution, the movement of each solute molecule is independent of the movements of the others: there is no co-operativity. Because more massive particles have lower velocities, the rate of diffusion decreases as the mass of the solute increases. Diffusion is very fast over very short, atomic distances but because it is a random walk of short jumps with frequent changes of direction it is very slow over large distances.

Bulk Flow or Convection

Bulk flow, otherwise termed convection, is the movement of a fluid where all components of the fluid tend to move together. For example in a river the water and all it carries moves by bulk flow. In pure bulk flow, movements against the flow can be ignored. Furthermore all components irrespective of their size within a small volume of the fluid, providing it is sufficiently far from all boundaries, will move at the same speed.

Instances and Relative Importance of Diffusion and Bulk Flow

When considering transport through extracellular fluid, cytoplasm or even the interior of a lipid bilayer, diffusion is the dominant process over short distances, i.e. less than a few microns and sometimes up to a few millimeters provided a concentration gradient exists. Bulk flow is the dominant process when transfers occur over long distances. Examples of diffusive movements include transfer of oxygen across a cell membrane and transfer of water across a lipid bilayer [42]. An example of bulk flow is blood flowing through the vasculature.

In most instances the actual process is neither just diffusion nor just bulk flow but something intermediate. A single process is actually occurring but it is often convenient to describe it as the sum of a movement due to diffusion superimposed on a movement due to flow of the fluid as a whole. Partly to emphasize that movements against the flow cannot be ignored, the flow component is often called convection rather than bulk flow.

Though convection is difficult to avoid in an unconfined space (even small differences in density resulting from differences in temperature are sufficient to produce eddies) large forces are required for flow to occur in confined spaces. This is illustrated by Poiseuille's equation showing that the pressure difference required, ΔP, to produce a flow, Q_V, through a tube with internal radius, r, increases with the inverse fourth power of the radius,

$$\Delta P = \text{constant} \times \frac{Q_V}{r^4}$$

It is this strong dependence on the size of the spaces available for flow which accounts for the very high resistance to flow through the interstitial matrix normally found between cells, but the more than 1000-fold lower resistance to flow in the same tissues when the matrix is expanded by oedema [43–45]. The resistance to flow in peripheral tissues appears to be vested in the nature of the gel within the interstitial spaces rather than the sizes of the spaces themselves. It has been estimated that in synovium [45] which has an extracellular space of about 20 % (within the range associated with grey matter in the brain [9, 46]) the actual resistance is about 5000-fold greater than if the spaces were filled with free fluid.

Transfers by Specific Transporters

In addition to diffusion or convection, transfer across membranes can often involve interactions with specific transporters, which include channels (e.g. Kv1.3 for K^+); cotransporters (e.g. NBCe1 for Na^+ and HCO_3^-) and exchangers (e.g. AE2 for Cl^- and HCO_3^-). Cotransporters and exchangers are examples of carriers. A more complicated carrier, the Na^+, K^+-ATPase couples the energy released by hydrolysis of ATP to force Na^+ and K^+ to move against their gradients of concentration and electrical potential. For that reason this type of transporter is often called a pump. The Na^+-pump is the key link between energy released from metabolism and the secretion processes in the choroid plexuses and at the blood-brain barrier.

Filtration, Secretion and Absorption

Net movement of fluid across barriers can be described as filtration or secretion. Either process in reverse is called absorption.

Filtration

Filtration is the production of a net volume flow across a barrier driven by hydrostatic pressure and/or osmotic pressure gradients. The energy source for the movement of water and solutes comes from outside the barrier, not from the processes occurring within the barrier.

Filtration across the walls of capillaries is often described using the language of non-equilibrium or irreversible thermodynamics. In this the net volume flow per unit area, the volume flux, is given by:

$$J_V = L_p\left(\Delta P - RT\sum_i(\sigma_i \Delta c_i)\right)$$

where L_p is the hydraulic permeability, ΔP the hydrostatic pressure difference across the membrane, R the universal gas constant, T the absolute temperature, Δc_i the concentration difference for the ith solute (e.g. Na^+, Cl^-, or serum albumin) and σ_i the corresponding reflection coefficient [47–49]. The reflection coefficient for a particular solute and a particular membrane is usually defined as the ratio of the osmotic pressure difference produced by that solute across that membrane to the osmotic pressure difference that would result if the solute were totally impermeant [47, 49]. The name "reflection coefficient" stems from its appearance in another equation from the same theory that applies when there are no solute gradients and the fluxes are driven solely by a hydrostatic pressure difference,

$$J_i = c_i(1-\sigma_i)J_V$$

where J_i is the flux of the ith solute present on both sides at concentration c_i. Thus if all of the solute is reflected $\sigma_i = 1$ and there is no flux of solute, while if none is reflected $\sigma_i = 0$ and the solute is transferred along with the water. One of the main results of the application of non-equilibrium thermodynamics to membrane transport is that the reflection coefficients are the same in the two equations given above [47, 48]. These equations come with two caveats: they have a firm basis only when the forces and flows are small enough that each flow is a linear function of the individual forces [50] and strictly the concentration differences should be differences in thermodynamic activities, which is important in quantitative work as the osmotic pressure of the proteins deviates markedly from the ideal value calculated from van't Hoff's law, $\Delta\pi = RT\Delta c$.

The Starling mechanism is a special case of filtration that accounts for fluid movement across the walls of peripheral capillaries. Starling [51] divided the solutes present in plasma and extracellular fluid at osmotically important concentrations into two groups: small molecules, the crystalloids like Na^+, Cl^- and glucose, which can easily cross peripheral microvessel walls, and all the rest, the colloids, like serum albumin which cross very slowly. Using this distinction he proposed that fluid movement across microvessel walls can be described as net filtration or reabsorption of a fluid composed of water and crystalloids driven by the differences of the hydrostatic pressure and the osmotic pressure of the colloids. The latter has been called variously the colloid osmotic pressure or the oncotic pressure. Starling's simplifications of the actual process cannot describe short-term transients when for instance the plasma NaCl concentration is abruptly changed but have otherwise been remarkably successful in describing fluid movements into and out of peripheral tissues (e.g. [49, 52, 53], see Endnote[a]).

In the language of non-equilibrium thermodynamics, Starling's approximation amounts to saying that σ_i is 1 for the colloids and $\sigma_i \Delta c_i$ approaches zero for all of the crystalloids. It has proven very difficult to determine the values of σ_i for small solutes and peripheral microvessels because permeation is so fast that equilibration between plasma and the interstitial fluid immediately outside the microvessels is reached during the measurement, i.e. $\Delta c_i \approx 0$[54]. The "true reflection coefficients" may be much greater (e.g. 0.5) than the "apparent reflection coefficients", 0.01 to 0.1, often quoted (for discussion see e.g. [54–56]).

Mechanistic descriptions of Filtration

The treatment of filtration using irreversible thermodynamics assumes that the energy sources for filtration are the hydrostatic and concentration gradients. It makes no attempt to elucidate the actual mechanisms involved other than to stipulate that they must explain the values of L_p and σ_i. Attempts at mechanistic descriptions have been complex and controversial and none has achieved widespread acceptance. For discussion of some of the difficulties see e.g. [53, 57, 58]. Bulat and Klarica [59], alongside their consideration of brain fluid movements, have proposed that in peripheral capillaries water moves down the resultant of the hydrostatic and total osmotic gradients and that the resulting change in total osmotic pressure rather than the colloid osmotic pressure within plasma limits the water fluxes and thus filtration. However, their proposal ignores the effect of Na^+ and Cl^- fluxes across the peripheral capillary walls that reduce the local solute concentration differences. In its present state this description is not an improvement on the Starling approximation.

Secretion and absorption

Secretion of fluid is a net volume flow into a region of interest, driven by energy derived from metabolism within a barrier (e.g. [5]). The energy is supplied as part of the transport of solutes; the water follows osmotically but see e.g. [60, 61] for consideration of the possibility of specific active water transport. Absorption or reabsorption of fluid could in principle be either filtration (including non-selective filtration, i.e. bulk flow) or secretion of fluid each directed back towards the blood. At least part of the absorption of CSF may be a non-selective flow of fluid carrying all its components through the arachnoid villi or the cribriform plate.

Uses of Tracers or Marker Substances and Lessons from the Periphery

Following the movement of tagged substances, i.e. tracers, has been a common method both for determining the characteristics of flow of the various brain fluids and for investigating transport processes across the choroid plexuses and the blood-brain barrier. The nature of the particular tracer used is crucial in the interpretation of results. For instance to measure flow, the marker must move only because there is net flow, must move at the same rate as the flow and must be easily detected quantitatively. Such markers have included inulin, dextrans and serum albumin. Other markers such as India ink particles and horseradish peroxidase have been very useful for tracing routes, but not for determining rates of flow. Tritiated water is not suitable for measuring rates of bulk flow because it is not constrained to stay within the fluid for which flow is being measured. (A well-known example is the flow of fluid through a capillary as described below). Furthermore, additions of tritiated water on just one or the other side of a barrier cannot be used to measure secretion or absorption because water moves so rapidly in both directions by diffusion. Net flow across a barrier occurs only if the fluxes of water and solutes in one direction exceed those in the other direction (Figure 4). It is possible for there to be large tracer fluxes while not having net flow. In that case the fluxes in the two directions are equal in magnitude.

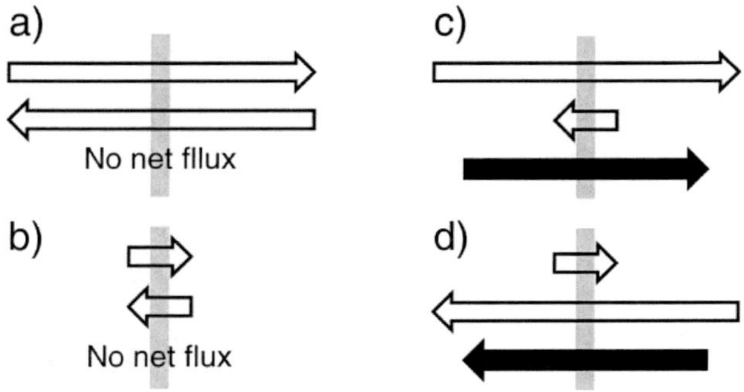

Figure 4: Fluxes inwards or outwards across a membrane compared with net flux. The membranes are indicated by the grey bars. In a) and b) the magnitudes of the fluxes are very different but in both cases the fluxes in the two directions are equal and hence there is no net flux. In c) and d) the magnitudes of the fluxes in the two directions are very different and there is a net flux in the direction indicated by the black arrow. Although the flux from left to right is the same in a) and c), because the fluxes in the opposite direction are not the same there is a net flux in c) but not in a). Hence net flux cannot be inferred from measurements of flux in a single direction.

To illustrate the importance of distinguishing the flux in one direction from net flux, one can consider the net flux of water across peripheral capillaries and venules of e.g. skin or muscle. The net flow can be determined by measuring water content, volume or weight changes in the tissues since almost all of the fluid is water. The result for non-fenestrated peripheral capillaries, which has stood the test of time, is that the net rate of filtration into the tissue is of the order of 0.1 % of the blood flow [62]. This filtration is driven by the Starling balance of forces, i.e. the net result of hydrostatic and colloid osmotic pressure differences (section 2.5.1). This net driving force can be expressed as a pressure difference or, using $\Delta\pi = RT\Delta c$ with $RT = 19$ mmHg/mM, as the concentration or osmolality difference Δc that would produce the same osmotic pressure, i.e. 19 mmHg is equivalent to 1 mM. The net driving force is very small, reduced by local concentration gradients in the regions of flow.

The water fluxes into and out of a tissue can be determined using radiotracers or heavy water. These fluxes are found to be between 1000- and 10000-fold greater than the net rate of filtration of water. Virtually all of the tritiated water molecules delivered by the blood cross the capillary wall before the blood leaves the tissue and thus transfer is limited by blood flow [55, 63, 64]. How is it that the net flux of tritiated water appears so large while the net flux of water is so small? The measurable net flux of tritiated water is driven by its own concentration gradient, which on a first pass through a tissue means effectively by its concentration in plasma since the concentration in the much larger volume of the tissue remains very small, i.e. the influx of tritiated water is almost equal to the net flux of tritiated water. Since, near enough, tritiated water and normal water behave the same, this means that influx of water, the inward unidirectional flux (Endnote[b]), is driven by its total concentration, approximately 55 M. However, the net flux of water, the filtration, which can be measured by volume change, is the difference between the influx and an almost equal efflux and thus it is driven by an equivalent concentration *difference* of water, which across a peripheral capillary wall is approximately 1 mM. Because the driving forces are so different, 55 M vs. 1 mM, it should not be surprising that the unidirectional flux is much larger than the net flux. Similar comparisons can be made for Na^+[55, 63, 64]. The molecular mechanisms for movements of Na^+ and water across the walls of peripheral microvessels look more like diffusion than flow, but nevertheless the net process is still filtration. The traditional [52] and current [49, 53] views on filtration and reabsorption in peripheral capillaries are described briefly in Endnote[c].

It is important to keep in mind that the terms filtration or reabsorption each describe a net process. They do not describe the movements in a single direction. No one now describes the process of fluid movement in peripheral capillaries as filtration of 99.9 % into the tissue with reabsorption of 99.8 % leaving 0.1 % in the tissue to be removed in the lymph flow. The difficulty is not with the numbers, as these are approximately the proportions of water molecules that move, but with the use of the words. These points are clear in the application of the principles to microvessels in the periphery. However, there are repeated instances of reports concerning fluid movements in the brain that have not been careful to distinguish between tracer fluxes and net fluxes of water and hence have incorrectly identified rates of secretion, filtration, absorption or flow of CSF and ISF (e.g. [65–72]).

Application of Basic Principles to Blood Vessels in the Brain

The same physical chemistry applies to brain microvessels as to peripheral capillaries, but there are important differences in the barrier properties of the vessel walls and in the transport pathways across them (Figure 5). In the periphery Na^+ and Cl^- cross the capillary walls easily in both directions and except possibly during relatively brief transients, e.g. when osmotic gradients are artificially imposed, their concentrations in the filtrate are similar to those in plasma. By contrast, at the blood-brain barrier, the low resistance pathways shared by water and hydrophilic solutes are absent, i.e. the paracellular route is blocked by tight-junctions. The blood-brain barrier has at least a 10 fold lower permeability to net flow of water than peripheral capillaries but the fluxes of water are still large enough that ca. 90 % of tracer water is extracted from the blood in a single pass [73–75]. Thus movements of water can occur relatively easily when gradients of total osmolality are imposed. Osmotic equilibrium is approached on a time scale of a few minutes [76–78] but equilibration half-times measured at various locations within the brain vary, presumably reflecting differences in the ratio of the surface area of microvessels to the local tissue volume [78]. Osmotically driven movement of water is the basis of the clinical use of hyperosmotic mannitol in plasma to draw water out of the brain to relieve elevated intracranial pressure. Net water movements in response to imposed osmotic differences are properly called filtration or absorption depending on the direction. They can be described (as in section 2.5.1) as flow driven by the resultant of hydrostatic and *total* osmotic pressures. This is not the same as the Starling mechanism concerned with fluxes of water and all small solutes across peripheral membranes. Starling's approximations only work when small solutes permeate rapidly (section 2.5) and this is not the case for the blood-brain barrier.

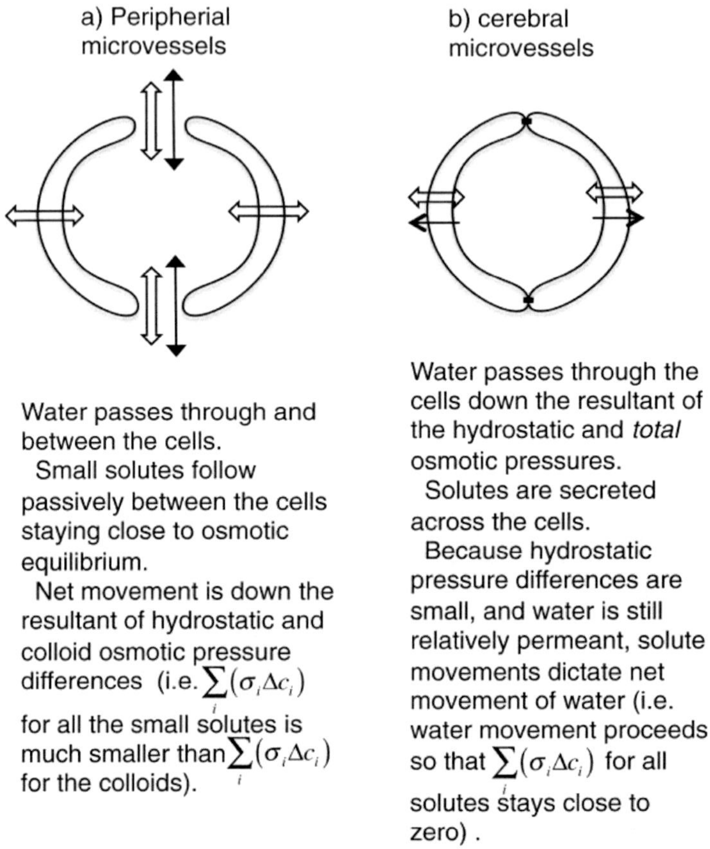

Figure 5: Comparisons of fluid transport across peripheral and cerebral microvessels. In the periphery a) small solutes cross the vessel walls via gaps between the cells. Small solute movement is rapid, therefore their concentration gradients are small and thus unlike the large solutes, the colloids, the small solutes do not oppose the movements of water. Thus except during brief transients, e.g. when osmotic gradients are artificially imposed, net fluid movement is governed by the hydrostatic and colloid osmotic pressure differences between blood and surrounding tissues. In the brain b) paracellular movement is limited by tight junctions. Thus small solutes cross the vessel walls only slowly and the direction and extent of their movement is determined by specific transporters. Hence they are as effective as the large solutes in producing osmotic gradients that dictate the extent of water movement into or out of the surrounding tissue.

The permeability of brain microvessels to Na^+ and Cl^- is more than 1000-fold lower than that of peripheral microvessels and the reflection coefficients of these and many other small solutes, e.g. sucrose and raffinose, are indistinguishable from 1 [7, 73]. This has very important implications for the "normal" movements of water and solutes in the absence of imposed osmotic differences. Any filtration driven by a difference in hydrostatic pressure will contain very little solute (section 2.5.1). Because even a small, e.g. 1 mM, concentration difference for NaCl across the barrier represents the same driving force as an unphysiologically large, e.g. 38 mmHg (= 56 cmH_2O), hydrostatic pressure difference, the rates of water and solute transport must be closely matched. Any possible increase in hydrostatic pressure difference will only produce a small flow of water that becomes self-limiting because it will reduce the osmolality in the interstitium. Furthermore net reabsorption of fluid driven by hydrostatic or colloid osmotic pressure gradients must leave the solutes behind and thus the reabsorption will be self-limiting. It can at most be very small compared with the rates at which CSF is produced.

Viewed the other way round, net solute transport across the blood-brain barrier will be matched by water flow. In other words, water follows salt. One could say that solutes are actively transported across the blood-brain barrier and water follows by filtration but the convention is to refer to the entire process as secretion of fluid. This has the important advantage that it emphasizes that the energy driving the process is derived from metabolism within the blood-brain barrier not from applied pressure gradients. *The rate of secretion of fluid across the blood-brain barrier is determined by the net rate of solute transport.* Furthermore, so long as water can move sufficiently rapidly between plasma and ISF that the osmolalities of these fluids are approximately equal, *the volume of ISF in the brain is determined by the solute content of the ISF and the osmolality of the plasma.*

Reabsorption across the brain microvessels of parenchymal fluid derived from the CSF has been proposed repeatedly [11, 65–70, 79–84] but in no instance have these proposals included a mechanism for reabsorption of the solutes. As explained above, the net pressures that can be produced, colloid osmotic or hydrostatic cannot drive net removal of fluid across the intact blood-brain barrier at a rate capable of dealing with excess fluid brought into the brain parenchyma (section 4.2.2 and Endnote[d]). This is not a new idea; this inability of the intact blood-brain barrier to reabsorb interstitial fluid has been argued before by many others [5, 7, 85–87]. Reabsorption could occur if the blood-brain barrier were to become permeable to small solutes, but this would amount to barrier failure and would mean it could no longer control ISF composition. The reason that removal of fluid across the intact blood-brain barrier cannot match the rate at which fluid is produced is that there is no mechanism for *net* transport of NaCl across the barrier in the direction from brain to blood. If salt could be transported, then water would follow.

The rate of secretion of Na^+, Cl^- and HCO_3^- is the primary determinant of the rate of fluid secretion into the parenchyma. Metabolically-produced water does not affect these rates and hence has little impact on fluid secretion [88]. The water derived from metabolism will either be absorbed across the blood-brain barrier or reduce the amount of water needed to accompany the secretion of solutes by the blood-brain barrier.

Diffusion and Convection within the Parenchyma

By analogy with peripheral tissues (section 2.3), the expectation is that movements of water and solutes in brain ISF will be primarily by diffusion rather than convection. Thus diffusion is the mechanism for movement of molecules between adjacent cells. On the distance scale of capillary to parenchymal cell for which estimates range from 10 μm to 50 μm [89–92] the dominant process still clearly is diffusion and this accounts for the important movements of O_2, CO_2, glucose and lactate. The sole exception may be local circulation of K^+ accompanied by water, a story tied up with the idea of glial K^+ buffering which is beyond the remit of this review. For distances up to a millimeter or more, diffusion still appears to be the most important process but additionally there is convection along fibre tracts in white matter [93] and other privileged pathways, which may include capillary basement membranes and the perivascular spaces along arteries and veins. These privileged routes will be considered in section 4. The evidence that diffusion is the dominant process within grey matter is considered immediately below. More extensive reviews of this subject and related matters are available [41, 46, 94–96].

The early studies on transport within the extracellular spaces of the parenchyma were aimed primarily at demonstrating that these spaces exist, a hot issue in the early 1960s. These studies were based on measured concentration profiles within the parenchyma of radiolabelled solutes chosen for their inability to enter cells or cross the blood-brain barrier [97, 98]. At various times after the labelled solutes were added to the ventricles (and kept at constant concentration by perfusion), the animals were killed, and parenchymal samples were taken at known distances from the ventricular surface. The concentration of solute found in the tissue adjacent to the perfused surface provided a measure of the volume occupied by the extracellular space while deeper samples provided information about diffusion. These values were then compared with those expected theoretically if transport were to be by diffusion:

$$\frac{C(x)}{C_0} = \text{erfc}\left(\frac{x}{2\sqrt{Dt}}\right)$$

where $C(x)$ is the concentration at distance x, C_0 is the concentration at the ventricular surface, t is the time after addition of the solute, D is the effective diffusion coefficient and erfc is the complementary error function. Hindrance to diffusion imposed by the tissue is described by the tortuosity, defined as the square root of the ratio D_0/D, where D_0 is the diffusion coefficient in a free medium. The comparison of experiment and theory provides two tests of whether or not solute movement is by diffusion by asking: "Do the data fit the shape of the theoretical curve?" and "Is the value calculated for the diffusion constant the same for different times?" It was found that the data obtained by this method pass both of these tests. However, Rall [38] has stated quoting C.S. Patlak that these tests could still be passed even if there were convection with a flow towards and across the ependyma of up to 1/4 of the CSF production into the ventricles. In other words the data do not exclude convection that has been inferred from experiments on efflux of test substances from the parenchyma (section 4.1).

More recent determinations of transport through interstitial spaces have been based on injection or iontophoresis of a test substance at a point in the parenchyma and detection in real-time of its spread through the tissue. In the Real Time Iontophoresis method using tetramethylammonium ions (TMA), these cations are iontophoresed from a microelectrode and their concentration measured a known distance away with an ion-selective microelectrode [95, 99]. This technique can be calibrated to give actual concentrations resulting from known amounts of TMA added in a very small region and enables both volume fraction and tortuosity to be calculated. In the Integrative Optical Imaging method, a fluorescent marker is pressure injected and the concentration profile determined from microscopic observation of the fluorescence as a function of position in the surroundings. This is more difficult to calibrate and only yields a measure of the effective diffusion coefficient and tortuosity for the marker, but it has the advantage that it allows checks on the symmetry of the spread away from the injection site [46, 95, 100]. For instance it can show that spread in white matter occurs preferentially parallel to the long fibres. The large amounts of data obtained using these techniques are consistent with diffusion as the mechanism of transport over short distances. Flow velocities around 10 μm min^{-1}, as inferred from tracer studies of solute flux (section 4.1 and [93]) would be too small to change significantly the calculated values of the diffusion parameters (Charles Nicholson and Tony Gardner-Medwin, personal communications) and even the larger flows inferred from some more recent measurements (see caveats in section 4.1.1) may be difficult to detect.

Convection may be more important as a mechanism for fluid movement in the brain under conditions where there is oedema associated with expanded extracellular spaces as can occur in white matter. By analogy with peripheral tissues, the resistance to convection may then be much smaller than in non-oedematous tissue (section 2.3). Reulen *et al.*[101–103] have considered convection in some detail in connection with the spread and resolution of oedema. Convection is important in the spread of drugs injected into the parenchyma in comparatively large volumes, so called convection-enhanced drug delivery [104–106] although the extent to which the flow passes through the interstitial spaces between cells in grey matter rather than being restricted to privileged pathways such as perivascular spaces has apparently not been determined.

Secretion by the Choroid Plexuses and the Blood-brain Barrier

The reader interested in the now historical argument about whether transport across the choroid plexus is secretion or filtration should consult reviews by Cserr [4], Davson and Segal [9] and Damkier *et al.*[15]. As discussed by Damkier *et al.*[15] the choroid plexus has all the hallmarks of a high capacity secretory epithelium whose product is isosmotic with plasma. The blood-brain barrier is a barrier at least partly because the endothelial cells lining the microvessels lack the low resistance paracellular pathways for small solutes that characterize microvessels in peripheral tissues (Figure 5). Instead the endothelial cells are sealed together by tight junctions that prevent (or at least greatly reduce) the passage of markers like horseradish peroxidase and lanthanum [32, 34] and raise the electrical resistance to levels similar to those in tight epithelia [107, 108]. As discussed in section 2.7, this low paracellular permeability means that sustained net movement of fluid across the blood-brain barrier requires transport of solutes to generate an osmotic gradient. Water then follows, partly by simple diffusion across the membranes of the endothelial cells, but possibly also via various transporters

[61, 109]. The mechanisms of ion transport at the blood-brain barrier have been reviewed by O'Donnell [18] and will be considered further in a sequel to this review.

FORMATION AND REMOVAL OF CSF AND ISF

CSF

In 1996 when Davson and Segal published their monumental review [9] the general view was that the rate of production of CSF was roughly 400 ml/day in humans and that of this perhaps 90 % arose as secretion by the choroid plexuses and 10 % by secretion of ISF across the blood-brain barrier and subsequent flow to join CSF. However, both the underlying assumption that all ISF flows into the CSF and the relative contribution of ISF to the secreted CSF have been and remain controversial.

Formation of CSF

If the choroid plexuses produce almost all the CSF, their removal should prevent the excessive fluid accumulation in the ventricles that occurs in hydrocephalus (section 4.2). Yet Milhorat [110–112] has stated quite firmly that choroidectomy does not work for the treatment of hydrocephalus and thus there must be sources of CSF in addition to the choroid plexuses. However, at least in the hands of Scarff [113], choroidectomy was found to be an effective therapy in a large proportion of patients with communicating hydrocephalus. One possible explanation of these discrepancies may be the extent to which total choroidectomy had been achieved. Regardless of the eventual resolution of the dispute over clinical utility, taking Milhorat's and Scarff's work together leaves little doubt of the basic physiological point, i.e. removal of the choroid plexuses reduces the rate of production of CSF but to nowhere near zero. The extensive series of choroid plexus cauterization treatments for hydrocephalus in children by Warf [114, 115] (see also [116]) is consistent with the choroid plexuses being a major, but not sole source of CSF even in children under one year (compare section 4.2.3).

Removal of CSF

As reviewed by Pollay [16, 117], CSF is removed by several routes including the arachnoid villi leading to venous sinuses, perineural pathways across the cribriform plate leading to nasal mucosa, and pathways at the roots of spinal nerves leading either to blood or lymph [8, 9, 118–127]. Large solutes like albumin that are removed from the brain via the cribriform plate enter lymph. By contrast, a large proportion of the lower molecular weight solutes and water leaving by the same route reach the blood directly across peripheral capillary walls in the nasal mucosa [120]. The relative rates at which CSF is removed by the various routes may vary markedly between species.

Measurement of CSF production rate

CSF production rate has been measured in several ways [9] of which three will be discussed here. These are a) collecting CSF as fast as it is formed; b) recording the rate of removal of injected markers into CSF; and c) measuring the dilution of markers perfused into the ventricles at a known rate. In the collection method a), CSF is removed through tubing or a needle held at a pressure sufficiently low that CSF exits solely by this route, it being assumed that all normal exit routes require a higher pressure. This was the method used in the influential study by Ekstedt [128, 129]. Caveat: It is no longer clear that this procedure can prevent exit of fluid from the brain by other routes. There may be large variations in the rate at which substances reach differing parts of the subarachnoid space when infused into the ventricles [130, 131]. Furthermore as discussed in section 4.1, it appears that some of the fluid secreted into the ISF in the forebrain returns to blood or lymph by routes that avoid mixing with CSF in the cisterna magna indicating that not all regions of the subarachnoid spaces are in free communication [119, 132]. These observations suggest that reducing pressure at the point of collection, e.g. cisterna magna or lumbar sac, may lead to collapse of part of the subarachnoid space and thus failure to transmit the reduction in pressure to other parts. Collecting CSF from some portions of the subarachnoid space may be a bit like trying to suck fluid through a flimsy straw. In the marker removal method b), a marker substance is injected into the cisterna magna and a complete collection of CSF is made at some later time. The rate of loss of the marker and the CSF production rate are calculated from the amount remaining. The calculation requires the assumption that CSF is well mixed throughout the period before collection and that all of the CSF can be collected. The method would still work if a portion of the CSF e.g. that in the spinal cord, were completely stagnant and were not reached by the marker and were not collected. Caveat: Studies such as those discussed in sections 4.1 and 4.3 indicate that CSF is not well mixed as required for these calculations. In the perfusion method c), fluid containing a known concentration of a marker is infused at a constant rate into a lateral ventricle and fluid is sampled by withdrawal from either the cisterna magna or lumbar sac. Assuming that the marker only leaves the CSF as part of fluid with the same concentration as in the sample, i.e. by bulk flow of well mixed CSF, and all of the secreted CSF is added to the mixture that is sampled, the secretion rate can be calculated as follows [133]:

$$\text{secretion rate} = \text{perfusion rate} \times \frac{C_{\text{perfusate}} - C_{\text{collected}}}{C_{\text{collected}}}$$

where each C is the concentration in the indicated fluid. Note that this method does not require total collection of CSF and thus the sample can be small and can, at least in principle, be taken without reducing intracranial pressure. Caveats: If CSF were secreted solely from blood into the ventricles from which it flowed into the cisterna magna and was then reabsorbed into blood or lymph from the subarachnoid spaces, the assumptions needed for this calculation to be correct would be plausible. However, as will be discussed in section 4, these are not safe assumptions. CSF is not well mixed and, as a result, the marker concentration in the sample does not accurately reflect those in the various outflows. At present there can be no "gold standard" method for measuring CSF production at least partly because there is no single compelling definition of what CSF production means, e.g. rate of fluid secretion into the ventricles or alternatively rate of fluid secretion into the CNS as a whole. The present estimates are likely to be reasonable values for the normal rate of secretion of fluid into the ventricles but may be underestimates of the total rate of fluid secretion into the brain as discussed in the following sections.

Measurement of ISF Production Rate

Estimates of the rate of ISF production can at least in principle be obtained from measurements of a) the rate of CSF production after destruction or removal of the choroid plexuses [134, 135], b) the rate of entry of fluid into the perfused cerebral aqueduct [136, 137] and most convincingly c) the rate of removal of markers injected into the parenchyma [132, 138, 139]. Caveats: There are drawbacks to each of these approaches. As Milhorat [112] was careful to emphasize, the first a) requires total removal of the choroid plexuses (or evidence about the contribution of the portions remaining) and measures a rate of fluid production after the fluid dynamics have been severely altered. The second method b) requires the assumption that ISF production in regions that drain towards the aqueduct is typical of production in all regions. In addition, the measurements are made after extensive recent surgical intervention. The third method c) requires that measurements be made in a number of locations (cf. [139, 140]) and assumes that the tracer substance is carried out of the parenchyma by convection, that there is no sieving of the markers and that the flow arises only by secretion of fluid across the blood-brain barrier.

Sieving refers to the retardation of movement for large solutes when fluid flows through a filter, a matrix or a gel. Flow through a matrix does not lead to flow of solutes of all sizes at the same rate, a defining characteristic of pure bulk flow, unless the sizes are all well below the typical distance between the fibres of the matrix [141]. The nature of the matrix, if any, along the rate limiting portions of the flow pathway for drainage of interstitial fluid (see section 4.1) is unknown. Thus estimating fluid flow rate from the rate at which large markers are removed from the tissue may underestimate the net rate of flow of water and small solutes. The evidence for bulk flow and for the absence of sieving in the parenchyma for solutes smaller than albumin is that the measured rates of removal of tracers of a range of sizes up to albumin were all the same [132]. However, the rate limiting stage of the flow was held to be along privileged pathways including perivascular (or Virchow-Robin) spaces, spaces between fibre tracts in white matter and spaces within the subependymal layer of the ventricles [7, 138]. The evidence obtained from removal of large markers is considered further in section 4.1 and the assumption that flow through the parenchyma arises only from the blood-brain barrier is considered in sections 4.3 and 5.

ONGOING APPROACHES TO THE INVESTIGATION OF BRAIN FLUID DYNAMICS

Studies of Movement of Substances and Routes of Outflow from the Brain Parenchyma

To discover whether the production of ISF contributes to the CSF, it is necessary to find out whether ISF drains into CSF before leaving the brain or whether it goes via some independent route. As described below, drainage of ISF appears to be partly to CSF as sampled in the cisterna magna and partly to cervical lymph nodes by routes that do not require the emerging ISF to mix with CSF in the cisterna magna.

Periarterial spaces as routes of efflux

When markers such as radiolabelled serum albumin or polyethylene glycols are injected into the caudate nucleus of one side of the forebrain, twice as much appears in the ipsilateral cervical lymph node as in the contralateral node in the neck, and less than 30 % reaches the cisterna magna [119, 132]. By contrast, if the same markers are injected into ventricular CSF a smaller proportion appears in the cervical lymph nodes but that which does so appears symmetrically on the two sides. These results have since been extended to other sites of injection revealing that the fraction of tracer reaching the cisterna magna varies depending on distance from the injection site [139, 140]. Furthermore, in experiments following the clearance of radiolabeled albumin, tracer concentration was found to be much higher in the walls of dissected arteries supplying the region where the tracer had been injected than in the surrounding CSF. For the caudate nucleus and internal capsule this prominently included the circle of Willis. Similarly, when horseradish peroxidase was injected (in the large volume of 30 μl) into the midbrain of a rat, it could be detected an hour later in or on the walls of branches of the posterior cerebral artery [139]. These findings were confirmed and extended [140] with the additional observation that the delay between the disappearance of tracer from the site of injection and its appearance in lymph was longer when the tracer was injected into the parenchyma than into CSF. Caveats: Two features of these studies have been challenged: the overall rate of removal and the assumption that injection of marker does not disturb the routes of outflow being examined. These experiments were all performed using barbiturate anaesthesia, but it has subsequently been shown that the half-life for efflux of ^{14}C-sucrose, which crosses the blood-brain barrier at a negligible rate, is about 4 h when the rats are awake or anaesthetized with ketamine and xylazine but about 25 h when they are anaesthetized with pentobarbital [142]. In ketamine/xylazine anaesthetized mice efflux of mannitol, another solute with low blood-brain barrier permeability, has a half-life similar to 4 h [143]: this confirms the result for sucrose. If the half-life for efflux of albumin is similarly faster in awake animals than in those anaesthetized with barbiturates, the estimates of the ISF production rate given by Cserr and co-workers [132, 138, 139] may need to be revised upwards, perhaps by as much as 6-fold (but see section 4.3 for alternative explanations for rapid removal of solutes). The rate of efflux of albumin clearly needs to be investigated further with attention being given to the state of anaesthesia.

Another serious concern is raised by the suggestion that the rates of infusion or volumes injected may have been sufficient to alter the mechanisms and routes of elimination. Iliff *et al.*[143] suggested that the observation of efflux along arteries might represent an artefact of "high local intraparenchymal pressure from the injection". Pressure-induced flow is used in convection-enhanced drug delivery [105]. However, even the lower end of the infusion rates employed, 0.5 μl min^{-1} to 20 μl min^{-1}, is 8-fold larger than the rate, 0.5 μl infused over 8 min, used by Szentistvanyi *et al.*[139]. Thus it is difficult to see how the infusions in the reported studies could have altered the efflux route from a putative physiological perivenous route (section 4.3) to the periarterial route observed.

Bradbury, Cserr and colleagues were of the opinion that flow out of the parenchyma occurred along periarterial spaces [144] and that where the arteries reached the subarachnoid space there was some form of barrier separating the periarterial and subarachnoid spaces. Of the structures described subsequently by Zhang *et al.*[23, 145] (Figure 1d) this barrier may be the sheath of pia mater enveloping the artery. Cserr and Patlak and Bradbury *et al.* incorporated this route of efflux into a general scheme in which ISF is secreted by the blood-brain barrier and then flows out of the parenchyma to CSF via periarterial spaces. The pial barrier delays but does not prevent passage of the albumin into CSF destined for the ipsilateral side of the cribriform plate and thus to the nasal mucosa [8, 119].

Further evidence supporting the existence of periarterial spaces as routes of efflux from the parenchyma has been provided using two-photon microscopy together with 3 kDa cascade blue dextran, viewing being through a cranial window [146]. Following injection into the parenchyma, the marker was detected in a space surrounding the arteries and in the extracellular matrix of the arterial smooth muscle layer but not along veins. The authors state (data not shown) that the rates of removal of fluorescence were similar for 3 kDa and 70 kDa markers but no removal was seen for a 2000 kDa marker. Fluorescence decreased more rapidly in the interstitium and periarterial space than in the smooth muscle layer. The authors suggest that this is because there are multiple routes for removal from the interstitium but only one for the smooth muscle layer. An alternative explanation for these observations may be that the dextran binds within the extracellular matrix of the smooth muscle layer.

Evaluation of the proposal that periarterial spaces provide an efflux route for markers

The idea that markers can move from the parenchyma via convection in periarterial spaces (Figure 6) and thence transfer into CSF has a number of plausible features. The convection inherent in this idea would always occur along pathways that should have a low resistance to flow. Similarly, transfer to and from CSF across the pial barrier should be rapid for small solutes [97, 143, 147–153] (but see section 4.3.1). It is slow, but still easily demonstrable, for macromolecules [32, 138, 143, 154]. The actual permeabilities for transport between CSF and parenchyma across the pia mater are not known because concentrations on both sides of the barrier have not been measured simultaneously. Some of the large solutes emerging from the parenchyma not only reach CSF but also find their way to the cisterna magna [119, 132]. This scheme suggests a plausible explanation for the observation that the proportion of substance reaching the cisterna magna is larger from injection sites that are closer to the cisterna magna.

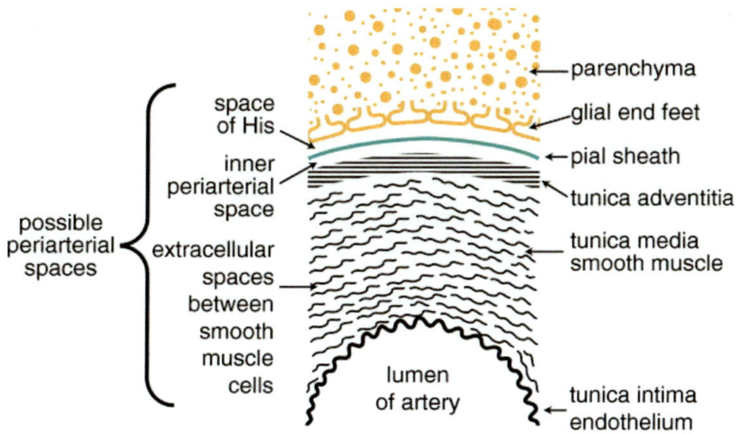

Figure 6: Diagram indicating positions within the cortical parenchyma of periarterial spaces that may allow fluid movement. The spaces shown correspond to those around the arteries in Figure 1d. The diagram has been formatted so that it may be compared with earlier published versions [1]. The diagram (not to scale) shows: the tunica intima, the endothelial lining of the lumen and a covering of elastic tissue; the tunica media, a smooth muscle layer; and the tunica adventitia, mainly connective tissue. Within these layers there are two possible free spaces. The inner, called the periarterial space by Zhang *et al.*[23] and labelled

the inner periarterial space in the diagram, is continuous with the periarterial space of the subarachnoid portion of the same artery. The outer that in contact with the brain parenchyma, is likely to be the space described by His [1]. In the view of Weller and associates (personal communication) both the inner periarterial space and the space of His are virtual spaces with no thickness and fluid movement occurs preferentially in the extracellular spaces of the smooth muscle layer. Other interpretations of the spaces are considered in section 4.3.4.

Rapid inflow via a periarterial space may also partly account for the early observation that horseradish peroxidase injected into CSF reaches the basement membranes of even the small microvessels in the cortex within a few minutes [79, 154–156] but at low concentration, such that at early times only very sensitive detection methods are capable of establishing its presence [157]. Rapid appearance of low concentrations deep in the cortex is the behaviour expected for relatively rapid transport over the large distance (millimeters) from the cortical surface, provided this is preceded or followed by slow penetration through a barrier. The rapid transport along the periarterial spaces could be provided by stirring or mixing in the perivascular spaces driven by the arterial pulse (compare [143, 154, 156, 158]). The barrier could be the pial coating of the arteries, which would precede the rapid transport process, or it might be the basement membranes or the layer of glial end-feet associated with the microvessels, either of which would follow the rapid process. There does not appear to be any data relating to the mechanism of transport for horseradish peroxidase through the subarachnoid space (section 4.3).

Whether veins also constitute an exit route for injected albumin from the parenchyma is not known. Most studies have not looked at this aspect, but in their fluorescent marker studies, Arbel-Ornath *et al.*[146] examined arteries and veins and saw fluorescence only along arteries.

Caveats: While efflux of solutes like albumin via periarterial spaces, CSF and the cribriform plate is an attractive proposal, it nevertheless raises a number of questions:

- Can albumin be detected in sufficient concentrations near the cribriform plate for the nasal route to be able to account for the removal of the markers? Does obstruction of the cribriform plate reduce transfer of markers injected into the parenchyma to cervical lymph nodes as it does [120] for markers injected into CSF (see also [159, 160])?
- Why have some markers (section 4.1.2) been difficult to detect in

the periarterial spaces? The present lack of such data in images of brain sections (e.g. [161]) is perhaps to be expected for two reasons: as the spaces are thought to have rapidly changing dimensions they may collapse during sample preparation and the concentrations in these spaces during efflux of markers may be low (section 4.3 for further discussion).

- Does the use of barbiturate anaesthesia alter the routes of efflux as well as reducing the rate (section 4.1.1)?

Extracellular spaces of the arterial smooth muscle layer as routes of efflux

Weller, Carare and colleagues have been pursuing an alternative hypothesis that flow out of the parenchyma occurs through the smooth muscle layer of the arterial walls (Figure 6). They have shown that fluorescently tagged 3-kDa dextran, 40-kDa ovalbumin and amyloid-β [153, 161–163] injected into the parenchyma become diffusely spread and associated with both the basement membranes of capillaries and the extracellular spaces separating the smooth muscle cells in arterial walls. Five minutes after injection into the parenchyma they found markers associated with the basement membranes. For tissue fixed after 30 min, the markers were still present diffusely in the parenchyma and were also found in the smooth muscle layer, but not within capillary basement membranes. For fixation after 3 h (for the dextran) or 24 h (for all markers), fluorescence was seen only in punctate form near the outside edges of the walls of arteries associated with macrophages. These data were interpreted as supporting clearance of the markers from the interstitium via bulk flow through microvascular basement membranes and the extracellular spaces between the smooth muscle cells of arterial walls [161].

Caveats: It is clear that markers can reach the extracellular spaces of the smooth muscle layer of the arteries from the parenchyma, but it is not clear that there can be significant flow along these layers. However, it would be unwise to dismiss movement via the smooth muscle layer as physically impossible without evidence. The data raise a number of questions.

- Why have the fluorescent markers investigated been concentrated in basement membranes and the smooth muscle layer? Is there some form of association or binding? Is there a sieving effect with retention of larger solutes with higher flow of water and smaller solutes than suggested by the movement of the markers?

- Can there be exchange between extracellular spaces of the arterial smooth muscle layer and nearby periarterial spaces? If so, efflux from the parenchyma might be via the periarterial spaces with the marker able to gain access to the smooth muscle from the adjacent periarterial space.
- Why are the fluorescent markers not found within the endothelial basement membranes of the arteries or within the outer basement membrane of the arterial wall [161, 164]?
- Why does fluorescent ovalbumin disappear from the capillary basement membranes while there is still diffuse fluorescence observed in the surrounding parenchyma [161]?
- What is the driving force for flow along arterial walls? Is it, as suggested, the arterial pulse changing compression in the wall [161, 165]? Is there any structural basis for the "valves" which appear to be needed to make compression drive movement preferentially in the outward direction [165, 166]?
- Can movement of markers along arteries from parenchyma to lymph node be measured? How long does it take for them to reach the subarachnoid space and subsequently the cervical lymph nodes?
- What fraction of an injected load of marker can be accounted for by that moving along arteries and reaching cervical lymph nodes?
- Does deposition of amyloid-β along presumed flow pathways obstruct fluid movement via these same pathways? What happens to water and small solutes?
- How in earlier experiments did radiolabelled human serum albumin injected into the parenchyma [119, 132, 139, 140] reach CSF in the cisterna magna (see earlier discussion)?

Even if the smooth muscle layer of arterial walls is not the main route for fluid transfer, it may still be important in clearance of amyloid-β and other proteins. This aspect has been the emphasis of recent work, e.g. [164, 167, 168]. It should be noted that the periarterial routes for transfer discussed in section 4.1.1 and in this section and indicated in Figure 6 are not mutually exclusive. It is possible that all pathways mediate transfer of fluid and solutes. It is also conceivable that net flow is in opposite directions in different spaces or even in both directions at different times within one. These possibilities are considered further in section 4.3.4.

Studies of CSF Flow and the Implications of the Flow Patterns for Sites and Rates of Production and Absorption of CSF and ISF

Measurement of CSF flow by phase contrast magnetic resonance imaging (PC-MRI)

Interest in patterns of CSF flow and what they reveal about the sites and rates of production and absorption of CSF have been rekindled by results obtained using PC-MRI and in particular measuring CSF flow through the cerebral aqueduct that connects the third and fourth ventricles. The development of PC-MRI has allowed real-time monitoring of blood flow in arteries and veins and, under favourable circumstances, also flow of CSF. MRI detects the magnetic dipoles of hydrogen atoms in small regions, called voxels, each of the order of 1 mm^3. Application of this technique is more difficult for measuring CSF movements than for measuring blood flow because CSF movement is slower and the regions of interest are often small, containing only a few voxels. Despite these limitations, results of PC-MRI have provided useful information about CSF flow.

Cyclic variations in CSF flow

From the results of PC-MRI and other techniques it is apparent that CSF flow in the cerebral aqueduct is not steady but rather varies over the cardiac cycle, driven by changes in blood volume in the brain vasculature [169]. In systole, more blood flows into the brain through arteries than leaves via veins while in diastole the reverse is true. Thus blood volume in the brain varies over the cardiac cycle and, because the skull is rigid and the inside volume is fixed, something else must compensate and this is achieved by shifting CSF from brain to spinal cord during systole and back during diastole. In normal people and experimental animals, a small part of this shift is movement of CSF in and out of the lateral and third ventricles via the cerebral aqueduct. Peak flow rates in each direction through the aqueduct of a normal subject are typically 10 times larger than the net average flow over the entire period. These fluid movements assist in mixing CSF in the ventricles and subarachnoid space.

Direction of net flow

Early PC-MRI results for normal subjects were interpreted as yielding calculated net flows through the aqueduct consistent with values of CSF production rate in the ventricles determined by other means [169–173]. However, later studies have challenged the traditional view that CSF is produced primarily in the ventricles and absorbed primarily in locations distant from the ventricles.

Changes in flow in hydrocephalus

Even before PC-MRI results became available there were reports showing that the standard story of a constant rate of CSF production in the ventricles and net flow from lateral and third ventricles outwards towards the subarachnoid space was not correct in patients with hydrocephalus, a condition where there is abnormal enlargement of the fluid-filled ventricles.

Non-communicating hydrocephalus

In patients with non-communicating hydrocephalus the normal exit route for CSF from the ventricles, the cerebral aqueduct, is blocked. In these patients the rate of increase in ventricular volume has been reported to be far less than the normal rate of CSF production [174] and this can be reproduced in animal models. Some even contend that, at least in the cat, blockage of the aqueduct produces no ventricular swelling at all [175, 176]. Set against these reports must be experimental results (for an early example see e.g. [177]) and clinical experience (e.g. [135]) indicating that blockage does produce ventricular swelling. Milhorat *et al.*[178] tested this explicitly in monkeys by inflating a balloon in the fourth ventricle so blocking the aqueduct. Swelling of the lateral and third ventricles was seen to be initially rapid, possibly corresponding to the normal rate of production of CSF, but then to become less rapid. Three alternative explanations may be considered for why the rate of swelling following blockage of the aqueduct can be substantially less than that generally believed to be the "normal" CSF production rate: firstly secretion of CSF may have become inhibited; secondly blockage may not have been achieved; and lastly alternative routes for removal of CSF from ventricles may have been created or become important. Agreement has been reached [5, 9] that the CSF production rate does not change or at least does not change enough to explain the above result (e.g. [179] and for an early study see [174]). Change of CSF production rate will not be considered further here. Possible alternative routes for fluid escape will be considered in section 4.2.4.

Communicating hydrocephalus and the possibility of reverse net flow

In communicating hydrocephalus, there is no impediment to CSF movement between ventricles and cisterna magna but there is still marked ventricular enlargement (by definition) and usually subarachnoid space contraction. PC-MRI measurements have shown greatly increased amplitudes of fluid flow through the aqueduct during the cardiac cycle, e.g 6-fold larger than normal, presumably because more CSF shifting from the cranium to the spinal cord must now come from the ventricles. At least one study comparing control subjects with patients with normal pressure hydrocephalus, found net flow in the expected direction and of the expected magnitude [180]. Some studies reporting large peak flows in each direction did not calculate a net flow [170, 181, 182] but there are now a number of reports of calculated net flows in the reverse direction, i.e. from fourth to third ventricle [69, 83, 84, 172, 183, 184].

Suggestions of reverse net flow through the cerebral aqueduct in communicating hydrocephalus predate evidence from PC-MRI. Milhorat summarizing early results concluded (p139 in [135]) "In communicating hydrocephalus the normal subarachnoid pathways are obstructed and the subarachnoid flow is frequently reversed and directed into the ventricular system. … Diagnostically, retrograde filling of the ventricular system is a characteristic cisternographic finding in communicating hydrocephalus". James and colleagues injected radiolabelled albumin into the cisterna magna of both dogs and monkeys with experimentally-induced communicating hydrocephalus and were able to see accumulation of the marker in the ventricles and periventricular regions, implying there must be net flow of fluid in the reversed direction, i.e. into the ventricles [185–188]. Such reverse net flow through the aqueduct implies both a major route for exit of CSF from the lateral and third ventricles other than the aqueduct and a major source of CSF other than the choroid plexuses (Figure 7). One proposal for an additional source is the cortical parenchyma, i.e. fluid secretion by the blood-brain barrier. Candidates for the route of outflow are less obvious. In conjunction with the ventricular enlargement there is periventricular oedema in white matter, which may allow abnormal routes for fluid removal from the ventricles [86].

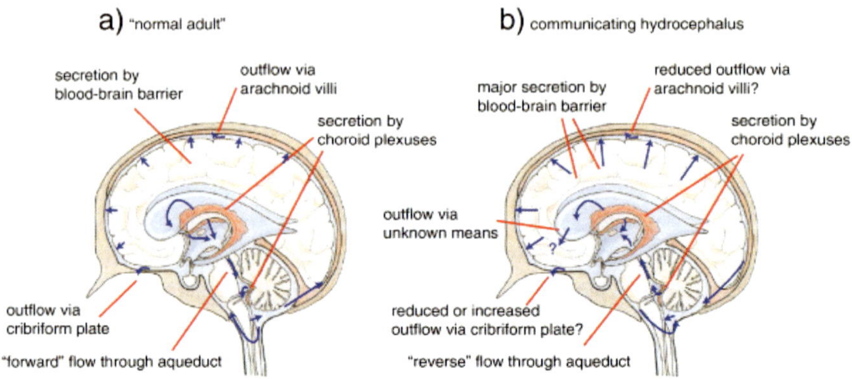

Figure 7: View of net CSF flow pattern in normal adult brain compared with that proposed for communicating hydrocephalus. a) In the normal adult, most of the CSF is secreted by the choroid-plexuses and is reabsorbed through the arachnoid villi or via the cribriform plate leading to the nasal mucosa. There is modest secretion across the blood-brain barrier into ISF most of which emerges into CSF across brain surfaces. The net flow through the cerebral aqueduct is from the third ventricle towards the fourth ventricle. b) In communicating hydrocephalus, the observation of reverse net flow through the aqueduct implies that formation of CSF occurs outside of the ventricles, probably by more extensive fluid secretion across the blood-brain barrier into the cortex and thence out into the CSF-containing spaces. Reverse net flow also requires some route for removal of CSF from the third and lateral ventricles. This route must accommodate both fluid secreted by the choroid plexuses located in these ventricles and fluid entering the third ventricle via the aqueduct. Possible routes are discussed in section 4.2.4. (The background image is the same as in Figure one-a taken from Strazielle *et al.* with permission [19]).

CSF flow in infants

Reverse flow in the aqueduct has also been reported by one group in children under two years; this being suggested as the normal condition for this age group [69, 84, 184]. There has also been a report of "early ventricular reflux" of radiopharmaceuticals in children under two years [189]. If net flow from fourth to third ventricle is accepted, "from where does it come, to where does it go" [69]? The blood-brain barrier can secrete fluid so the suggestion that the choroid plexuses are not the main source of CSF until two years is a hypothesis that should be tested. However, no evidence using any technique other than PC-MRI has been presented that would indicate a change in formation rate at or about the age of two. Using an infusion technique to determine the relation between intracranial pressure and rate of addition of fluid to CSF, Blomquist *et al.* found 588 ml/day as the control rate of CSF production, with no clear variation in age from two months to about ten years of age [190]. This is towards the high end of estimates of the normal adult rate. There is greater difficulty in understanding the fate of the CSF as discussed below.

Possible alternative routes for CSF outflow from the ventricles

If net flow of CSF through the aqueduct is towards the lateral and third ventricles, then where does CSF leave these ventricles? Three possibilities need to be considered: a) there is another conduit for outward flow; b) CSF passes into parenchyma where it mixes with ISF and finally enters microvessels, i.e. it crosses the blood-brain barrier; and c) CSF flows into the parenchyma, mixes with ISF, and finally exits by other routes as discussed in section 4.1.

- Possible alternatives to the cerebral aqueduct as a conduit for flow of CSF out of the central ventricles have been discussed by Ghersi-Egea, Nagaraja, Fenstermacher and colleagues [130, 131]. These putative conduits are in regions where there is only a thin wall separating the ventricles from subarachnoid space [3]. However, there has been little support from any other groups for the idea that these regions do provide a pathway for flow. Nevertheless a search for such a conduit would be appropriate as there could be benefits from its presence as is seen in the treatment of hydrocephalus when an outflow route is produced surgically by third ventriculostomy (e.g. [114]). A further alternative route proposed by Milhorat [191] for flow out of the ventricles in hydrocephalic infants is via the choroid plexuses. This would require net flow to occur across the epithelium towards the blood, possibly via the paracellular route. However, Wislocki and Putnam found no movement of dyes via the choroid plexus in hydrocephalus induced by injection of lamp black into the cisterna magna [192] and Eisenberg, McLennan and Welch [193] found in kaolin-induced hydrocephalus that there was no reabsorption of fluid from the ventricles and thus perforce none via the choroid plexuses. However, Welch and Sadler had earlier found free diffusion of solutes as large as inulin from ventricles to blood but at a low rate as if across only a small fraction of the area of the choroid plexuses [194]. There was also an additional component of transport as if ventricular fluid were being moved across the choroid plexus by pinocytosis. The rate at which fluid was transferred by this route was later estimated to be at about 1/10th the normal rate of secretion [5]. In summary, while the balance of evidence is against a role for the choroid plexuses in net flow from ventricle to blood, the idea has not been conclusively excluded and requires further investigation.

- With regard to the second possible exit route for CSF, as already discussed in section 2.7, CSF absorption by penetration into the parenchyma and subsequent absorption into brain capillaries driven by colloid osmotic or hydrostatic pressures cannot be sustained for more than a few minutes so long as the blood-brain barrier is intact. It seems unlikely that this route is important for infants because there is no evidence of a major difference in the properties of their blood-brain barrier compared to older children. In hydrocephalus, alteration of the blood-brain barrier should be considered, but no evidence appears to have been produced that such alteration provides a route for the net removal of fluid from regions of the parenchyma. Linninger *et al.*[82] and Penn *et al.*[83] concluded that the blood-brain barrier was an important site of fluid removal from the parenchyma, but while their analysis supports the conclusion that there is some route of removal, the conclusion that this route is across the blood-brain barrier is the result of circular logic as their model for fluid movement considered only the blood-brain barrier as a possibility.
- Large scale flows of water in and out of parenchyma can be produced osmotically (section 2.7) and large fluid flows can also occur in regions of oedema. Thus in hydrocephalus, where there is periventricular oedema in white matter, it may well be the case that there are pathways for flow through the brain to alternative routes of exit such as those considered in section 4.1. Such a flow of fluid from ventricles into the parenchyma and thence to perivascular spaces was proposed as early as 1921 by Wislocki and Putnam [192]. By contrast, flow of the magnitude proposed through non-oedematous parenchyma from ventricles to the subarachnoid space would be without precedent. But "without precedent" only means that more direct evidence is required before the idea can be accepted.

Irrespective of the ultimate site of reabsorption of the fluid, if there is a marked increase in fluid transfer through the parenchyma there should be a marked change in the fate of markers injected into it. These are experiments that need to be done.

Caveats on PC-MRI results for net flow through the aqueduct

The idea that in communicating hydrocephalus and in infants there is a cortical source of CSF and a sink for fluid in the ventricles via the parenchyma (or possibly the choroid plexuses) is based primarily on the validity of results for reverse net flow in the aqueduct. There are a number of features in the PC-MRI experiments, usually interpreted using proprietary software, which may not have been adequately tested and/or calibrated (for a recent review see [182]). Here are some of the pitfalls.

Calculating net flows from large flows in opposite directions suffers from the classic problem of subtracting large numbers to determine small differences; the signal can easily get lost in the noise. Experimentally there are large variations in the values of the net flows reported by various groups. Before the idea that reverse net flow occurs is finally accepted the reasons for these variations/discrepancies need to be better understood. At present the results for children are suggestive but not compelling. In communicating hydrocephalus the results from PC-MRI are supported by those from other techniques, notably cisternography and quantitative autoradiography [187, 198] which show that in these patients, but not in normal subjects, there is movement of radiolabelled albumin from cisterna magna to ventricles with accumulation in the ventricles and/or periventricular tissue. Furthermore in both communicating and non-communicating hydrocephalus there is oedema in the parenchyma adjacent to the ventricles, which may reduce resistance to fluid flow into the tissue from the ventricles. This might provide a route of fluid movement ultimately either to CSF in the subarachnoid space and thence to blood or lymph or more directly to sites outside the skull via perivascular or perineural spaces (section 4.1). It is also possible that the hydrocephalic state is associated with damage to or alterations in the properties of the blood-brain barrier or the choroid plexuses. There is a real need for experiments in which results from PC-MRI and from other techniques are compared directly in the same animals or patients.

Recent Studies on Perivascular Routes for Entry into and Exit from the Cortex

Proposals that substances enter and leave the parenchyma via perivascular spaces are as old as studies on the formation and absorption of brain extracellular fluids. Some aspects have been considered in earlier sections and earlier work has been reviewed elsewhere [9, 10]. It is clear that substances can move via vascular-associated routes. This section considers recent evidence obtained in rats and mice that has led to the well-publicised (e.g. [199] and [200]) glymphatic circulation proposal [201]: " … CSF passes through the para-arterial space that surrounds arteries … and into the interstitial space … vectorial convective fluxes drive waste products away from the arteries and toward the veins. ISF and its constituents then enter the para-venous space. As ISF exits the brain through the paravenous route, it reaches lymphatic vessels in the neck, and eventually returns its contents to the systemic circulation".

The basis of the glymphatic circulation proposal: evidence from fluorescence imaging studies

Routes of transfer within mouse brain have been characterized by Iliff *et al.* using imaging of a number of different fluorescent markers [143]. For events near the cortical surface they used two-photon microscopy through a cranial window. For events occurring more than about 200 μm below the surface, movements were inferred from fluorescence imaging of brain sections. Iliff *et al.* found that when the 3 kDa labelled dextran, TR-d3, was included in closed-circuit ventriculo-cisternal perfusion of mice, it penetrated slowly into the parenchyma consistent with diffusion, just as expected from earlier results (section 2.8). By contrast when injected into the cisterna magna allowing access to the subarachnoid space, it distributed quickly to the cortical surfaces of the brain and within thirty minutes was seen prominently along the course of the arteries penetrating the cortex and also within the interstitium. Likewise 45 kDa ovalbumin-647, OA-647, also injected into the cisterna magna, followed the arteries to the region of observation and into the cortex but was not seen in the interstitium until later. Within the cortex, it was seen both in the periarterial space and within the smooth muscle layer of the arteries.

In a subsequent publication Iliff *et al.* emphasized that when the 3 kDA TR-d3 but also larger markers, 40 kDa dextran TR-d40 and 2000 kDa dextran

FITC-d2000, were injected into the cisterna magna they each reached the region of observation along the walls of arteries [202]. Initially, however, they were not seen within the bulk of the CSF. The time-lapse images for TR-d40 (in the supplementary material of [143]) show particularly clearly that this marker followed the track of the arteries within the subarachnoid space but then spread out from these into the CSF. Such movement along arteries is consistent with the data obtained using MRI [203]. The route by which the markers gained access to the periarterial space from CSF in the subarachnoid cisternae is not known [202, 203]. The data show that the markers travelled rapidly along the arteries but also that, at least for markers of 45 kDa or less, they were able to move between the periarterial space and CSF, a manoeuvre which requires crossing the pial sheath (section 4.1). The idea that injected materials become associated with arterial walls has also been put forward by Rennels *et al.* for horseradish peroxidase [154, 156] and by Ghersi-Egea, Nagaraja, Fenstermacher and colleagues for sucrose [130, 131]. This retention presumably indicates a relatively low permeability of the pial sheath of the arteries to sucrose.

Caveat: 10 μl is a large volume to infuse into the CSF of a mouse and it is possible that the infusion may alter pressures and flows with the brain. Iliff *et al.*[202] note that in experiments under isoflurane anaesthesia, infusions like those used in their earlier work [143, 203, 204] “resulted in the non-physiological reflux of CSF tracer from the cisternal spaces into the fourth, third, and lateral ventricles” and they showed that even infusion at 1 μl min^{-1} for 10 min was associated with a temporary increase in intracranial pressure (ICP). They concluded that “Given that rapid paravascular CSF influx continues long after the normalization of these shifts in ICP, the paravascular CSF fluxes observed ....appear to represent physiological fluxes and not artefacts of changes in ICP resulting from CSF tracer infusion”. It is however conceivable that the increase in pressure might produce changes that outlive the increased pressure in the pial sheath of arteries traversing the CSF cisterns or even in the pial sheath of arteries within the cortex.

As found by Iliff *et al.* movement of fluorescent markers from periarterial spaces into the parenchyma depended upon their size [143]. Both 3 kDa TR-d3 and 45 kDa OA-647 spread into the parenchyma from both the cortical surface and the periarterial spaces surrounding the penetrating arteries, but 2000 kDa FITC-d2000 did not appear to spread at all even though it was clearly visualized in the periarterial space. These fluorescence data were not calibrated to allow quantitation of concentrations. However, from the comparison of the widespread and prominent distribution of 3 kDa TR-d3 seen using very sensitive fluorescence detection and the much less widespread distribution of 1 kDa gadolinium-DTPA detected using a less sensitive MRI-based method, Iliff *et al.* concluded that the concentrations of both markers in the parenchyma were much lower than those in the periarterial space near the cortical surface [203]. It is difficult to be quantitative about the transfer process from periarterial space to interstitium. In the case of TR-d3 the fluorescence signals from the periarterial regions may be saturated and reveal little about the concentrations while gadolinium-DTPA cannot be seen in the parenchyma.

The above fluorescence and MRI data are consistent with earlier findings concerning cortical penetration of horseradish peroxidase [79, 154–156] (cf. section 4.1) showing very rapid delivery of small quantities of markers into the parenchyma. From the speed of delivery over distances of millimeters it is apparent that transport along periarterial spaces cannot be just via diffusion (one point on which all are agreed [157]), i.e. both earlier and current data provide strong evidence for a convective component to the inward movement. These data are also consistent with studies on access of drugs to the parenchyma from CSF [158]. As reviewed by Papisov *et al.*, the evidence shows that drugs administered via the CSF are subject to "pulsation assisted translocation … into the perivascular space" [205]. The fluorescence results of Iliff *et al.* confirm and extend these findings and in addition, demonstrate that along the cortical surface movement of markers appears to be within a periarterial sheath, rather than just diffusely through the subarachnoid space [143].

Having reached the parenchyma, OA-647 (45 kDa) was seen in or on the walls of veins in the cortex 1 h but not 10 min after it was added to the cisterna magna. Iliff *et al.* interpret this as showing that the marker leaves the parenchyma via the veins [143]. A similar time sequence has been seen with another probe, rhod-2, consistent with this idea of progression [206]. The observed sequence, near arteries, near microvessels then near veins, is a major pillar of the glymphatic circulation proposal: fluid flow inwards along the arteries, then through the interstitium and thence outwards along the veins [143].

Quantification of influx and efflux using radiotracers

Although fluorescence measurements are well suited for identifying pathways and are the most sensitive detection method that can be used in real time, they are difficult to make quantitative. Radiotracers have also been used by Iliff *et al.* to examine time courses for movement of markers [143]. They obtained data for uptake of mannitol (182 Da) and dextran-10 (10 kDa) injected into the cisterna magna and for retention of mannitol and amyloid $\beta_{1\text{-}40}$ injected into the parenchyma. Their uptake curves for mannitol were interpreted as showing a single component with half-time of about 15-20 min. Their time course for retention of mannitol had a half-life of a few hours, perhaps similar to the half-life of about 4 h found for sucrose by Groothuis *et al.* as discussed in section 4.1. Caveats: by simple inspection of the sample uptake traces of Iliff *et al.,* it appears much more likely that the uptake of mannitol has two components, one very rapid, perhaps into the periarterial spaces, and another much slower, possibly into the parenchyma. Subsequent data reported for mannitol by Xie *et al.*, (supplementary material in [207]), confirm that there are at least two components of the uptake. The earlier data comparing the uptakes of mannitol and dextran-10 are not extensive enough to allow any conclusion about the relative rates of uptake.

Influences of aquaporin 4 (AQP4) located on astrocyte end-feet on perivascular flow

It is interesting to consider how perivascular fluid movement might be influenced by the presence of AQP4 located on glial end-feet enveloping blood vessels within the parenchyma [208]. Water and solutes moving between perivascular spaces and parenchyma must either be transported into the end-feet or go through the narrow spaces between them. AQP4 is abundantly expressed in the end-foot membrane facing towards the vasculature and thus is in a position to facilitate water transport via a transcellular route. It was found in AQP4 knock-out mice that entry of the smaller markers 3 kDa TR-d3 and 45 kDa OA-647 into the parenchyma from the periarterial space was greatly slowed [143]. This was interpreted by Iliff *et al.* as showing that by removing a pathway for water flow from periarterial space into interstitium the rate of glymphatic circulation was reduced. But reduction in glymphatic circulation should involve not only penetration into the parenchyma but also flow down the periarterial space (Figure 8) and yet data from Iliff *et al.* for the larger marker FITC-d2000 argues against such a reduction in periarterial convection: absence of AQP4 did not appear to affect penetration of the marker deep into the periarterial spaces. Thus the observation that AQP4 knock-out reduces movement into the parenchyma without reducing convection in the periarterial space presents the glymphatic circulation proposal with a substantial challenge.

It indicates that convection in some manner assists movement along the periarterial spaces without producing a large net flow in one direction.

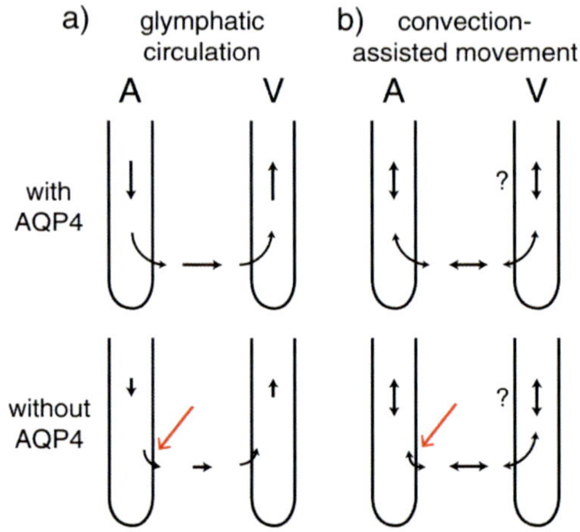

Figure 8: Effect of absence of AQP4 on fluid flow along perivascular routes. In both a) the glymphatic circulation proposal and b) the convection-assisted movement hypotheses described in Figure 9, the absence of AQP4 results in reduced movement of fluid from the periarterial spaces into the parenchyma (red arrow). In the glymphatic circulation proposal this would produce a build-up of fluid in the periarterial space, which would reduce flow into that space and would also remove the source of fluid destined for perivenous outflow. If instead, as in the convection-assisted movement hypotheses, fluid flow can occur in both directions in the periarterial spaces with little net flow, absence of AQP4 may not affect fluid flow along the periarterial space, a result more consistent with the data. A indicates the periarterial spaces, V the perivenous spaces. The question marks indicate that convection-assisted movement may or may not also occur in the perivenous spaces.

Reassessment of the evidence: alternatives to the glymphatic proposal

Caveats: Although the glymphatic proposal provides a basis for understanding much of the data, it does not explain the lack of effect of AQP4 knock-out on periarterial flux of the large dextran TR-d2000 discussed in the preceding section or the data for efflux via arterial routes discussed in section 4.1. So, are there alternatives to the glymphatic circulation proposal? In part, yes (Figure 9). Convection in the periarterial spaces within the cortex driven by cyclic changes in arterial pressure may produce either mixing

(Figure 9b) or flows in opposite directions in different spaces of the arterial wall (Figures 6 and 9c). In either of these alternatives, convection assists movement into and out of the cortex without necessarily producing the net flow of fluid into the cortex as required by the glymphatic circulation proposal (Figure 9a). The back and forth periarterial convection resulting from the cardiac cycle is rapid on a time scale of seconds. Thus movement of a marker observed on a time scale of minutes would be averaged such that back and forth movements would not be detected only its net movement down its concentration gradient.

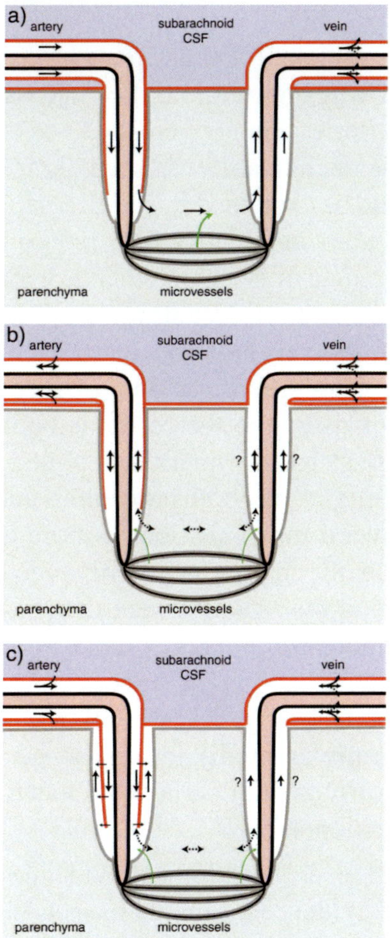

Figure 9: Possible schemes to explain rapid influx of markers via periarterial spaces: a) the glymphatic proposal based on Figure five of Iliff *et al.* [[143]], b) stirring or mixing and c) layered flow. In a) and c) there is preferential influx via the space between the arterial wall and the pial sheath (the inner periarterial

space in Figure 6) while in b) convection back and forth, speeds up the rate of transfer of markers in both directions. Red lines represent pial membranes, grey lines the layer of glial end-feet or glia limitans, solid arrows are fluxes of markers carried or assisted by convection, dashed arrows are either by diffusion or assisted by convection and green arrows have been added as a reminder that fluid secreted by the blood-brain barrier contributes to the fluid in the parenchyma. The location of the pial barriers is based on Zhang *et al.*[23] (Figures 1d and and6).6). In **c)** the influx of fluid via the periarterial spaces may inflate the space of His providing a route for the return of fluid to the subpial space at the cortical surface and then to CSF in the subarachnoid space.

Can the alternatives explain Iliff *et al.* 's data [143]? The key observations are that when a marker was injected into the cisterna magna it was seen to enter the cortex along arteries and not veins whereas when the same marker was injected into the parenchyma, it was seen along veins but not arteries. These observations could be explained by the alternative hypotheses in terms of concentration differences and relative flow rates along arteries and veins. As shown in Iliff *et al.* 's data in the supplementary material [143] discussed earlier in this section, a marker injected into the cisterna magna would be delivered rapidly and at relatively high concentration to the periarterial spaces at the cortical surface but not the surrounding CSF or the perivenous spaces. Hence movement of a marker into the cortex is bound to be preferentially along arteries. By contrast when a marker is added to the parenchyma, a significant fraction of it might enter both perivenous and periarterial pathways but with mixing or outward movement being more effective in periarterial than in perivenous paths, the marker would be removed more rapidly from the periarterial spaces and the amounts seen there would be less. Thus the observations with the fluorescent markers, taken alone, can be rationalized. However neither the glymphatic circulation proposal nor the convection-assisted movement hypotheses can easily reconcile the observation that the fluorescent markers are preferentially located in perivenous spaces at late times with the data for preferential periarterial location and efflux of other markers discussed in section 4.1.

From both the more recent histological evidence ([23, 144], Figure 1c) and from the functional evidence discussed in sections 4.1 and 4.3, it can be concluded that the spaces or pathways involved in periarterial flow within the parenchyma are not in free communication with CSF in the subarachnoid space but are separated from it by at least the pial sheath or the pia mater. This separation is, however, challenged by the older observation that colloidal carbon gains access to a periarterial space from subarachnoid CSF [1]. There would appear to be three candidates for periarterial flow pathways within cortical parenchyma (Figure 6). One is the extracellular space in the smooth muscle layer as proposed by Weller and Carare (section 4.1). Another is the space between the smooth muscle layer and the pial sheath which Zhang *et al.* call the periarterial space [23] and is probably the Virchow-Robin space of the older literature (see [1] for discussion). The third is the space of His [1] that may exist immediately adjacent to the glial end-feet. Brinker *et al.* [17] label this outer space as the Virchow-Robin space. Because the space of His is continuous with the space below the pia mater at the cortical surface (Figure 1c) it could be called the perivascular subpial space.

It is tempting to speculate (see Figure 9c) that fluid arriving within the pial sheath of arteries in the subarachnoid space might be able to move inwards along the (inner) periarterial space between the smooth muscle and the sheath, much as proposed by Iliff, Nedergaard and colleagues. However, rather than this fluid continuing into the parenchyma (as in Figure 9a) it might inflate the space of His (an outer periarterial space) allowing fluid to return to the subpial space at the cortical surface and thence cross the pia mater to the subarachnoid CSF. The idea that the periarterial spaces may vary in calibre depending on local pressures is inherent in the idea that arterial pulsation drives convection. Such variability may be part of the explanation of the observation by Iliff *et al.* that there was far less inward movement of markers when the dura was not intact [143].

The alternative hypotheses, based on mixing or layers of flow as shown in Figure 9b and c, appear to explain the fluorescence data at least qualitatively. Furthermore, unlike the glymphatic circulation proposal, they can also accommodate the data considered in section 4.1 for periarterial efflux of markers (though not simultaneously preferential perivenous outflow of the fluorescent markers). The alternatives also provide an explanation for the lack of effect of AQP4 knockout on the movements of large markers into the periarterial space as they do not require net fluid movement from the periarterial space into the parenchyma.

The two alternative hypotheses indicated and the glymphatic circulation proposal are not mutually exclusive and furthermore there may be other scenarios. At present there is conflicting evidence for outward flow along veins and little evidence for *net* flow (as opposed to net marker flux down a concentration gradient) inwards along arteries. There is no evidence that the net flows, i.e. those inwards along the arteries, those proceeding through the parenchyma and those going outwards along the veins are of similar magnitude as required by the glymphatic circulation proposal. The implications of any of the schemes in Figure 9 which bring fluid deep into the cortex and then remove it, are that they would greatly shorten the diffusion distance from CSF to points within the parenchyma and hence the time taken for solutes to penetrate into or be cleared from regions of the cortex. The principle of operation in 9b and 9c is the same as that of a microdialysis probe which allows diffusional exchange between the interstitial fluid and the fluid within the probe.

Studies Concerning the Influence of Sleep and Anaesthesia on Perivascular Fluid Flow and Interstitial Fluid Volume

The glymphatic circulation proposal has been used to interpret experimental results on the effects of sleep and anaesthesia on perivascular fluid flow and interstitial volume in the parenchyma [143, 201].

Changes in flow and volume inferred from rates of delivery of fluorescent markers

Xie *et al*. found that in awake mice the entry into the parenchyma of either of the 3 kDa fluorescent markers, FITC-dextran or TR-dextran, added via the cisterna magna was reduced compared to that seen in asleep or anaesthetized animals [207]. They interpreted their results as demonstrating the interstitial fluid volume to be less in the awake state. Some of their data, however, suggest another explanation. The data show that in the awake state the markers appear along the arteries in the subarachnoid space to a much smaller extent (or much later?) thus reduced penetration into the parenchyma would follow simply because the concentration in the periarterial space is lower. If this is the correct explanation, the observation that the amount reaching the interstitium was reduced reveals nothing about the interstitial spaces in the parenchyma. Although Xie *et al.* reported examples of data for concentrations measured simultaneously in the periarterial space and in a surrounding annulus of parenchyma, they did not report ratios of these concentrations [207]. If these ratios could be obtained they would provide information about the process of penetration of the markers into the interstitium and might allow conclusions about changes in interstitial volume.

Interstitial fluid volume changes inferred from TMA iontophoresis measurements

Evidence that the interstitial fluid volume in the cortical parenchyma may be different in the awake state has been obtained [207] using the TMA iontophoresis method [46, 99] discussed in section 2.8. If there is no bulk flow and diffusion is isotropic (the same in all directions), the data from this method can be interpreted to yield an interstitial volume fraction and tortuosity. Xie *et al.* observed that the volume fraction estimated by the TMA method was higher if the rats were asleep than if they were awake. The transitions from the asleep to awake or awake to anaesthetized states were rapid, e.g. taking 15 min, (see supplementary material in [207]) and the inferred volume fraction changes were large, from 23 % to 14 % and 14 % to 23 % respectively. The experimental data are for regions within 300 μm of the cortical surface, but they are discussed by Xie *et al.* as if they apply to the entire cerebral cortex.

Caveat: As these changes are surprisingly large and fast it is necessary to ask if there are any confounding factors in the TMA measurements. Possible factors to consider include fluid flow, the structure of the extracellular space and voltage gradients with the associated patterns of current flow. For instance if there were fluid flow in the direction from the injection electrode to the measuring electrode in the awake but not in the sleeping mice this might account for a higher concentration of TMA reaching the recording electrode in the awake mice without any volume change. Alternatively a similar result might be seen if there were more flow in the opposite direction in the sleeping mice. However, to explain the data by such mechanisms would involve larger flows than are currently known to occur (Charles Nicholson and Tony Gardner-Medwin, personal communications) and would further require that these flows vary with behavioural state. (It should be noted, that the estimates of flows in the literature may be underestimates (sections 4.1 and 4.3.2) as they were based on measurements of outflow of markers from the cortex during barbiturate anaesthesia.) Changes of interstitial space configuration (for example opening or closing of diffusion blockages) seem also unlikely to account for the data, since they would be expected to affect the calculated tortuosity, which was not seen to vary significantly [207]. Thus it is hard at present to see a clear basis for challenging the inference of an altered space fraction for the regions within 300 μm of the cortical surface where the measurements were carried out (see Endnote[e]).

In a Physiological Society poster, Gardner-Medwin noted that in rats there is evidence against a large increase in ISF volume at the onset of sleep [209]. If there were to be an increase in volume the impedance of the tissue should decrease, but the only changes in impedance observed have been an increase associated with onset of rapid eye movement sleep and decrease on return to "normal" sleep [210, 211]. He also proposed for discussion several ideas based on movements of ions and water during slow wave sleep but at present additional experiments are needed to establish what effects these might have on TMA diffusion dynamics. Brinker *et al.* have focussed attention on possible factors other than altered glymphatic circulation that may have altered the width of interstitial spaces in animals exposed to the very artificial recording conditions, including changes in arterial blood pressure, venous blood pressure, stress hormone levels, and blood gases [17].

Interstitial fluid volume changes inferred from radiotracer measurements

Xie *et al.* also used radiotracers to compare differences in interstitial fluid volume in awake, asleep or anaesthetized mice [207]. For this purpose they injected either radiolabelled inulin or amyloid-β into the cortex and followed the decrease in concentration over time. The half-time for the decrease was shorter for amyloid-β than for inulin probably reflecting the fact that amyloid-β is transported across the blood-brain barrier. The half-times for decrease for both substances were longer in awake mice compared to asleep or anaesthetized mice. Xie *et al.* interpreted these data (as with their fluorescence data, section 4.4.1) as implying smaller interstitial fluid volume and hence smaller rates of transfer within the interstitium in the awake mice. (For inulin the rate constants were roughly 0.008 min^{-1} awake compared to 0.016 min^{-1} asleep and 0.019 min^{-1} anaesthetized while for amyloid-β they were approximately 0.021 min^{-1}, 0.05 min^{-1} and 0.045 min^{-1}). Caveats: Although the changes in rate constants may reflect changes in interstitial fluid volume, many other things may have changed as well.

A CURRENT VIEW OF REGULATION OF THE EXTRACELLULAR FLUIDS OF THE BRAIN AND THEIR CONSTANT RENEWAL

The composition of the extracellular fluids and the volume they occupy clearly have to be regulated to allow effective functioning of cells in the brain. This regulation is achieved by the presence of barriers that prevent free exchange between CNS and blood and mechanisms for production, distribution and reabsorption of fluid hence providing regular fluid turnover. The application of modern techniques to the study of the processes involved in brain fluid turnover has yielded new and important information. But in interpreting this information it is important not to neglect the basic anatomy and physiology of transport and flow. The underlying principle of control of total extracellular fluid volume in the brain is simple. Fluid is secreted into the brain at a rate that is relatively insensitive to changes in intracranial pressure (reviewed in [5, 9]). Fluid is returned from the brain to the blood at a rate which increases with intracranial pressure. The rate of fluid return increases with the extracellular fluid volume because increased volume increases intracranial pressure. At steady-state, when extracellular volume and intracranial pressure are constant, they take on whatever values balance inflow and outflow. This process has been considered extensively in earlier reviews [9, 16].

Regulation of the composition of extracellular fluid (section 1.4) depends primarily on the transport properties of the blood-brain barrier. This review has been concerned only with the evidence that the net process is a secretion and not with either the mechanisms or the importance of short-term local control. The net process must be secretion rather than filtration because the blood-brain barrier is an effective barrier. While hydrostatic pressures can drive water fluxes across the blood-brain barrier, unless these are accompanied by altered transport of solutes they will soon lead to concentration changes that produce osmotic pressures that oppose the net water fluxes (section 2.7). Viewed the other way round, solute fluxes, via specific transporters at the blood-brain barrier will produce osmotic changes that will force water to follow. These principles were established many years ago [7, 73] but some of their consequences have not been sufficiently widely appreciated. Importantly, there cannot be significant net reabsorption of fluid across the intact blood-brain barrier into the blood unless there is a mechanism for producing net movement of NaCl across the barrier in the direction from brain to blood and no such mechanism has been found or proposed.

Constant turnover of extracellular fluids has been widely viewed as necessary to provide a non-selective means to remove wastes for which there are no specific transporters. Continual secretion of “clean” fluid into the brain and reabsorption of whatever fluid is present by non-selective pressure-driven mechanisms achieves this end. Secretion of this “clean” fluid is partly the function of the choroid plexuses and partly the function of the blood-brain barrier. Perhaps surprisingly the relative contributions by each of these structures are still not known as their secretion rates are difficult to determine (sections 3.1.3 and 3.2). ISF flow which has been assumed to reflect blood-brain barrier secretion rate has been estimated from the rate of removal of markers from the brain but these measurements are called into question on two grounds: firstly they were performed with animals under barbiturate anaesthesia which may have inhibited both secretion and flow (sections 4.1 and 4.3.2) and secondly influx of CSF into the parenchyma via perivascular routes may contribute to ISF flow (section 4.3), which then does not reflect solely blood-brain barrier secretion. The reabsorption of fluid from brain eventually to blood occurs by a number of mechanisms (reviewed in [16]), all of which are relatively non-selective and increase in rate with intracranial pressure.

The broader outlines of brain fluid regulation were largely established in the older work. Much of more recent work has uncovered more complex fluid movements within the brain that have implications for the functions and regulation of the extracellular fluids. The results from PC-MRI measurement of CSF flow rates (section 4.2) leave no doubt that movement of CSF is much more extensive and variable than the slow steady flow envisaged in the classical account, i.e. secretion by the choroid plexuses and a sedate steady flow to exit sites. Results of some PC-MRI studies imply that *net* flow through the cerebral aqueduct is reversed so that the net flow is into rather than out of the third and lateral ventricles. Reversed flow has been proposed as the normal direction for net flow in children under two (in one study) and as the flow pattern in communicating hydrocephalus (in a number of studies). Reversal of flow has important implications, i.e. it requires that there is an important source of CSF other than the choroid plexuses and some means for removing fluid from the third and lateral ventricles other than via the cerebral aqueduct. In communicating hydrocephalus the route of removal may be created as a pathological consequence of the disorder but in children no plausible route has been proposed. Fluorescent tracer analysis has shown that flow can occur into the parenchyma along periarterial spaces. Whether or not this represents *net* fluid flow into the cortex and whether there is subsequent flow through the interstitium between arteriolar and venular ends of the microcirculation and *net* flow out of the cortex via perivenous routes, described as the glymphatic circulation, remain to be established (section 4.3).

CONCLUSION

Fluid is secreted into CSF by the choroid plexuses and into ISF at the blood-brain barrier. These secretions are driven by metabolic energy coupled to transport by the ubiquitous Na^+, K^+-ATPase, the Na^+-pump. Fluid is returned to lymph and blood by a number of routes prominently including perineural pathways leading to the olfactory epithelium and the arachnoid villi. Fluid cannot be reabsorbed across the blood-brain barrier so long as the barrier remains intact. The relative rates of secretion by the choroid plexuses and at the blood-brain barrier remain uncertain. Similarly more data are required to be sure of the sources of fluid involved in turnover of ISF. Previous estimates of total CSF production rate including those made by the ventriculo-cisternal perfusion method may not include a large fraction of the contribution of ISF secretion at the blood-brain barrier. Earlier determinations of ISF production rates from measurements of the rate of efflux of markers from brain parenchyma may have underestimated the rate of fluid turnover by as much as 6-fold because experiments were performed under barbiturate anaesthesia, which may have suppressed both secretion and turnover. However, calculations of the rate of secretion at the blood-brain barrier made using measurements of rates of efflux of markers from the parenchyma may overestimate the secretion because fluid turnover mediating the marker efflux may include perivascular influx of CSF as well as blood-brain barrier secretion. If the observation of reversed net CSF flow through the aqueduct using phase contrast magnetic resonance imaging is substantiated, this would provide strong evidence that the blood-brain barrier is able to secrete fluid at a substantial rate compared to that of the choroid plexuses. Finally data showing net influx of CSF into the parenchyma via periarterial routes may indicate that local convection is occurring in periarterial spaces both within the subarachnoid spaces and within the cortex with or without associated flow through the parenchyma. The field is wide open for further experiments!

Endnotes

[a]The Starling mechanism does not assume that proteins cannot cross peripheral capillary walls, only that they cross sufficiently slowly that they can exert almost their full osmotic pressure. The filtered fluid has concentrations of small solutes that are virtually the same as those in plasma, but the concentration of proteins is about 1/4 of that in plasma. At any point in time there is approximately the same amount of plasma protein

in the tissues, equal to (average interstitial concentration of protein) × (total interstitial fluid volume), as in the plasma (plasma protein concentration in plasma) × (plasma volume). However, because the interstitial volume is about 4-fold bigger than the plasma volume there is still a marked difference in concentration and hence a marked colloid osmotic pressure across the membrane. In the central nervous system the microvascular permeability to protein is very much less than in the periphery and the plasma protein concentration in brain interstitium is very small.

[b]A unidirectional flux of water across a membrane, barrier or region is the flux of water molecules that start on one side and end up on the other. A unidirectional flux is normally measured using a tracer, e.g. for water often tritiated or deuterated water. Tracer molecules are added to one side of a membrane at a known fraction of the total number of water molecules on that side (including the tracer). The unidirectional flux of water is then calculated as the net flux of tracer divided by the fraction. The net flux of water is the difference between the unidirectional fluxes in the two directions.

For bulk flow as with the flow of blood along the length of a blood vessel, the unidirectional flux of water in the opposite direction to the flow is negligible and the unidirectional flux in the direction of flow is the same as the net flux. By contrast for pure diffusion down a gradient from a higher to a lower concentration, the ratio of the unidirectional flux in the same direction to the net flux is (higher concentration)/(difference in concentration) while the ratio of the unidirectional flux in the opposite direction to the net flux is (lower concentration)/(difference in concentration). For the actual process occurring across peripheral microvascular walls, the ratio of the back flux to the net flux is intermediate between the values for bulk flow and diffusion but much closer to that for diffusion. To put it another way, the movement of molecules against the flow in an artery can be ignored, but not the movements against the net flow across a capillary wall. The back flux of water is then almost as large as the forward flux.

[c]In most textbooks the story is told that in peripheral tissues there is filtration (a net movement out of the blood) of perhaps 1 % of the blood volume arriving in the tissue at the arterial ends of capillaries but that as a consequence of the drop in hydrostatic pressure along the capillaries there is reabsorption (a net inward movement) at the venous end of 90 % of that filtered leaving the 0.1 % of blood flow to become lymph. While this story uses the terms filtration and reabsorption correctly, it has been known for many years that it is rarely a correct description largely because it was based

on outdated estimates of the interstitial forces and ignored changes in the hydrostatic and colloid osmotic pressures within the tissues and within the endothelial junctions. In those tissues with no other input of fluid, e.g. skin and muscle, filtration rather than absorption normally occurs over the entire length of the microvessels, capillaries and venules, with the overall amount filtered balanced by lymph flow. By contrast in tissues that have an input of a fluid with substantially less protein than plasma, e.g. the intestines during absorption of a meal or the cortex of the kidneys, there can be absorption of most of the added fluid into the vasculature while the rest carries interstitial protein into lymph. A similar process is thought to be occurring in the absorption of CSF that reaches the nasal mucosa (section 4.1).

[d]The inability of the intact blood-brain barrier to reabsorb interstitial fluid has been noted repeatedly. However from the frequency with which such reabsorption has been suggested, the point appears not to have been universally accepted. It may therefore be useful to provide a simple numerical calculation that emphasizes the difficulties. Imagine that CSF and ISF production containing 150 mM NaCl is 0.35 ml min^{-1} and that all of the reabsorption is occurring by pressure driven flow across microvessels in the parenchyma. If these have the normal properties of the blood-brain barrier, the reabsorbate will be nearly pure water, which by the hypothesis is being removed from the brain at 0.35 ml min^{-1}. The net effect is that salt is being added by fluid production but is not being removed by fluid reabsorption. How long will it take to accumulate enough salt in the brain to stop the net reabsorption of water? The rate of addition of salt is 0.35 ml min^{-1} $\times$ 0.15 mmol ml^{-1} ~ 0.0475 mmol min^{-1}. If the total water volume of the brain is taken to be 1 litre this amounts to a rate of change of concentration of 0.0475 mM min^{-1} which is a rate of increase of osmolality of 0.095 mOsmolal min^{-1}. In other words each minute the salt build-up can offset a pressure of 0.095 mOsmolal * 19 mmHg $mOsmolal^{-1}$ = 0.9 mmHg. In other words, *unless there is a means for removing the salt,* within a few minutes the osmolality of the brain will increase to offset any realistic hydrostatic (or colloid osmotic) pressure *difference* that could be driving the reabsorption. The difficulty with postulating pressure driven filtration of fluid across the blood-brain barrier either into the parenchyma or out of it is explaining how one moves the salt.

[e]The transitions from asleep to awake or awake to anaesthetized states reported by Xie *et al.* [207] are rapid, e.g. taking 15 min, and the inferred volume changes are large, from 23 % to 14 % and 14 % to 23 % respectively. They require addition to or removal from the interstitial spaces of both water

and solutes as the fluid must remain nearly isosmotic with CSF and cells. It is difficult to imagine how the volume of the interstitium can be changed by so large a fraction so quickly unless the change is very localized.

To gain some appreciation of the size of the effects proposed consider the increase in ISF volume when the awake mice were anaesthetized. The same percentage change in ISF volume in humans if it occurred over the entire brain would require addition of about 90 ml of fluid containing about 300 mOsmoles of solutes per litre. These solutes cannot be K^+ and Cl^- released from cells for the simple reason that, if maintained throughout a period of sleep, the increase in extracellular K^+ concentration (to about 60 mM) together with the associated depolarization and sequelae would kill the cells. The solutes have to be primarily Na^+ and Cl^-. Indeed it would be very interesting to know if the NaCl content of brain extracellular fluid is different in the awake and asleep states as the volume is approximately proportional to this content.

The only possible mechanisms for NaCl to be added to the interstitial fluid are efflux from cells, fluxes across the blood-brain barrier or flow from CSF. The volume of the cells is much larger than the interstitial volume and thus even though the intracellular Na^+ and Cl^- concentrations are small the cells may contain just about enough NaCl that if all were effluxed accompanied by water it might account for the shift. However, it is difficult to imagine how this could be made to occur. Secretion at the blood-brain barrier even if it were at the full rate needed for "normal" production of CSF (about 20 ml h^{-1}) would be much too slow and there is no known mechanism for subsequent absorption of fluid by this route upon awakening. Furthermore introduction of fluid into the brain from the blood would be expected to increase intracranial pressure. That leaves exchanges with CSF. Can NaCl and water, about half of that in the entire CSF, be moved into the parenchyma from CSF in so short a time?

ACKNOWLEDGEMENTS

We would like to thank Charles Nicholson and Tony Gardner-Medwin for helpful discussions on the TMA method and the interpretation of results and for suggesting changes to sections 2.8 and 4.4.2, TG-M for a preprint of his poster on possible explanations of the TMA results and for pointing out to us the extremely rapid nature of the volume changes being inferred by Xie *et al.*[207] and the work on impedance changes by Ranck *et al.*[210],[211] in rats, Rodney Levick for advice on fluid movements into and through peripheral tissues and for suggested changes to sections 2.2 through 2.7, and Stefan Piechnik for initial help with the pitfalls of PC-MRI.

Authors' Contributions

SBH carried out the literature search and prepared the figures. SBH and MAB wrote the manuscript. Both authors have read and approved the final version of the manuscript.

REFERENCES

1. Woollam DHM, Millen JW. Perivascular spaces of the mammalian central nervous system. *Biol Rev Camb Philos Soc.* 1954;29:251–283. doi: 10.1111/j.1469-185X.1954.tb00597.x.
2. Millen J, Woollam D. *The Anatomy of the Cerebrospinal Fluid.* London: Oxford University Press; 1962.
3. Hayman LA, Weathers SW, Kirkpatrick JB. Atlas of cerebrospinal fluid spaces. In: Hayman LA, Hinck VC, editors. *Clinical Brain Imaging: Normal Structure and Functional Anatomy.* St. Louis: Mosby-Year Book; 1992. pp. 306–328.
4. Cserr HF. Physiology of choroid plexus. *Physiol Rev.* 1971;51:273–311.
5. Welch K. The principles of physiology of the cerebrospinal fluid in relation to hydrocephalus including normal pressure hydrocephalus. *Adv Neurol.* 1975;13:247–332.
6. Bradbury MWB. *The Concept of a Blood-Brain Barrier.* Chichester: Wiley; 1979.
7. Cserr HF, Patlak CS. Regulation of brain volume under isosmotic and anisosmotic conditions. In: Gilles R, Hoffmann EK, Bolis L, editors. *Advances in Comparative and Environmental Physiology.* Heidelberg: Springer; 1991. pp. 61–80.
8. Cserr HF, Patlak CS. Secretion and bulk flow of interstitial fluid. In: Bradbury MWB, editor. *Physiology and Pharmacology of the Blood-Brain Barrier.* Berlin: Springer-Verlag; 1992. pp. 245–261.
9. Davson H, Segal MB. *Physiology of the CSF and Blood-brain Barriers.* Boca Raton: CRC Press; 1996.
10. Abbott NJ. Evidence for bulk flow of brain interstitial fluid: significance for physiology and pathology. *Neurochem Int.* 2004;45:545–552. doi: 10.1016/j.neuint.2003.11.006.
11. Johanson CE, Duncan JA, III, Klinge PM, Brinker T, Stopa EG, Silverberg GD. Multiplicity of cerebrospinal fluid functions: new challenges in health and disease. *Cerebrospinal Fluid Res.* 2008;5(10):1–32.
12. Saunders NR, Ek CJ, Habgood MD, Dziegielewska KM. Barriers in the brain: a renaissance? *Trends Neurosci.* 2008;31:279–286. doi: 10.1016/j.tins.2008.03.003.

13. Liddelow SA. Fluids and barriers of the CNS: a historical viewpoint. *Fluids Barriers CNS.* 2011;8:2. doi: 10.1186/2045-8118-8-2.
14. Abbott NJ. Blood-brain barrier structure and function and the challenges for CNS drug delivery. *J Inherit Metab Dis.* 2013;36:437–449. doi: 10.1007/s10545-013-9608-0.
15. Damkier HH, Brown PD, Praetorius J. Cerebrospinal fluid secretion by the choroid plexus. *Physiol Rev.* 2013;93:1847–1892. doi: 10.1152/physrev.00004.2013.
16. Pollay M. The function and structure of the cerebrospinal fluid outflow system. *Cerebrospinal Fluid Res.* 2010;7:9. doi: 10.1186/1743-8454-7-9.
17. Brinker T, Stopa EG, Morrison J, Klinge PM. A new look at cerebrospinal fluid circulation. *Fluids Barriers CNS.* 2014;11:10. doi: 10.1186/2045-8118-11-10.
18. O'Donnell ME. Ion and water transport across the blood-brain barrier. In: Alvarez-Leefmans FJ, Delpire E, editors. *Physiology and Pathology of Chloride Transporters and Channels in the Nervous System: From Molecules to Diseases.* Amsterdam: Elsevier Science; 2009. pp. 585–606.
19. Strazielle N, Ghersi-Egea JF. Physiology of blood-brain interfaces in relation to brain disposition of small compounds and macromolecules. *Mol Pharm.* 2013;10:1473–1491. doi: 10.1021/mp300518e.
20. Engelhardt B, Sorokin L. The blood-brain and the blood-cerebrospinal fluid barriers: function and dysfunction. *Semin Immunopathol.* 2009;31:497–511. doi: 10.1007/s00281-009-0177-0.
21. Luissint A-C, Artus C, Glacial F, Ganeshamoorthy K, Couraud P-O. Tight junctions at the blood brain barrier: physiological architecture and disease-associated dysregulation. *Fluids Barriers CNS.* 2012;9:23. doi: 10.1186/2045-8118-9-23.
22. Woollam DH, Millen JW. The perivascular spaces of the mammalian central nervous system and their relation to the perineuronal and subarachnoid spaces. *J Anat.* 1955;89:193–200.
23. Zhang ET, Inman CBE, Weller RO. Interrelationships of the pia mater and the perivascular (Virchow-Robin) spaces in the human cerebrum. *J Anat.* 1990;170:111–123.

24. Mokri B. The Monro-Kellie hypothesis: applications in CSF volume depletion. *Neurology.* 2001;56:1746–1748. doi: 10.1212/WNL.56.12.1746.
25. Lehtinen MK, Bjornsson CS, Dymecki SM, Gilbertson RJ, Holtzman DM, Monuki ES. The choroid plexus and cerebrospinal fluid: emerging roles in development, disease, and therapy. *J Neurosci.* 2013;33:17553–17559. doi: 10.1523/JNEUROSCI.3258-13.2013.
26. Hansen AJ. Effect of anoxia on ion distribution in the brain. *Physiol Rev.* 1985;65:101–148.
27. Jones HC, Keep RF. The control of potassium concentration in the cerebrospinal-fluid and brain interstitial fluid of developing rats. *J Physiol.* 1987;383:441–453. doi: 10.1113/jphysiol.1987.sp016419.
28. Bito LZ, Davson H. Local variations in cerebrospinal fluid composition and its relationship to the composition of the extracellular fluid of the cortex. *Exp Neurol.* 1966;14:264–280. doi: 10.1016/0014-4886(66)90114-2.
29. Bradbury MW, Davson H. The transport of potassium between blood, cerebrospinal fluid and brain. *J Physiol.* 1965;181:151–174. doi: 10.1113/jphysiol.1965.sp007752.
30. Oldendorf WH. The blood-brain barrier. *Exp Eye Res.* 1977;25(Suppl):177–190. doi: 10.1016/S0014-4835(77)80016-X.
31. Merritt HH, Fremont-Smith F. *The Cerebrospinal Fluid.* Philadelphia, London: WB Saunders Co; 1937.
32. Brightman MW, Reese TS. Junctions between intimately apposed cell membranes in the vertebrate brain. *J Cell Biol.* 1969;40:648–677. doi: 10.1083/jcb.40.3.648.
33. Brightman MW. Physiology and Pharmacology of the Blood-Brain Barrier. In: Bradbury MWB, editor. *Ultrastructure of Brain Endothelium.* Berlin: Springer-Verlag; 1992. pp. 1–22.
34. Reese TS, Karnovsky MJ. Fine structural localization of a blood-brain barrier to exogenous peroxidase. *J Cell Biol.* 1967;34:207–217. doi: 10.1083/jcb.34.1.207.
35. Oldendorf WH, Cornford ME, Brown WJ. Large apparent work capability of blood-brain-barrier - study of mitochondrial content of capillary endothelial cells in brain and other tissues of rat. *Ann Neurol.* 1977;1:409–417. doi: 10.1002/ana.410010502.

36. Nabeshima S, Reese TS, Landis DM, Brightman MW. Junctions in the meninges and marginal glia. *J Comp Neurol.* 1975;164:127–169. doi: 10.1002/cne.901640202.
37. Becker NH, Novikoff AB, Zimmerman HM. Fine structure observations of the uptake of intravenously injected peroxidase by the rat choroid plexus. *J Histochem Cytochem.* 1967;15:160–165. doi: 10.1177/15.3.160.
38. Rall DP. Transport through the ependymal linings. In: Lajtha A, Ford DH, editors. *Progress in Brain Research.* Amsterdam: Elsevier; 1968. pp. 159–172.
39. Curran RE, Mosher MB, Owens ES, Fenstermacher JD. Cerebrospinal fluid production rates determined by simultaneous albumin and inulin perfusion. *Exp Neurol.* 1970;29:546–553. doi: 10.1016/0014-4886(70)90079-8.
40. Saunders NR, Knott GW, Dziegielewska KM. Barriers in the immature brain. *Cell Mol Neurobiol.* 2000;20:29–40. doi: 10.1023/A:1006991809927.
41. Wolak DJ, Thorne RG. Diffusion of macromolecules in the brain: implications for drug delivery. *Mol Pharm.* 2013;10:1492–1504. doi: 10.1021/mp300495e.
42. Fettiplace R, Haydon DA. Water permeability of lipid membranes. *Physiol Rev.* 1980;60:510–550.
43. Guyton AC, Granger HJ, Taylor AE. Interstitial fluid pressure. *Physiol Rev.* 1971;51:527–563.
44. Granger HJ, Laine GA, Barnes GE, Lewis RE. Dynamics and control of transmicrovascular fluid exchange. In: Staub NC, Taylor AE, editors. *Edema.* New York: Raven; 1984. pp. 189–228.
45. Levick JR. Flow through interstitium and other fibrous matrices. *Q J Exp Physiol.* 1987;72:409–437. doi: 10.1113/expphysiol.1987.sp003085.
46. Sykova E, Nicholson C. Diffusion in brain extracellular space. *Physiol Rev.* 2008;88:1277–1340. doi: 10.1152/physrev.00027.2007.
47. Staverman AJ. The theory of measurement of osmotic pressure. *Recl Trav Chim Pays Bas.* 1951;70:344–352. doi: 10.1002/recl.19510700409.
48. Kedem O, Katchalsky A. Thermodynamic analysis of the permeability of biological membranes to non-electrolytes. *Biochim Biophys Acta.* 1958;27:229–246. doi: 10.1016/0006-3002(58)90330-5.

49. Michel CC. Fluid movement through capillary walls. In: Renkin EM, Michel CC, Geiger SR, editors. *Handbook of Physiology Section 2 The Cardiovascular System Volume 4 Part 1 Microcirculation.* Bethesda, MD: American Physiological Society; 1984. pp. 375–409.
50. Staverman AJ. Non-equilibrium thermodyanamics of membrane processes. *Trans Faraday Soc.* 1952;48:176–185. doi: 10.1039/tf9524800176.
51. Starling EH. On the absorption of fluids from the connective tissue spaces. *J Physiol.* 1896;19:312–326. doi: 10.1113/jphysiol.1896.sp000596.
52. Landis EM, Pappenheimer JR. Exchange of substances through the capillary walls. In: Hamilton WF, editor. *Handbook of Physiology, Section 2, Circulation.* Washington, D.C: American Physiological Society; 1963. pp. 961–1034.
53. Levick JR, Michel CC. Microvascular fluid exchange and the revised Starling principle. *Cardiovasc Res.* 2010;87:198–210. doi: 10.1093/cvr/cvq062.
54. Pappenheimer JR. Osmotic reflection coefficients in capillary membrane. In: Crone C, Lassen NA, editors. *Capillary Permeability: the Transfer of Molecules and Ions between Capillary Blood and Tissue.* Copenhagen, New York: Munksgaard, Academic; 1970. pp. 278–286.
55. Dl Y, Alvarez OA. Water sodium and thiourea transcapillary diffusion in dog heart. *Am J Physiol.* 1967;213:308–314.
56. Wolf MB, Watson PD. Measurement of osmotic reflection coefficient for small molecules in cat hindlimbs. *Am J Physiol.* 1989;256:H282–H290.
57. Hill A. Osmosis. *Q Rev Biophys.* 1979;12:67–99. doi: 10.1017/S0033583500002602.
58. Kim KS, Davis IS, Macpherson PA, Pedley TJ, Hill AE. Osmosis in small pores: a molecular dynamics study of the mechanism of solvent transport. *Proc R Soc Lond A Math Phys Sci.* 2005;461:273–296. doi: 10.1098/rspa.2004.1374.
59. Bulat M, Klarica M. Fluid filtration and reabsorption across microvascular walls: control by oncotic or osmotic pressure? *Periodicum Biologorum.* 2005;107:147–152.

60. Loo DDF, Wright EM, Zeuthen T. Water pumps. *J Physiol.* 2002;542:53–60. doi: 10.1113/jphysiol.2002.018713.
61. MacAulay N, Hamann S, Zeuthen T. Water transport in the brain: role of cotransporters. *Neuroscience.* 2004;129:1031–1044. doi: 10.1016/j.neuroscience.2004.06.045.
62. Pappenheimer JR, Soto-Rivera A. Effective osmotic pressure of the plasma proteins and other quantities associated with the capillary circulation in the hindlimbs of cats and dogs. *Am J Physiol.* 1948;152:471–491.
63. Flexner LB, Cowie DB, Vosburgh GJ. Studies on capillary permeability with tracer substances. *Cold Spring Harb Symp Quant Biol.* 1948;13:88–98. doi: 10.1101/SQB.1948.013.01.015.
64. Pappenheimer JR. Passage of molecules through capillary walls. *Physiol Rev.* 1953;33:387–423.
65. Bulat M, Lupret V, Oreskovic D, Klarica M. Transventricular and transpial absorption of cerebrospinal fluid into cerebral microvessels. *Coll Antropol.* 2008;32(Suppl 1):43–50.
66. Bulat M, Klarica M. Recent insights into a new hydrodynamics of the cerebrospinal fluid. *Brain Res Rev.* 2011;65:99–112. doi: 10.1016/j.brainresrev.2010.08.002.
67. Oreskovic D, Klarica M. The formation of cerebrospinal fluid: nearly a hundred years of interpretations and misinterpretations. *Brain Res Rev.* 2010;64:241–262. doi: 10.1016/j.brainresrev.2010.04.006.
68. Oreskovic D, Klarica M. Development of hydrocephalus and classical hypothesis of cerebrospinal fluid hydrodynamics: facts and illusions. *Prog Neurobiol.* 2011;94:238–258. doi: 10.1016/j.pneurobio.2011.05.005.
69. Bateman GA, Brown KM. The measurement of CSF flow through the aqueduct in normal and hydrocephalic children: from where does it come, to where does it go? *Child's Nerv Syst.* 2012;28:55–63. doi: 10.1007/s00381-011-1617-4.
70. Klarica M, Mise B, Vladic A, Rados M, Oreskovic D. "Compensated hyperosmolarity" of cerebrospinal fluid and the development of hydrocephalus. *Neuroscience.* 2013;248:278–289. doi: 10.1016/j.neuroscience.2013.06.022.
71. Chikly B, Quaghebeur J. Reassessing cerebrospinal fluid (CSF) hydrodynamics: a literature review presenting a novel hypothesis for

CSF physiology. *J Bodyw Mov Ther.* 2013;17:344–354. doi: 10.1016/j.jbmt.2013.02.002.

72. Igarashi H, Tsujita M, Kwee IL, Nakada T. Water influx into cerebrospinal fluid is primarily controlled by aquaporin-4, not by aquaporin-1: O-17 JJVCPE MRI study in knockout mice. *Neuroreport.* 2014;25:39–43.
73. Fenstermacher JD, Johnson JA. Filtration and reflection coefficients of the rabbit blood-brain barrier. *Am J Physiol.* 1966;211:341–346.
74. Fenstermacher JD, Patlak CS. The movements of water and solutes in the brains of mammals. In: Pappius HM, Feindel W, editors. *Dynamics of Cerebral Edema.* Berlin: Springer-Verlag; 1976. pp. 87–94.
75. Paulson OB, Hertz MM, Bolwig TG, Lassen NA. Filtration and diffusion of water across blood-brain-barrier in man. *Microvasc Res.* 1977;13:113–123. doi: 10.1016/0026-2862(77)90120-0.
76. Sweet WH, Selverstone B, Soloway S, Stetten D., Jr Studies of formation, flow and absorption of cerebrospinal fluid. II. Studies with heavy water in the normal man. *Surg Forum.* 1950;92:376–381.
77. Sweet WH, Brownell GL, Scholl JA, Bowsher DR, Benda P, Stickley EE. The formation, flow and absorption of cerebrospinal fluid - newer concepts based on studies with isotopes. *Res Publ Assoc Res Nerv Ment Dis.* 1954;34:101–159.
78. Bering EA., Jr Water exchange of central nervous system and cerebrospinal fluid. *J Neurosurg.* 1952;9:275–287. doi: 10.3171/jns.1952.9.3.0275.
79. Borison HL, Borison R, McCarthy LE. Brain stem penetration by horseradish peroxidase from the cerebrospinal fluid spaces in the cat. *Exp Neurol.* 1980;69:271–289. doi: 10.1016/0014-4886(80)90211-3.
80. Greitz D. Radiological assessment of hydrocephalus: new theories and implications for therapy. *Neurosurg Rev.* 2004;27:145–165.
81. Greitz D. Paradigm shift in hydrocephalus research in legacy of Dandy's pioneering work: rationale for third ventriculostomy in communicating hydrocephalus. *Child's Nerv Syst.* 2007;23:487–489. doi: 10.1007/s00381-007-0303-z.
82. Linninger AA, Xenos M, Sweetman B, Ponkshe S, Guo X, Penn RD. A mathematical model of blood, cerebrospinal fluid and brain dynamics. *J Math Biol.* 2009;59:729–759. doi: 10.1007/s00285-009-0250-2.

83. Penn RD, Basati S, Sweetman B, Guo X, Linninger A. Ventricle wall movements and cerebrospinal fluid flow in hydrocephalus. *J Neurosurg.* 2011;115:159–164. doi: 10.3171/2010.12.JNS10926.

84. Bateman GA, Napier BD. External hydrocephalus in infants: six cases with MR venogram and flow quantification correlation. *Child's Nerv Syst.* 2011;27:2087–2096. doi: 10.1007/s00381-011-1549-z.

85. Davson H. Dynamic aspects of cerebrospinal fluid. *Dev Med Child Neurol Suppl.* 1972;27:1–16.

86. McComb JG. Recent research into the nature of cerebrospinal-fluid formation and absorption. 8. *J Neurosurg.* 1983;59:369–383. doi: 10.3171/jns.1983.59.3.0369.

87. Kimelberg HK. Water homeostasis in the brain: basic concepts. *Neuroscience.* 2004;129:851–860. doi: 10.1016/j.neuroscience.2004.07.033.

88. Bradbury MW, Cserr HF. Drainage of cerebrospinal fluid and brain interstitial fluid into cervical lymphatics. In: Johnston MG, editor. *Experimental Biology of the Lymphatic Circulation.* Amsterdam, Oxford: Elsevier; 1985. pp. 355–394.

89. Gjedde A, Diemer NH. Double-tracer study of the fine regional blood-brain glucose transfer in the rat by computer-assisted autoradiography. *J Cereb Blood Flow Metab.* 1985;5:282–289. doi: 10.1038/jcbfm.1985.36.

90. Klein B, Kuschinsky W, Schrock H, Vetterlein F. Interdependency of local capillary density, blood flow, and metabolism in rat brains. *Am J Physiol.* 1986;251:H1333–H1340.

91. Gross PM, Sposito NM, Pettersen SE, Panton DG, Fenstermacher JD. Topography of capillary density, glucose metabolism, and microvascular function within the rat inferior colliculus. *J Cereb Blood Flow Metab.* 1987;7:154–160. doi: 10.1038/jcbfm.1987.38.

92. Schlageter KE, Molnar P, Lapin GD, Groothuis DR. Microvessel organization and structure in experimental brain tumors: microvessel populations with distinctive structural and functional properties. *Microvasc Res.* 1999;58:312–328. doi: 10.1006/mvre.1999.2188.

93. Rosenberg GA, Kyner WT, Estrada E. Bulk flow of brain interstitial fluid under normal and hyperosmolar conditions. *Am J Physiol.* 1980;238:F42–F49.

94. Fenstermacher J, Kaye T. Drug diffusion within the brain. *Ann N Y Acad Sci.* 1988;531:29–39. doi: 10.1111/j.1749-6632.1988.tb31809.x.
95. Nicholson C, Sykova E. Extracellular space structure revealed by diffusion analysis. *Trends Neurosci.* 1998;21:207–215. doi: 10.1016/S0166-2236(98)01261-2.
96. Verkman AS. Diffusion in the extracellular space in brain and tumors. *Phys Biol.* 2013;10:045003. doi: 10.1088/1478-3975/10/4/045003.
97. Rall DP, Oppelt WW, Patlak CS. Extracellular space of brain as determined by diffusion of inulin from the ventricular system. *Life Sci.* 1962;1:43–48. doi: 10.1016/0024-3205(62)90104-2.
98. Patlak CS, Fenstermacher JD. Measurements of dog blood-brain transfer constants by ventriculocisternal perfusion. *Am J Physiol.* 1975;229:877–884.
99. Nicholson C, Phillips JM. Ion diffusion modified by tortuosity and volume fraction in the extracellular micro-environment of the rat cerebellum. *J Physiol.* 1981;321:225–257. doi: 10.1113/jphysiol.1981.sp013981.
100. Nicholson C, Tao L. Hindered diffusion of high molecular weight compounds in brain extracellular microenvironment measured with integrative optical imaging. *Biophys J.* 1993;65:2277–2290. doi: 10.1016/S0006-3495(93)81324-9.
101. Reulen HJ, Graham R, Spatz M, Klatzo I. Role of pressure gradients and bulk flow in dynamics of vasogenic brain edema. *J Neurosurg.* 1977;46:24–35. doi: 10.3171/jns.1977.46.1.0024.
102. Reulen HJ, Tsuyumu M, Tack A, Fenske AR, Prioleau GR. Clearance of edema fluid into cerebrospinal fluid. A mechanism for resolution of vasogenic brain edema. *J Neurosurg.* 1978;48:754–764. doi: 10.3171/jns.1978.48.5.0754.
103. Reulen HJ. Bulk flow and diffusion revisited, and clinical applications. In: Czernicki Z, editor. *Brain Edema XIV.* Vienna: Springer-Verlag; 2010. pp. 3–13.
104. Bobo RH, Laske DW, Akbasak A, Morrison PF, Dedrick RL, Oldfield EH. Convection-enhanced delivery of macromolecules in the brain. *Proc Natl Acad Sci U S A.* 1994;91:2076–2080. doi: 10.1073/pnas.91.6.2076.

105. Morrison PF, Laske DW, Bobo H, Oldfield EH, Dedrick RL. High-flow microinfusion: tissue penetration and pharmacodynamics. *Am J Physiol.* 1994;266:R292–R305.

106. Chen XM, Sarntinoranont M. Biphasic finite element model of solute transport for direct infusion into nervous tissue. *Ann Biomed Eng.* 2007;35:2145–2158. doi: 10.1007/s10439-007-9371-1.

107. Crone C, Olesen SP. Electrical-resistance of brain microvascular endothelium. *Brain Res.* 1982;241:49–55. doi: 10.1016/0006-8993(82)91227-6.

108. Butt AM, Jones HC, Abbott NJ. Electrical resistance across the blood-brain barrier in anaesthetized rats: a developmental study. *J Physiol.* 1990;429:47–62. doi: 10.1113/jphysiol.1990.sp018243.

109. MacAulay N, Hamann S, Zeuthen T. Chloride transporters as water pumps: elements in a new model of epithelial water transport. In: Alvarez-Leefmans FJ, Delpire E, editors. *Physiology and Pathology of Chloride Transporters and Channels in the Nervous System: From Molecules to Diseases.* Amsterdam: Elsevier Science; 2009. pp. 547–568.

110. Milhorat TH. Failure of choroid plexectomy as treatment for hydrocephalus. *Surg Gynecol Obstet.* 1974;139:505–508.

111. Milhorat TH. Third circulation revisited. *J Neurosurg.* 1975;42:628–645. doi: 10.3171/jns.1975.42.6.0628.

112. Milhorat TH. *Cerebrospinal Fluid and the Brain Edemas.* New York: Neuroscience Society of New York; 1987.

113. Scarff JE. Treatment of nonobstructive (communicating) hydrocephalus by endoscopic cauterization of choroid plexuses. *J Neurosurg.* 1970;33:1–18. doi: 10.3171/jns.1970.33.1.0001.

114. Warf BC. Comparison of endoscopic third ventriculostomy alone and combined with choroid plexus cauterization in infants younger than 1 year of age: a prospective study in 550 African children. *J Neurosurg.* 2005;103:475–481.

115. Warf BC. The impact of combined endoscopic third ventriculostomy and choroid plexus cauterization on the management of pediatric hydrocephalus in developing countries. *World Neurosurg.* 2013;79(S23):e13–e15.

116. Zhu XL, Di Rocco C. Choroid plexus coagulation for hydrocephalus not due to CSF overproduction: a review. *Child's Nerv Syst.* 2013;29:35–42. doi: 10.1007/s00381-012-1960-0.

117. Pollay M. Overview of the CSF dual outflow system. In: Aygok GA, Rekate HL, editors. *Hydrocephalus.* Vienna: Springer-Verlag; 2012. pp. 47–50.

118. Bradbury MW, Cole DF. The role of the lymphatic system in drainage of cerebrospinal fluid and aqueous humour. *J Physiol.* 1980;299:353–365. doi: 10.1113/jphysiol.1980.sp013129.

119. Bradbury MW, Cserr HF, Westrop RJ. Drainage of cerebral interstitial fluid into deep cervical lymph of the rabbit. *Am J Physiol.* 1981;240:F329–F336.

120. Bradbury MWB, Westrop RJ. Factors influencing exit of substances from cerebrospinal-fluid into deep cervical lymph of the rabbit. *J Physiol.* 1983;339:519–534. doi: 10.1113/jphysiol.1983.sp014731.

121. Kida S, Pantazis A, Weller RO. CSF drains directly from the subarachnoid space into nasal lymphatics in the rat - anatomy, histology and immunological significance. *Neuropathol Appl Neurobiol.* 1993;19:480–488. doi: 10.1111/j.1365-2990.1993.tb00476.x.

122. Johnston M, Papaiconomou C. Cerebrospinal fluid transport: a lymphatic perspective. *News Physiol Sci.* 2002;17:227–230.

123. Papaiconomou C, Zakharov A, Azizi N, Djenic J, Johnston M. Reassessment of the pathways responsible for cerebrospinal fluid absorption in the neonate. *Child's Nerv Syst.* 2004;20:29–36. doi: 10.1007/s00381-003-0840-z.

124. Johnston M, Zakharov A, Papaiconomou C, Salmasi G, Armstrong D. Evidence of connections between cerebrospinal fluid and nasal lymphatic vessels in humans, non-human primates and other mammalian species. *Cerebrospinal Fluid Res.* 2004;1:2. doi: 10.1186/1743-8454-1-2.

125. Johnston M, Zakharov A, Koh L, Armstrong D. Subarachnoid injection of Microfil reveals connections between cerebrospinal fluid and nasal lymphatics in the non-human primate. *Neuropathol Appl Neurobiol.* 2005;31:632–640. doi: 10.1111/j.1365-2990.2005.00679.x.

126. Nagra G, Koh L, Zakharov A, Armstrong D, Johnston M. Quantification of cerebrospinal fluid transport across the cribriform plate into lymphatics in rats. *Am J Physiol.* 2006;291:R1383–R1389.

127. Murtha LA, Yang Q, Parsons MW, Levi CR, Beard DJ, Spratt NJ, McLeod DD. Cerebrospinal fluid is drained primarily via the spinal canal and olfactory route in young and aged spontaneously hypertensive rats. *Fluids Barriers CNS.* 2014;11:12. doi: 10.1186/2045-8118-11-12.

128. Ekstedt J. CSF hydrodynamic studies in man. 1. Method of constant pressure CSF infusion. *J Neurol Neurosurg Psychiatry.* 1977;40:105–119. doi: 10.1136/jnnp.40.2.105.

129. Ekstedt J. CSF hydrodynamic studies in man. 2. Normal hydrodynamic variables related to CSF pressure and flow. *J Neurol Neurosurg Psychiatry.* 1978;41:345–353. doi: 10.1136/jnnp.41.4.345.

130. Ghersi-Egea JF, Finnegan W, Chen JL, Fenstermacher JD. Rapid distribution of intraventricularly administered sucrose into cerebrospinal fluid cisterns via subarachnoid velae in rat. *Neuroscience.* 1996;75:1271–1288. doi: 10.1016/0306-4522(96)00281-3.

131. Nagaraja TN, Patel P, Gorski M, Gorevic PD, Patlak CS, Fenstermacher JD. In normal rat, intraventricularly administered insulin-like growth factor-1 is rapidly cleared from CSF with limited distribution into brain. *Cerebrospinal Fluid Res.* 2005;2:5. doi: 10.1186/1743-8454-2-5.

132. Cserr HF, Cooper DN, Suri PK, Patlak CS. Efflux of radiolabeled polyethylene glycols and albumin from rat brain. *Am J Physiol.* 1981;240:F319–F328.

133. Heisey SR, Held D, Pappenheimer JR. Bulk flow and diffusion in the cerebrospinal fluid system of the goat. *Am J Physiol.* 1962;203:775–781.

134. Milhorat TH. Choroid plexus and cerebrospinal fluid production. *Science.* 1969;166:1514–1516. doi: 10.1126/science.166.3912.1514.

135. Milhorat TH. *Hydrocephalus and the Cerebrospinal Fluid.* Baltimore: Williams & Wilkins; 1972.

136. Pollay M, Curl F. Secretion of cerebrospinal fluid by the ventricular ependyma of the rabbit. *Am J Physiol.* 1967;213:1031–1038.

137. Curl FD, Pollay M. Transport of water and electrolytes between brain and ventricular fluid in the rabbit. *Exp Neurol.* 1968;20:558–574. doi: 10.1016/0014-4886(68)90109-X.

138. Cserr HF, Cooper DN, Milhorat TH. Flow of cerebral interstitial fluid as indicated by removal of extracellular markers from rat caudate-

nucleus. *Exp Eye Res.* 1977;25:461–473. doi: 10.1016/S0014-4835(77)80041-9.

139. Szentistvanyi I, Patlak CS, Ellis RA, Cserr HF. Drainage of interstitial fluid from different regions of rat brain. *Am J Physiol.* 1984;246:F835–F844.

140. Yamada S, DePasquale M, Patlak CS, Cserr HF. Albumin outflow into deep cervical lymph from different regions of rabbit brain. *Am J Physiol.* 1991;261:H1197–H1204.

141. Curry F-RE, Renkin EM, Michel CC, Geiger SR. *Handbook of Physiology Section 2 The Cardiovascular System. Volume 4 Part 1 Microcirculation.* Bethesda, MD: American Physiological Society; 1984. Mechanics and thermodynamics of transcapillary exchange; pp. 309–374.

142. Groothuis DR, Vavra MW, Schlageter KE, Kang EW-Y, Itskovich AC, Hertzler S, Allen CV, Lipton HL. Efflux of drugs and solutes from brain: the interactive roles of diffusional transcapillary transport, bulk flow and capillary transporters. *J Cereb Blood Flow Metab.* 2007;27:43–56. doi: 10.1038/sj.jcbfm.9600315.

143. Iliff JJ, Wang M, Liao Y, Plogg BA, Peng W, Gundersen GA, Benveniste H, Vates GE, Deane R, Goldman SA, Nagelhus EA, Nedergaard M. A paravascular pathway facilitates CSF flow through the brain parenchyma and the clearance of interstitial solutes, including amyloid beta. *Sci Transl Med.* 2012;4:147ra111. doi: 10.1126/scitranslmed.3003748.

144. Hutchings M, Weller RO. Anatomical relationships of the pia mater to cerebral blood vessels in man. *J Neurosurg.* 1986;65:316–325. doi: 10.3171/jns.1986.65.3.0316.

145. Zhang ET, Richards HK, Kida S, Weller RO. Directional and compartmentalized drainage of interstitial fluid and cerebrospinal-fluid from the rat-brain. *Acta Neuropathol.* 1992;83:233–239. doi: 10.1007/BF00296784.

146. Arbel-Ornath M, Hudry E, Eikermann-Haerter K, Hou S, Gregory JL, Zhao LZ, Betensky RA, Frosch MP, Greenberg SM, Bacskai BJ. Interstitial fluid drainage is impaired in ischemic stroke and Alzheimer's disease mouse models. *Acta Neuropathol.* 2013;126:353–364. doi: 10.1007/s00401-013-1145-2.

147. Fenstermacher J. Comment after paper by Katzman. In: Coxon RV, editor. *Symposium on the Blood-Brain Barrier Wates Foundation; Oxford.* Oxford: Truex; 1970. p. 166.

148. Levin VA, Fenstermacher JD, Patlak CS. Sucrose and inulin space measurements of cerebral cortex in four mammalian species. *Am J Physiol.* 1970;219:1528–1533.

149. Levin E, Sisson WB. The penetration of radiolabeled substances into rabbit brain from subarachnoid space. *Brain Res.* 1972;41:145–153. doi: 10.1016/0006-8993(72)90622-1.

150. Milhorat TH. Some observations on circulation of phenosulfonpthalein in cerebrospinal fluid - normal flow and flow in hydrocephalus. *J Neurosurg.* 1970;32:522–528. doi: 10.3171/jns.1970.32.5.0522.

151. Pape LG, Katzman R. K^{42} distribution in brain during simultaneous ventriculocisternal and subarachnoid perfusion. *Brain Res.* 1972;38:49–69. doi: 10.1016/0006-8993(72)90589-6.

152. Bradbury MW, Segal MB, Wilson J. Transport of potassium at the blood-brain barrier. *J Physiol.* 1972;221:617–632. doi: 10.1113/jphysiol.1972.sp009771.

153. Weller RO, Djuanda E, Yow H-Y, Carare RO. Lymphatic drainage of the brain and the pathophysiology of neurological disease. *Acta Neuropathol.* 2009;117:1–14. doi: 10.1007/s00401-008-0457-0.

154. Rennels ML, Gregory TF, Blaumanis OR, Fujimoto K, Grady PA. Evidence for a paravascular fluid circulation in the mammalian central nervous system, provided by the rapid distribution of tracer protein throughout the brain from the subarachnoid space. *Brain Res.* 1985;326:47–63. doi: 10.1016/0006-8993(85)91383-6.

155. Wagner HJ, Pilgrim C, Brandl J. Penetration and removal of horseradish peroxidase injected into the cerebrospinal fluid: role of cerebral perivascular spaces, endothelium and microglia. *Acta Neuropathol.* 1974;27:299–315. doi: 10.1007/BF00690695.

156. Rennels ML, Blaumanis OR, Grady PA. Rapid solute transport throughout the brain via paravascular fluid pathways. *Adv Neurol.* 1990;52:431–439.

157. Ichimura T, Fraser PA, Cserr HF. Distribution of extracellular tracers in perivascular spaces of the rat brain. *Brain Res.* 1991;545:103–113. doi: 10.1016/0006-8993(91)91275-6.

158. Hadaczek P, Yamashita Y, Mirek H, Tamas L, Bohn MC, Noble C, Park JW, Bankiewicz K. The "perivascular pump" driven by arterial pulsation is a powerful mechanism for the distribution of therapeutic molecules within the brain. *Mol Ther.* 2006;14:69–78. doi: 10.1016/j.ymthe.2006.02.018.

159. Mollanji R, Bozanovic-Sosic R, Silver I, Li B, Kim C, Midha R, Johnston M. Intracranial pressure accommodation is impaired by blocking pathways leading to extracranial lymphatics. *Am J Physiol.* 2001;280:R1573–R1581.
160. Mollanji R, Bozanovic-Sosic R, Zakharov A, Makarian L, Johnston MG. Blocking cerebrospinal fluid absorption through the cribriform plate increases resting intracranial pressure. *Am J Physiol.* 2002;282:R1593–R1599.
161. Carare RO, Bernardes-Silva M, Newman TA, Page AM, Nicoll JAR, Perry VH, Weller RO. Solutes, but not cells, drain from the brain parenchyma along basement membranes of capillaries and arteries: significance for cerebral amyloid angiopathy and neuroimmunology. *Neuropathol Appl Neurobiol.* 2008;34:131–144. doi: 10.1111/j.1365-2990.2007.00926.x.
162. Weller RO, Massey A, Newman TA, Hutchings M, Kuo YM, Roher AE. Cerebral amyloid angiopathy: amyloid beta accumulates in putative interstitial fluid drainage pathways in Alzheimer's disease. *Am J Pathol.* 1998;153:725–733. doi: 10.1016/S0002-9440(10)65616-7.
163. Preston SD, Steart PV, Wilkinson A, Nicoll JAR, Weller RO. Capillary and arterial cerebral amyloid angiopathy in Alzheimer's disease: defining the perivascular route for the elimination of amyloid beta from the human brain. *Neuropathol Appl Neurobiol.* 2003;29:106–117. doi: 10.1046/j.1365-2990.2003.00424.x.
164. Carare RO, Hawkes CA, Jeffrey M, Kalaria RN, Weller RO. Review: Cerebral amyloid angiopathy, prion angiopathy, CADASIL and the spectrum of protein elimination failure angiopathies (PEFA) in neurodegenerative disease with a focus on therapy. *Neuropathol Appl Neurobiol.* 2013;39:593–611. doi: 10.1111/nan.12042.
165. Schley D, Carare-Nnadi R, Please CP, Perry VH, Weller RO. Mechanisms to explain the reverse perivascular transport of solutes out of the brain. *J Theor Biol.* 2006;238:962–974. doi: 10.1016/j.jtbi.2005.07.005.
166. Weller RO, Subash M, Preston SD, Mazanti I, Carare RO. Perivascular drainage of amyloid-beta peptides from the brain and its failure in cerebral amyloid angiopathy and Alzheimer's disease. *Brain Pathol.* 2008;18:253–266. doi: 10.1111/j.1750-3639.2008.00133.x.
167. Hawkes CA, Sullivan PM, Hands S, Weller RO, Nicoll JAR, Carare RO. Disruption of arterial perivascular drainage of amyloid-beta from

the brains of mice expressing the human APOE epsilon 4 allele. *PLoS One.* 2012;7:e41636. doi: 10.1371/journal.pone.0041636.

168. Hawkes CA, Gatherer M, Sharp MM, Dorr A, Yuen HM, Kalaria R, Weller RO, Carare RO. Regional differences in the morphological and functional effects of aging on cerebral basement membranes and perivascular drainage of amyloid- from the mouse brain. *Aging Cell.* 2013;12:224–236. doi: 10.1111/acel.12045.
169. Feinberg DA, Mark AS. Human brain motion and cerebrospinal fluid circulation demonstrated with MR velocity imaging. *Radiology.* 1987;163:793–799. doi: 10.1148/radiology.163.3.3575734.
170. Enzmann DR, Pelc NJ. Normal flow patterns of intracranial and spinal cerebrospinal fluid defined with phase-contrast cine MR imaging. *Radiology.* 1991;178:467–474. doi: 10.1148/radiology.178.2.1987610.
171. Enzmann DR, Pelc NJ. Cerebrospinal fluid flow measured by phase-contrast cine MR. *AJNR Am J Neuroradiol.* 1993;14:1301–1307.
172. Baledent O, Gondry-Jouet C, Meyer M-E, De Marco G, Le Gars D, Henry-Feugeas M-C, Idy-Peretti I. Relationship between cerebrospinal fluid and blood dynamics in healthy volunteers and patients with communicating hydrocephalus. *Invest Radiol.* 2004;39:45–55. doi: 10.1097/01.rli.0000100892.87214.49.
173. Piechnik SK, Summers PE, Jezzard P, Byrne JV. Magnetic resonance measurement of blood and CSF flow rates with phase contrast - normal values, repeatability and CO(2) reactivity. In: Manley G, Hemphill C, Stiver S, editors. *Intracranial Pressure and Brain Monitoring XIII: Mechanisms and Treatment.* Vienna: Springer; 2008. pp. 263–270.
174. Bering EA, Jr, Sato O. Hydrocephalus: changes in formation and absorption of cerebrospinal fluid within the cerebral ventricles. *J Neurosurg.* 1963;20:1050–1063. doi: 10.3171/jns.1963.20.12.1050.
175. Oreskovic D, Klarica M, Vukic M. The formation and circulation of cerebrospinal fluid inside the cat brain ventricles: a fact or an illusion? *Neurosci Lett.* 2002;327:103–106. doi: 10.1016/S0304-3940(02)00395-6.
176. Klarica M, Oreskovic D, Bozic B, Vukic M, Butkovic V, Bulat M. New experimental model of acute aqueductal blockage in cats: effects on cerebrospinal fluid pressure and the size of brain

ventricles. *Neuroscience.* 2009;158:1397–1405. doi: 10.1016/j.neuroscience.2008.11.041.

177. Dandy WE, Blackfan KD. Internal hydrocephalus - an experimental, clinical and pathological study. *Am J Dis Child.* 1914;8:406–482. doi: 10.1001/archpedi.1914.02180010416002.
178. Milhorat TH, Clark RG, Hammock MK. Experimental hydrocephalus. 2. Gross pathological findings in acute and subacute obstructive hydrocephalus in the dog and monkey. *J Neurosurg.* 1970;32:390–399. doi: 10.3171/jns.1970.32.4.0390.
179. James AE, Novak G, Bahr AL, Burns B. Production of cerebrospinal-fluid in experimental communicating hydrocephalus. *Exp Brain Res.* 1977;27:553–557. doi: 10.1007/BF00239043.
180. Gideon P, Stahlberg F, Thomsen C, Gjerris F, Sorensen PS, Henriksen O. Cerebrospinal fluid flow and production in patients with normal pressure hydrocephalus studied by MRI. *Neuroradiology.* 1994;36:210–215. doi: 10.1007/BF00588133.
181. Nitz WR, Bradley WG, Jr, Watanabe AS, Lee RR, Burgoyne B, O'Sullivan RM, Herbst MD. Flow dynamics of cerebrospinal fluid: assessment with phase-contrast velocity MR imaging performed with retrospective cardiac gating. *Radiology.* 1992;183:395–405. doi: 10.1148/radiology.183.2.1561340.
182. Baledent O. Imaging of the cerebrospinal fluid circulation. In: Rigamonti D, editor. *Adult Hydrocephalus.* Cambridge: Cambridge University Press; 2014. pp. 121–138.
183. Kim DS, Choi JU, Huh R, Yun PH, Kim DI. Quantitative assessment of cerebrospinal fluid hydrodynamics using a phase-contrast cine MR image in hydrocephalus. *Child's Nerv Syst.* 1999;15:461–467. doi: 10.1007/s003810050440.
184. Bateman GA. Correction to "External hydrocephalus in infants: six cases with MR venogram and flow quantification correlation" *Child's Nerv Syst.* 2011;27:2033–2034. doi: 10.1007/s00381-011-1610-y.
185. Strecker EP, James AE. Evaluation of cerebrospinal-fluid flow and absorption - clinical and experimental studies. *Neuroradiology.* 1973;6:200–205. doi: 10.1007/BF00335324.
186. James AE, Jr, Strecker EP, Sperber E, Flor WJ, Merz T, Burns B. An alternative pathway of cerebrospinal fluid absorption in communicating hydrocephalus. Transependymal movement. *Radiology.* 1974;111:143–146. doi: 10.1148/111.1.143.

187. Strecker EP, Kelley JET, Merz T, James AE. Transventricular albumin absorption in communicating hydrocephalus - semiquantitative analysis of periventricular extracellular-space utilizing autoradiography. *Arch Psychiatr Nervenkr.* 1974;218:369–377. doi: 10.1007/BF00342579.

188. James AE, Flor WJ, Novak GR, Strecker EP, Burns B, Epstein M. Experimental hydrocephalus. *Exp Eye Res.* 1977;25(Suppl):435–459. doi: 10.1016/S0014-4835(77)80040-7.

189. Maki Y, Kokubo Y, Nose T, Yoshii Y. Some characteristic findings of isotope cisternograms in children. *J Neurosurg.* 1976;45:56–59. doi: 10.3171/jns.1976.45.1.0056.

190. Blomquist HK, Sundin S, Ekstedt J. Cerebrospinal fluid hydrodynamic studies in children. *J Neurol Neurosurg Psychiatry.* 1986;49:536–548. doi: 10.1136/jnnp.49.5.536.

191. Milhorat TH, Mosher MB, Hammock MK, Murphy CF. Evidence for choroid-plexus absorption in hydrocephalus. *N Engl J Med.* 1970;283:286–289. doi: 10.1056/NEJM197008062830604.

192. Wislocki GB, Putnam TJ. Absorption from the ventricles in experimentally produced internal hydrocephalus. *Am J Anat.* 1921;29:313–320. doi: 10.1002/aja.1000290302.

193. Eisenberg HM, McLennan JE, Welch K. Ventricular perfusion in cats with kaolin-induced hydrocephalus. *J Neurosurg.* 1974;41:20–28. doi: 10.3171/jns.1974.41.1.0020.

194. Welch K, Sadler K. Permeability of the choroid plexus of the rabbit to several solutes. *Am J Physiol.* 1966;210:652–660.

195. Wåhlin A, Ambarki K, Hauksson J, Birgander R, Malm J, Eklund A. Phase contrast MRI quantification of pulsatile volumes of brain arteries, veins, and cerebrospinal fluids compartments: repeatability and physiological interactions. *J Magn Reson Imaging.* 2012;35:1055–1062. doi: 10.1002/jmri.23527.

196. Greitz D, Wirestam R, Franck A, Nordell B, Thomsen C, Stahlberg F. Pulsatile brain movement and associated hydrodynamics studied by magnetic resonance phase imaging. The Monro-Kellie doctrine revisited. *Neuroradiology.* 1992;34:370–380. doi: 10.1007/BF00596493.

197. Piechnik SK, Jezzard P, Byrne JV, Summers PE. Physiological component in background flow velocity in MR phase contrast measurements. *Proc Intl Soc Mag Reson Med.* 2008;16:2235.

198. James AE, Jr, Flor WJ, Merz T, Strecker EP, Burns B. A pathophysiologic mechanism for ventricular entry of radiopharmaceutical and possible relation to chronic communicating hydrocephalus. *Am J Roentgenol Radium Ther.* 1974;122:38–43. doi: 10.2214/ajr.122.1.38.

199. Glymphatic systemhttp://en.wikipedia.org/wiki/Glymphatic_system

200. Konnikova M. *New York Times.* New York: New York Times Company; 2014. Goodnight. Sleep clean.

201. Nedergaard M. Neuroscience. Garbage truck of the brain. *Science.* 2013;340:1529–1530. doi: 10.1126/science.1240514.

202. Iliff JJ, Wang MH, Zeppenfeld DM, Venkataraman A, Plog BA, Liao YH, Deane R, Nedergaard M. Cerebral arterial pulsation drives paravascular CSF-interstitial fluid exchange in the murine brain. *J Neurosci.* 2013;33:18190–18199. doi: 10.1523/JNEUROSCI.1592-13.2013.

203. Iliff JJ, Lee H, Yu M, Feng T, Logan J, Nedergaard M, Benveniste H. Brain-wide pathway for waste clearance captured by contrast-enhanced MRI. *J Clin Invest.* 2013;123:1299–1309. doi: 10.1172/JCI67677.

204. Yang LJ, Kress BT, Weber HJ, Thiyagarajan M, Wang BZ, Deane R, Benveniste H, Iliff JJ, Nedergaard M. Evaluating glymphatic pathway function utilizing clinically relevant intrathecal infusion of CSF tracer. *J Transl Med.* 2013;11:107. doi: 10.1186/1479-5876-11-107.

205. Papisov MI, Belov VV, Gannon KS. Physiology of the intrathecal bolus: the leptomeningeal route for macromolecule and particle delivery to CNS. *Mol Pharm.* 2013;10:1522–1532. doi: 10.1021/mp300474m.

206. Thrane VR, Thrane AS, Plog BA, Thiyagarajan M, Iliff JJ, Deane R, Nagelhus EA, Nedergaard M. Paravascular microcirculation facilitates rapid lipid transport and astrocyte signaling in the brain. *Sci Rep.* 2013;3:2582. doi: 10.1038/srep02582.

207. Xie L, Kang H, Xu Q, Chen MJ, Liao Y, Thiyagarajan M, O'Donnell J, Christensen DJ, Nicholson C, Iliff JJ, Takano T, Deane R, Nedergaard M. Sleep drives metabolite clearance from the adult brain. *Science.* 2013;342:373–377. doi: 10.1126/science.1241224.

208. Papadopoulos MC, Verkman AS. Aquaporin water channels in the nervous system. *Nat Rev Neurosci.* 2013;14:265–277. doi: 10.1038/nrn3468.

209. The challenge presented by altered brain interstitial fluid dynamics during slow wave sleephttp://www.ucl.ac.uk/~ucgbarg/res/Phys2014_poster.pdf

210. Ranck JB., Jr Electrical impedance in the subicular area of rats during paradoxical sleep. *Exp Neurol.* 1966;16:416–437. doi: 10.1016/0014-4886(66)90107-5.

211. Ranck JB., Jr Electrical impedance changes in many sites of brain in paradoxical sleep, anesthesia, and activity. *Exp Neurol.* 1970;27:454–475. doi: 10.1016/0014-4886(70)90107-X.

Chapter 9

Effects of Passive Hydrotherapy WATSU (WaterShiatsu) in the Third Trimester of Pregnancy: Results of a Controlled Pilot Study

Agnes M. Schitter[1], Marko Nedeljkovic[1], Heiner Baur[2], Johannes Fleckenstein[1], and Luigi Raio[3]

[1] Department of TCM/Acupuncture, Institute of Complementary Medicine IKOM, University of Bern, Imhoof-Pavillon, Inselspital, 3010 Bern, Switzerland.
[2] Department of Health, Bern University of Applied Sciences, Murtenstrasse 10, 3008 Bern, Switzerland.
[3] Department of Obstetrics and Gynecology Inselspital, University Hospital of Bern, Effingerstrasse 102, 3010 Bern, Switzerland

Citation: Effects of Passive Hydrotherapy WATSU (WaterShiatsu) in the Third Trimester of Pregnancy: Results of a Controlled Pilot Study. Agnes M. Schitter, Marko Nedeljkovic, Heiner Baur, Johannes Fleckenstein, and Luigi Raio. Hindawi Publishing Corporation Evidence-Based Complementary and Alternative Medicine Volume 2015, Article ID 437650, 10 pages http://dx.doi.org/10.1155/2015/437650.

ABSTRACT

Background. WATSU (WaterShiatsu) is a complementary therapeutic treatment method comprising passive stretches and massage techniques administered in 35°C warm water. Pregnant women claim safe methods to reduce pain, stress, and fatigue. Therefore, we conducted a pilot study evaluating the effects of WATSU on pregnancy-related complaints in third trimester pregnant women. *Methods*. Nine healthy pregnant women at gestational week ≥34 were included in an intervention group (receiving WATSU) and compared to eight women in a passive control group (receiving no treatment). WATSU was performed on days 1 and 4 of the study, accompanied by ultrasound examinations. Outcomes include physiological and psychometric as well as qualitative data. Participants in the control group completed questionnaires only. *Results*. WATSU was found to significantly lower participants' levels of stress and pain and to improve their mental health-related quality of life and mood. In comparison to the passive control group, participants in the intervention group reported reduction in perceived stress from day 1 to day 8 ($P = 0.036$, Cohen›s $f = 0.57$). Qualitative data indicate that WATSU was appreciated as enjoyable and deeply relaxing. No negative side effects were reported. *Conclusion*. Our findings support the notion that WATSU yields therapeutic benefits for pregnant women and warrant further research. This study has been registered at ClinicalTrials.gov: NCT01708018.

INTRODUCTION

Several researchers propose that, during pregnancy, maternal mental and physical wellbeing are transferred to the fetus resulting in epigenetic changes implicating consequences for a lifetime [1–4]. Hence, with respect to the child's future development, severe maternal stress and distress cannot be considered to be of transient nature [5]. Moreover, maternal effect and attitude during pregnancy are also predictive factors for postnatal depressive symptoms [6], which in turn are likely to affect the child's wellbeing.

WATSU (an acronym based on WAter and shiaTSU) is a body-based method [7] comprising buoyancy, passive stretches, and massage techniques, including massage and palpation of acupuncture points that is administered in warm water. WATSU has been described as applicable during pregnancy [8], where it is claimed to reduce pregnancy-related low back pain, to relax hypertonic muscles including those of the uterus, to improve the overall sense of wellbeing, and to deepen the relationship of the mother with her unborn child.

In the 1980s, WATSU was created by massage-therapist Harold Dull, whose personal affinity to water led him to practice Masunaga-Shiatsu (also known as Zen-Shiatsu) in water heated to roughly skin temperature (308.15 Kelvin; 35°C, 95°F). He used Shiatsu massage of acupuncture points, joint mobilization, and tissue stretches and added gentle massage techniques to harmonize the energy flow (qi, 氣) according to the principles of Traditional Chinese Medicine (TCM), from which Shiatsu originates [9]. The increased clinical implementation of WATSU in interdisciplinary treatment settings such as rehabilitation facilities indicates a growing acceptance of this body-based complementary therapeutic intervention; it is used as a component in multimodal treatment settings focusing on posttraumatic stress disorder, and anxiety [10, 11], chronic pain and fibromyalgia [11, 12], stress-related illnesses [13], depression [11, 14], and sexual dysfunction [15]. WATSU has also been recommended as a treatment for patients with hemiparesis, multiple sclerosis, cerebral palsy, and spinal cord injury [16, 17]. Considering physiological effects on the cardiopulmonary system due to physical exposure to hydrostatic pressure, certain cardiac conditions, for example, chronic heart failure [18] and respiratory impairments such as cystic fibrosis [19] can be regarded as potential indications for this treatment.

However, clinical trials investigating therapeutic effects of WATSU are still scarce. To our knowledge, to date, only three small scale trials on WATSU have been published [12, 16, 20].

Regarding pregnancy, gymnastics in water were observed to have pain-relieving effects [21, 22], as were massage [23] and birth in water [24]. Active hydrotherapy during pregnancy is considered to be beneficial and safe [25], while the effects of passive hydrotherapy during pregnancy so far have not been the subject of scientific investigation. Therefore, we designed a pilot study to examine safety of WATSU as well as a broad range of issues to identify those worthwhile pursuing in future research, that is, potential therapeutic effects of WATSU on self-reported stress, pregnancy-related pain, mood and quality of life, amount of amniotic fluid, blood flow characteristics, spontaneous course of breech presentations, and prospects of external cephalic versions in third trimester pregnant women. Additionally, qualitative data reflecting participants' perception of the intervention were assessed and analyzed.

METHODS

Study Design

We conducted a controlled clinical pilot study at the Department of Obstetrics and Gynecology at Bern University Hospital investigating the effects of WATSU on pregnant women with pregnancy-related complaints at week ≥34 of gestation. The research protocol was formally approved by the Ethics Committee of the Canton of Bern, Switzerland, and carried out according to the Helsinki Declaration. Recruitment was carried out from May 2012 to May 2014 by medical staff at Bern University Hospital, as well as through advertisements via the midwifery association of the Canton of Bern. Potential participants were provided with complete written and verbal information about the study and written consent was obtained prior to participation. Interested healthy women with a singleton pregnancy in the 34th or greater week of gestation underwent a telephone-screening. Exclusion criteria were any pathological findings during pregnancy, neurological deficits resulting from low back pain, WATSU-treatment within the past four weeks, and poor language skills. Women reporting a breech presentation were included in the study if they did not plan external cervical version. Participants in the intervention group received standardized WATSU treatments at days 1 and 4, accompanied by ultrasound examinations before and after WATSU treatment and on day 8. Participants were allocated to the passive control group if they refused to undergo intervention or ultrasound examinations, or if they lived too far from the site of the intervention. If allocated to the passive control group, participants did receive neither WATSU nor any alternative treatment on the part of the study. All participants were free to maintain additional medical and/or therapeutic treatments during the study. Assessments took place on days 1 (baseline), 4, and 8 (follow-up). Participants were not financially compensated; participants in the control group were offered a free WATSU-treatment after study completion.

The study has been registered at ClinicalTrials.gov: NCT01708018.

Intervention

WATSU treatments were performed by four specialized therapists with ten to sixteen years of professional experience and certification by the Swiss Aquatic Bodywork Association (Netzwerk für Aquatische Körperarbeit (NAKA)). Each participant in the intervention group received a standardized WATSU-treatment on the first and fourth day of study participation, administered by the same therapist. Participants were free to choose a female or a male therapist. Sessions lasted 60 minutes and took place in the morning from 9:00 a.m. to 10:00 a.m. in a therapy pool located at the University Hospital of Bern filled with 35°C fresh water. The administered motion sequence "WATSU-Transition-Flow" [9] was adapted for women in their third trimester of pregnancy and was followed closely. During the WATSU-treatment, participants' abdomens were not touched. Deviations from the standardized treatment protocol were documented for each treatment session. Each session started with a brief verbal description of the procedure and ended with the opportunity for the client to give verbal feedback. During WATSU treatment, participants rested in a supine position, predominantly being supported at the back of their head and at their pelvis or knees by their therapist's forearms. To unburden participants' lower backs, floating devices were attached to their thighs. In the course of treatment, therapist and participant were in continuous physical contact with distances varying from a full arm's length to cradled positions. The participant was slowly floated back and forth through the water in large circular patterns, generated by the therapist's rotation around her/his own body axis. Following the session, participants were asked to drink 500 mL of water to compensate for body fluid loss due to increased diuresis [26, 27]. If indicated and with participants' consent, external cephalic version (ECV) was performed on day 9 of the study. Participants in the passive control group received neither WATSU nor an alternative treatment in the context of this trial.

Outcome Measures

Sociodemographic data (age, height, weight, prior deliveries, week of gestation, and fetal position) and baseline data (perceived stress, pain, and quality of life) were assessed on day 1.

Baseline Assessment on Day 1 and Follow-Up Assessment on Day 8

Included the Perceived Stress Scale (PSS) and theMedical Outcomes Study 36-Item Short Form Health Survey (SF-36). Both questionnaires referred to participants' self-reported condition in the past week.

Perceived Stress Scale (PSS). This 10-item questionnaire assesses participants' cognitive evaluation of stress perception [28]. Participants estimate how unpredictable, uncontrollable, and overloaded they perceive their lives to be on a 5-point scale ranging from "never" to "very often." Good internal consistency is reported (Cronbach's $\alpha = 0.87$) [29].

Medical Outcomes Study 36-Item Short Form Health Survey (SF-36). Health-related quality of life was assessed using a questionnaire comprising eight subscales: "physical functioning," "role physical," "bodily pain," "role emotional," "vitality," "mental health," "general health perception," and "social functioning." The SF-36 shows good internal consistencies with Cronbach's $\alpha > 0.70$ in all subscales except "general health perception" (0.57) and "social functioning" (0.69) [30].

Psychometric Assessment on Days 1 and 4 Immediately before and after WATSU Treatments

Included Visual Analogue Scales (VAS) assessing pain and stress and the Multidimensional Mood Questionnaire (MDMQ). These instruments referred to participants' self-reported actual condition.

Stress and Pain Related Visual Analogue Scales (VAS). Visual Analogue Scales have a length of 100 mm, with 0 mm indicating no stress, respectively, no pain at all, and 100 mm representing maximal perception of stress, respectively, high pain. VAS scales proved to be valid and reliable [31, 32].

Multidimensional Mood Questionnaire (MDMQ). The validated mood questionnaire with good internal consistencies (Cronbach's α between 0.73 and 0.89) assesses treatment-related changes in self-reported mood. It consists of a list of 12 adjectives that address current mood, calmness, and alertness (e.g., "happy," "nervous," and "awake") ranked on a 5-point scale ranging from "not at all" to "very much." The sum score ranges from 12 to 60 with higher scores indicating better mood [33].

Ultrasound Examination

All ultrasound examinations were carried out twenty minutes before and after WATSU treatment, as well as in the morning of day 8 by gynecologists using Voluson E8 Expert, GE Medical Systems (Zipf, Austria), with curved array transducer of 5–8 MHz. Ultrasound measurements included examination of fetal position and assessment of amniotic fluid volume, measuring the single deepest amniotic fluid pocket (SDP) [34, 35]. Doppler examination of the umbilical artery and both uterine arteries was performed. Blood flow characteristics were assessed measuring the pulsatility index (PI). Along with this assessment, the tonus of the uterus was palpated by the examining gynecologist.

Qualitative Outcome Measures

Participants in the intervention group were asked to fill out a qualitative questionnaire answering the following questions right after their second WATSU-treatment: "How was your experience being treated with WATSU?" "Which changes did you notice in response to your WATSU treatment?" "Which aspects of your WATSU treatment were less pleasant for you?" and "Do you have any suggestions for improvement?"

Data Analysis

Quantitative data were analyzed based on intention to treat. For missing data, the last value was carried forward. Quantitative data analyses were conducted using SPSS (version 19) statistical software package for Windows (IBM SPSS Statistics, Somers, NY, USA). Prior to statistical analyses, all data were tested for homogeneity of variance and normal distribution employing Levene and Kolmogorov-Smirnov tests. Between-group differences in sociodemographic characteristics, baseline values, and mean change values of outcome measures from baseline to follow-up assessment were analyzed using Mann-Whitney *U* test for continuous data, while the evaluation of categorical data was based on visual inspection. Analyses of outcome measures within the intervention group were performed using Wilcoxon test. All analyses were two tailed, with the level of significance set at $P < 0.05$ with 95% confidence interval. All continuous data are presented as mean value ± standard deviation (SD). For the purpose of international comparability, outcome values of the SF-36 main scales were standardized by employing the weighting coefficient for US population and transformed into percentages [30]. Effect size parameters (f) were derived from partial η^2 values and were reported based on the following effect size conventions: f: 0.10 = small, 0.25 = medium, and 0.40 = large [36].

Narrative questionnaire data reflecting participants' experiences with WATSU were systematically organized into analytical units, which were inductively classified into thematic subcategories and main categories according to the Mayring-triangulation-model [37]. All generated categories were quantitatively described by indicating frequency of mentions in absolute and percentage values. Qualitative data reflecting context-related information were narratively summarized.

RESULTS

Seventeen of the 26 recruited women were included with nine participants being allocated to the intervention group and eight to the control group. Three participants in the control group were lost to follow-up (for details see Figure 1). Finally, data from all seventeen study participants were analyzed according to the intention-to-treat principle. No adverse events were reported, and all of the WATSU-treatments were carried out as scheduled.

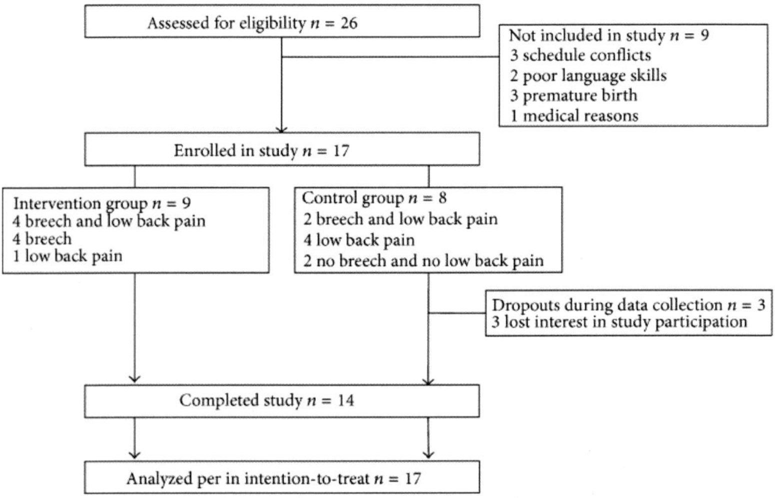

Figure 1: Consort flow diagram.

Group and baseline characteristics did not differ significantly between the two study groups (see Table 1). Study participants were between 27 and 40 years of age (32.0 ± 3.2) and 69% of them were primiparous.

Table 1: Group and baseline characteristics

Characteristics[a]	**Intervention group (n = 9)**	**Control group (n = 8)**	**Complete sample (n = 17)**	P[b]
Age (years)	31.3 (±2.7)	32.8 (±3.7)	32.0 (±3.2)	0.56
BMI (kg/m^2)	27.5 (±3.4)	25.5 (±2.3)	26.6 (±3.1)	0.31
Primiparous (yes/no)	6/3	7/1	13/4	n/a
Week of gestation	36.3 (±0.7)	35.8 (±1.2)	36.1 (±1.0)	0.43

Breech presentation (yes/no)	8/1	2/6	9/8	n/a
PSS score	14.8 (±5.0)	14.8 (±11.7)	14.8 (±8.5)	0.81
SF-36 main scales				
(i) physical component	47.0 (±6.8)	40.4 (±10.7)	43.9 (±9.2)	0.18
(ii) mental component	46.4 (±11.6)	52.8 (±6.9)	49.4 (±9.9)	0.12

[a]All continuous data are presented as mean (±SD).

[b] *P* values refer to Mann-Whitney *U* test.

n/a, not applicable; BMI, body mass index; PSS, Perceived Stress Scale; SF-36, Medical Outcomes Study 36-Item Short-Form Health Survey.

Quantitative Results

As presented in Table 2, analyses within the intervention group revealed a significant improvement in mental health-related quality of life (SF-36: $P = 0.018$) and a significant reduction in perceived stress (PSS: $P = 0.027$) from baseline to follow-up assessment.

Table 2: Analyses of medium-term changes in psychometric outcome measures within the intervention group.

Variables[a]	**At baseline (day 1)**	**At follow-up (day 8)**	**Δ value**	***P*** [b]
PSS score[c]	14.8 (±5.0)	12.7 (±5.3)	−2.1 (±2.2)	0.027*
SF-36 main scales[d]				
(i) Physical component	47.0 (±6.8)	45.3 (±11.0)	−1.7 (±6.7)	0.735
(ii) Mental component	46.4 (±11.6)	50.0 (±12.3)	+3.3 (±4.0)	0.018*
SF-36 sub scales[d]				
(i) Physical function	58.9 (±11.4)	61.1 (±18.7)	+2.2 (±10.9)	0.527
(ii) Role physical	72.2 (±49.1)	66.7 (±43.3)	−5.6 (±48.1)	1.00
(iii) Bodily pain	72.7 (±49.2)	70.4 (±26.3)	−2.2 (±4.1)	0.157
(iv) General health	81.3 (±10.8)	79.7 (±17.4)	−2.4 (±13.8)	0.786
(v) Vitality	50.6 (±18.3)	55.6 (±18.3)	+5.0 (±9.0)	0.114
(vi) Social function	76.4 (±22.1)	79.2 (±21.7)	+2.8 (±8.3)	0.317
(vii) Role emotional	63.0 (±38.9)	70.4 (±38.9)	+7.4 (±22.2)	0.317
(viii) Mental health	67.1 (±16.9)	72.0 (±18.6)	+4.9 (±8.4)	0.125

[a]All continuous data are presented as mean (±SD).

[b] *P* values refer to Wilcoxon test. [*] $P > 0.05$; CI 95%.
[c]PSS, Perceived Stress Scale; positive mean change values represent an increase in perceived stress.
[d]SF-36, Medical Outcomes Study 36-Item Short-Form Health Survey; positive mean change values represent an increase in health related quality of life.

Analyses examining immediate effects of the two WATSU-treatments on participants' level of stress, pain, and mood consistently showed significant improvements in all outcome measures (see Table 3; for individualized stress curves see Figure 2).

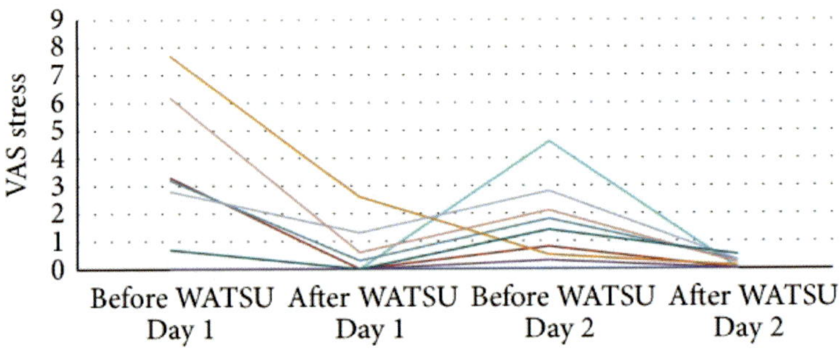

Figure 2: Changes in perceived stress measured by VAS (Visual Analog Scale; higher scores indicate increased stress, maximum score: 10).

Table 3: Analyses of short-term changes in psychometric outcome measures within the intervention group

Variables[a]	Before WATSU	After WATSU	Δ value	*P* [b]
VAS stressfulness (mm)[c]				
(i) At day 1	27 (±28)	5 (±9)	−21 (±22)	0.028*
(ii) At day 4	16 (±15)	2 (±2)	−14 (±14)	0.012*
VAS pain (mm)[c]				
(i) At day 1	9 (±12)	2 (±3)	−8 (±9)	0.028*
(ii) At day 4	12 (±10)	1 (±1)	−11 (±11)	0.012*
MDMQ mood scale score[d]				
(i) At day 1	16.9 (±2.8)	18.7 (±2.6)	+1.8 (±1.9)	0.042*
(ii) At day 4	14.7 (±6.8)	19.3 (±1.0)	+4.7 (±6.2)	0.027*

MDMQ alertness scale score[d]				
(i) At day 1	13.9 (±3.2)	16.0 (±2.4)	+2.1 (±3.6)	0.122
(ii) At day 4	10.7 (±5.0)	16.3 (±2.4)	+5.7 (±4.7)	0.007*
MDMQ calmness scale score[d]				
(i) At day 1	13.7 (±3.0)	18.0 (±2.9)	+4.3 (±3.0)	0.012*
(ii) At day 4	12.8 (±5.6)	18.3 (±1.7)	+5.6 (±5.8)	0.015*

[a]All continuous data are presented as mean (±SD).

[b] P values refer to Wilcoxon test. * $P > 0.05$; CI 95%.

[c]VAS, Visual Analog Scale; positive mean change values represent an increase in actual stress (actual pain).

[d]MDMQ, Multidimensional-Mood-Questionnaire; positive mean change values represent an increase in mood, alertness, and calmness.

In contrast to the control group, participants in the intervention group reported significant decreases in perceived stress from day 1 to day 8 whereas no significant group differences were found for mean change values of SF-36 main scales (see Table 4). Short term effects of WATSU on stress, pain, and mood that were observed within the intervention group were also present in comparison to the control group (see Table 5).

Table 4: Analyses of medium-term changes in outcome measures between the intervention and the control group

Variables[a]	**Δ values**		**P [b]**	**Cohen's f**
	From baseline (day 1) to follow-up (day 8)			
	Intervention group	**Control group**		
PSS score[c]	−2.1 (±2.2)	+.9 (±3.2)	0.036*	0.57
SF-36 main scales[d]				
(i) Physical component	−1.6 (±6.6)	+.2 (±6.4)	0.500	
(ii) Mental component	+3.3 (±4.0)	+3.1 (±3.6)	0.923	

[a]All continuous data are presented as mean (SD).

[b] P values refer to Mann-Whitney test. * $P > 0.05$; CI 95%.

[c]PSS, Perceived Stress Scale; positive mean change values represent an increase in perceived stress.

[d]SF-36, Medical Outcomes Study 36-Item Short-Form Health Survey; positive mean change values represent an increase in health related quality of life.

Table 5: Analyses of short-term changes in outcome measures between the intervention and the control group

Variables[a]	Δ values		P^{b}	Cohen's *f*
	Intervention group before and after WATSU	**Control group before and after 2 hours waiting period**		
VAS stressfulness (mm)[c]				
(i) At day 1	−21 (±22)	−3 (±11)	0.090	
(ii) At day 4	−14 (±14)	+1 (±12)	0.021*	0.61
VAS pain (mm)[c]				
(i) At day 1	−8 (±9)	−1 (±4)	0.037*	0.51
(ii) At day 4	−11 (±11)	+3 (±10)	0.005*	0.72
MDMQ mood scale score[d]				
(i) At day 1	+1.8 (±1.9)	+0.5 (±2.51)	0.175	
(ii) At day 4	+4.7 (±6.2)	−0.1 (±1.13)	0.013*	0.56
MDMQ alertness scale score[d]				
(i) At day 1	+2.1 (±3.6)	+0.1 (±1.2)	0.200	
(ii) At day 4	+5.7 (±4.7)	+0.4 (±1.2)	0.001*	0.79
MDMQ calmness scale score[d]				
(i) At day 1	4.6 (±3.0)	−0.3 (±1.9)	0.002*	0.96
(ii) At day 4	5.6 (±5.8)	−0.5 (±2.2)	0.008*	0.71

[a]All continuous data are presented as mean (SD).

[b] *P* values refer to Mann-Whitney test. * $P > 0.05$; CI 95%.

[c]VAS, Visual Analog Scale; positive mean change values represent an increase in actual stress (actual pain).

[d]MDMQ, Multidimensional-Mood-Questionnaire; positive mean change values represent an increase in mood, alertness, and calmness.

Data from ultrasound examinations in the intervention group revealed normal amounts of amniotic fluid with respect to the gestational age, that is, >2 cm and <8 cm at the single deepest amniotic fluid pocket (SDP) [38, 39] at baseline, before and after each WATSU treatment, and at follow-up (day

8) for all participants of the intervention group. Four participants with low baseline values (≤3.8 cm, mean 3.6 cm ± 0.1 cm) showed an average increase in amniotic fluid to 5.9 cm ± 1.5 cm on day 8. For one participant, the values remained practically unchanged (from 5.5 cm to 5.6 cm), and in four cases that presented values ≥5.6 cm (6.8 cm ± 0.8 cm) at baseline, the amniotic fluid value decreased to a mean value of 5.02 cm ± 0.7 cm (see Figure 3).

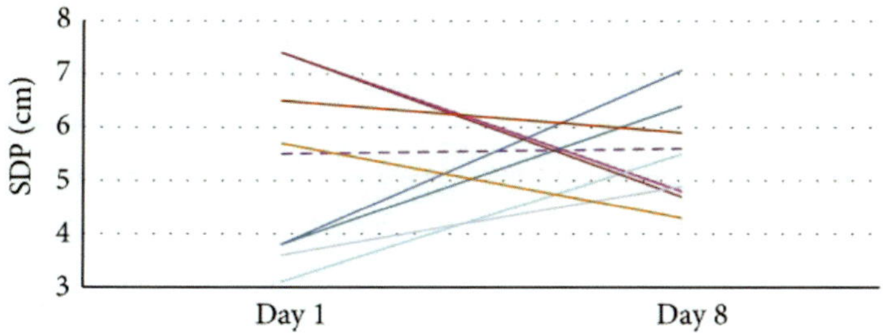

Figure 3: Changes in the amount of amniotic fluid in centimeter in the intervention group measured at the SDP (single deepest amniotic fluid pocket).

Similarly, the pulsatility indexes (PI) obtained from the umbilical artery and from both maternal vessels were within normal ranges for the respective gestational age and no significant changes were observed comparing pre- and posttreatment values (P values ≥ 0.23). No signs of adverse reactions were detectable in ultrasound measures at any point in the study.

Due to optimal tonus of the uterus, four of the participants in the intervention group underwent cephalic version on day 9 of the study, with two attempts being successful. Of the eight fetuses in breech position in the intervention group, one spontaneously presented in cephalic position on day 4. The two fetuses in breech in the control group remained in this position up to day 8 and one attempt for cephalic version failed.

Qualitative Results

A total of 57 comments reflecting subjective perceptions of the intervention were obtained from nine participants and were classified into main and subcategories as presented in Table 6.

Table 6: Results from analysis of qualitative data reflecting participants' perception of the intervention

Categories	Mentions (n = 60)		Sample phrases
	In absolute values	In % values	
(1) Overall impression	12	21.1	
(1.1) Overall relaxation	8	14.0	*"Total relaxation!"*
(1.2) Overall positive impression	4	7.0	*"Very pleasant!"*
(2) Body related	10	17.5	
(2.1) Physical relaxation	4	7.0	*"I think my belly relaxed too."*
(2.2) Mobilization	2	3.5	*"My lower back and the right side of the trunk are loosened."*
(2.3) Transfer to daily life	2	3.5	*"Walking became easier."*
(2.4) Weightlessness	1	1.8	*"It is great to feel light and weightless once in a while during pregnancy!"*
(2.5) Reduction of pain	1	1.8	*"Less pain."*
(3) Mind related	15	26.3	
(3.1) Perception	6	10.5	*"No distractions since ears were underneath the surface."*
(3.2) Peace of mind	4	7.0	*"Free of thoughts and fears."*
(3.3) Learning	3	5.3	*"Ideal preparation for birth."*
(3.5) Focus on breech position	2	3.5	*"My child moved a lot but did not turn."*
(4) Child related	10	17.5	
(4.1) Child relaxed	2	3.5	*"After a while, my child relaxed with me."*
(4.2) Child active	5	8.8	*"I felt my child moving a lot."*
(4.3) Interpretation of child's feeling	1	1.8	*"The baby liked it too; she was very quiet."*
(4.4) Contact to child	2	3.5	*"I think it is a good way to come close to the child—both are in water."*
(5) Further suggestions	10	17.5	
(5.1) Improvements	4	7.0	*"Start WATSU earlier in pregnancy."*
(6.1) Emphasizing satisfaction	6	10.5	*"There was nothing I disliked; everything was just great."*

Relaxation was most frequently reported as a major overall impression of the WATSU-treatments but was also distinctly mentioned by participants as an effect of treatment on their physical and mental states. Participants shared observations regarding body-related changes, for example, regained mobility and flexibility and decreased pain. They empathically reflected on how this experience might be perceived by their unborn children and on the nature of jointly experiencing an aquatic environment. Some of the women expressed gratitude for having been able to participate in the study and emphasized their impressions of having learned something about relaxation and surrendering that they perceived as a very helpful preparation for the upcoming birth. The intervention was referred to using positive descriptions such as "*like a dream.*" Mind-related aspects represented the majority (26.3%) of comments. All participants described WATSU as an experience they had fully enjoyed, and the opportunity to suggest improvements was instead used to emphasize satisfaction by six participants. The four stated suggestions concerning technical and process-related aspects of the study involved timing issues such as delays in ultrasound examination as well as a lack of clarity about where to hand in feedback-forms. One person mentioned a WATSU treatment-specific observation (having felt uncomfortable with hip flexion), and one woman suggested starting WATSU treatment earlier in pregnancy.

DISCUSSION

Active aquatic therapies have been studied extensively and proven to be beneficial for pregnant women [25]. Since passive hydrotherapy was not yet scientifically investigated with respect to its effects on pregnant women, we conducted a controlled pilot study to investigate the effects of WATSU on women in their ≥34th week of gestation, as well as to evaluate which of the investigated parameters would be suitable outcome measurements for further large-scale clinical trials.

As the impacts of maternal stress on unborn children can be severe [1–5], interventions that effectively reduce stress during pregnancy are desirable. Significant short-term improvements measured with VAS and MDMQ and significant medium-term decreases in perceived stress (PSS), with respective increases in the mental component of SF-36 main scale, indicate an optimized situation concerning the emotional wellbeing of the participants in the intervention group. Components of WATSU including gentle touch are believed to act in a stress reducing manner [40, 41]. In addition, the therapist's thoroughly compassionate attitude allows her/him to enter a parasympathetic state that the patient is nonverbally encouraged to join in [42]. During immersion, patients experience decreased heart rates [18] that are organically anticipating and promoting a parasympathetic state of relaxation. Reduced hypothalamic-pituitary-adrenal axis activity, that is, lower plasma cortisol levels along with increased mental and physical relaxation in context with immersion, has previously been reported [43, 44]. A potential mode of action promoting emotional wellbeing might be the activation of afferent C-tactile fibers during immersion. It has been proposed that these fibers transmit slow gentle touch—analogous to bypassing water—that has been observed to activate emotional brain areas [45, 46].

Short-term effects on pain were significant in our study (see Tables 3 and 5). This is of clinical relevance as epidemiological data indicate that two out of three pregnant women suffer pregnancy-related low back pain [22]. During pregnancy, it is recommended to refrain from analgesics [47], and thus nonpharmacologic alternatives are desirable. Prior research found that merely being immersed in warm water decreases pain [44, 48]. Passive hydrotherapy offers in addition unique possibilities of weightless mobilization in a quasi gravity-free environment with reduced joint compression forces. This in turn creates turbulence and currents causing a sensory overflow, which is regarded as one mechanism in pain reduction following immersion [17]. According to expert opinion, WATSU might be alleviating excessive muscle tone and pain due to rotational movements of the trunk and gentle rocking of the whole body, leading to dampened muscle tone as a side effect of vestibular system activation [49].

Neither the amniotic fluid volume nor umbilical blood flow characteristics were significantly altered after WATSU treatment. Although the sample size is small and therefore generalizability of the results has to be taken carefully, no adverse maternal and fetal reactions have been observed. The examining gynecologists perceived the tone of the uterus as being lower after WATSU treatment, encouraging them to emphatically suggest external cephalic version in four cases. Their observations might be evaluated objectively by electromyogram (EMG) in future studies. Breech presentation is a complication that occurs in only 3-4% of pregnancies but is a very real threat to the mother's as well as the child's life [50]. External cephalic version to correct the child's position is being suggested although associated with risks for mother and child [51, 52]. The success rate of external cephalic version attempts is strongly dependent on uterine tension (OR 18; 95% CI, 12–29) [53] and to a lesser extent also on the amount of amniotic fluid [54].

Only 10% of the amniotic fluid is regulated by transmembranous processes while intra-amniotic interactions—mainly fetal urine production, swallowing, and fluid secretions from the respiratory tract—regulate 90% [55, 56]. However, four participants showed slight increases in SDP measurements from day 1 to day 8. We also measured decreases in amniotic fluid over the period of seven days, which stands in contrast to the findings described in the literature [55, 57]. Since no formalized guidelines for the management of polyhydramnion exist to date [38], the decrease of amniotic fluid due to immersion might be a finding warranting further exploration. While short-term changes in the amount of amniotic fluid due to immersion were observed to be equalized within 30 minutes after immersion [58], the changes observed in our study took place over a longer period of time and would therefore suggest the involvement of respective regulatory processes.

Perceived relaxation and pleasantness seem to be among the most obvious features of passive hydrotherapy and were mentioned by participants in this study with great consistency in qualitative feedback. Interestingly, WATSU, despite its massage character, was attributed to mental effects most frequently. It appears to have been perceived by our participants rather as a mind-body-intervention, particularly suitable to inviting serenity as well as broadened awareness and mindfulness. Comments suggest usefulness of WATSU not only at the third trimester of pregnancy, but also in earlier stages, for example, potentially supporting women in overcoming ambivalence with respect to their motherhood.

LIMITATIONS

The following limitations of this study need to be addressed. The original study design implied ultrasound examinations on the control group as well to assess the natural course of changes in the amount of amniotic fluid [59]. In fact, only one woman in the control group agreed to follow this procedure. Also blinding of any party but the statistician is a challenge.

Most of the mothers-to-be in our intervention group learned about the diagnosis of breech presentation a short time before their enrollment and were thus suddenly being confronted with decisions about whether or not to undergo external cephalic version or cesarean section. In this respect, they differed considerably from participants in the control group, who represented a more general population. It is therefore unclear, to what extent specifically the medium-term effects of emotional recovery can be attributed to the treatment.

A small sample size may lead to the identification of large, clinically relevant effects; however, selection bias and overly weighted outliers cannot be ruled out. Hence, caution is appropriate when interpreting the present results. Due to the small sample size, we were not able to definitively interpret all of our findings scientifically, particularly the impact of WATSU treatment on spontaneous versions of children in breech position.

In qualitative data, saturation might have been achieved in terms of categories; however, since several aspects were mentioned only once, there might have been additional new statements concerning the experience, had more individuals participated in the survey.

CONCLUSIONS

To our knowledge, this is the first clinical study investigating the effects of passive hydrotherapy during pregnancy. It demonstrates significant benefits of WATSU with respect to stress, pain, mood, and mental health-related quality of life. Apparently, lowered tone of the uterus encouraged attempts of cephalic versions when children were in breech position. The treatment was described as very agreeable by the participants and appears to be a safe intervention.

These findings support the notion that WATSU yields therapeutic benefits for pregnant women; therefore, its integration into interdisciplinary treatment approaches should be considered. On the basis of this pilot study, larger trials should be established to further investigate and confirm the impact of the observed effects.

ACKNOWLEDGMENTS

This study was sponsored by Bern University Hospital and funded by the Edith Maryon Foundation and the Reiner Foundation. The funding sources had no role in the study design, nor were they involved in the collection, analysis, and interpretation of data. Likewise, they took part neither in preparation of the paper, nor in the decision to submit the paper for publication. The authors wish to thank the Department of Physiotherapy at Bern University Hospital for access to the therapy pool, Karin Dauwalder, Krista Dick, Yvonne Müller-Rohrer, and Christian Rothenbühler for administering WATSU, and Lorenz Radlinger as well as Marc Baumann, MD, for their contributions to the study. We thank Monica Van Niel, OTR/L, for proofreading the paper.

CONFLICT OF INTERESTS

Agnes M. Schitter is chairwoman of the Swiss Aquatic Bodyworkers' Association, NAKA, and member of the supervisory board of the Institute for Aquatic Bodywork IAKA AG, Bern, Switzerland. These honorary engagements do not alter the author's adherence to publication ethics. The other authors declare that they have no conflict of interests regarding the publication of this paper.

REFERENCES

1. Monk C., Spicer J., Champagne F. A. Linking prenatal maternal adversity to developmental outcomes in infants: the role of epigenetic pathways. *Development and Psychopathology.* 2012;24(4):1361–1376. doi: 10.1017/S0954579412000764.
2. Urech C., Fink N. S., Hoesli I., Wilhelm F. H., Bitzer J., Alder J. Effects of relaxation on psychobiological wellbeing during pregnancy: a randomized controlled trial. *Psychoneuroendocrinology.* 2010;35(9):1348–1355. doi: 10.1016/j.psyneuen.2010.03.008.
3. Kivlighan K. T., DiPietro J. A., Costigan K. A., Laudenslager M. L. Diurnal rhythm of cortisol during late pregnancy: associations with maternal psychological well-being and fetal growth. *Psychoneuroendocrinology.* 2008;33(9):1225–1235. doi: 10.1016/j.psyneuen.2008.06.008.
4. Smythies J., Edelstein L., Ramachandran V. Molecular mechanisms for the inheritance of acquired characteristics—exosomes, microRNA shuttling, fear and stress: Lamarck resurrected? *Frontiers in Genetics.* 2014;5:p. 133. doi: 10.3389/fgene.2014.00133.
5. Entringer S., Buss C., Kumsta R., Hellhammer D. H., Wadhwa P. D., Wüst S. Prenatal psychosocial stress exposure is associated with subsequent working memory performance in young women. *Behavioral Neuroscience.* 2009;123(4):886–893. doi: 10.1037/a0016265.
6. Bos S. C., Macedo A., Marques M., et al. Is positive affect in pregnancy protective of postpartum depression? *Revista Brasileira de Psiquiatria.* 2013;35(1):5–12. doi: 10.1016/j.rbp.2011.11.002.
7. NCCAM. *What is Complementary and Alternative Medicine?* National Center for Complementary and Alternative Medicine, National Institutes of Health; 2008.
8. Marafon G. Tecniques of body mediation from alternative medicine—WATSU. *Sexologies.* 2008;17(S1):p. 29. doi: 10.1016/S1158-1360(08)72584-2.
9. Dull H. *Watsu: Freeing the Body in Water.* Vol. 11. Victoria, Canada: Trafford Publishing; 2004.
10. Scaer R. *The Body Bears the Burden: Trauma, Dissociation, and Disease.* Binghamton, NY, USA: Haworth Medical; 2001.

11. Smeeding S. J. W., Bradshaw D. H., Kumpfer K., Trevithick S., Stoddard G. J. Outcome evaluation of the Veterans affairs Salt Lake City integrative health clinic for chronic pain and stress-related depression, anxiety, and post-traumatic stress disorder. *Journal of Alternative and Complementary Medicine.* 2010;16(8):823–835. doi: 10.1089/acm.2009.0510.
12. Faull K. A pilot study of the comparative effectiveness of two water-based treatments for fibromyalgia syndrome: watsu and Aix massage. *Journal of Bodywork and Movement Therapies.* 2005;9(3):202–210. doi: 10.1016/j.jbmt.2004.12.001.
13. Cinica Holistica Engadina, http://www.clinica-holistica.com/index.php?page=587.
14. Maczkowiak S., Hölter G., Otten H. WATSU—the effect of differently accentuated movement therapy interventions on clinically depressive patients. *Bewegungstherapie und Gesundheitssport.* 2007;23(2):58–64. doi: 10.1055/s-2007-960607.
15. Marafon G. Tecniques of body mediation from alternative medicine-WATSU. *Sexologies.* 2008;17(1):p. 29. doi: 10.1016/S1158-1360(08)72584-2.
16. Chon S. C., Oh D. W., Shim J. H. Watsu approach for improving spasticity and ambulatory function in hemiparetic patients with stroke. *Physiotherapy Research International.* 2009;14(2):128–136. doi: 10.1002/pri.421.
17. Becker B. E. Aquatic therapy: scientific foundations and clinical rehabilitation applications. *PM & R.* 2009;1(9):859–872. doi: 10.1016/j.pmrj.2009.05.017.
18. Cider A., Sveälv B. G., Täng M. S., Schaufelberger M., Andersson B. Immersion in warm water induces improvement in cardiac function in patients with chronic heart failure. *European Journal of Heart Failure.* 2006;8(3):308–313. doi: 10.1016/j.ejheart.2005.08.001.
19. Jarad N. A., Powell T., Smith E. Evaluation of a novel sputum clearance technique—hydro-acoustic therapy (HAT) in adult patients with cystic fibrosis: a feasibility study. *Chronic Respiratory Disease.* 2010;7(4):217–227. doi: 10.1177/1479972310376082.
20. Maczkowiak S., Hölter G., Otten H. WATSU—the effect of differently accentuated movement therapy interventions on clinically depressive patients (WATSU—zur Wirksamkeit unterschiedlich

akzentuierter bewegungstherapeutischer Interventionen bei klinisch depressiven Patienten) *B&G Bewegungstherapie und Gesundheitssport.* 2007;23(2):58–64.

21. Waller B., Lambeck J., Daly D. Therapeutic aquatic exercise in the treatment of low back pain: a systematic review. *Clinical Rehabilitation.* 2009;23(1):3–14. doi: 10.1177/0269215508097856.
22. Pennick V. E., Young G. Interventions for preventing and treating pelvic and back pain in pregnancy. *Cochrane Database of Systematic Reviews.* 2007;(2)CD001139
23. Hansen M. *If Massage Therapy Is Applied in a Hydrotherapy Pool; Then the Discomforts of Pregnancy during Second/Third Trimester Will Be Relieved.* MTABC; 2011.
24. Cluett E. R., Nikodem V. C., McCandlish R. E., Burns E. E. Immersion in water in pregnancy, labour and birth. *The Cochrane Database of Systematic Reviews.* 2004;(2)CD000111
25. Katz V. L. Exercise in water during pregnancy. *Clinical Obstetrics and Gynecology.* 2003;46(2):432–441. doi: 10.1097/00003081-200306000-00022.
26. Hope A., Aanderud L., Aakvaag A. Dehydration and body fluid-regulating hormones during sweating in warm (38°c) fresh- and seawater immersion. *Journal of Applied Physiology.* 2001;91(4):1529–1534.
27. Sherer D. M., Cullen J. B. H., Thompson H. O., Woods J. R., Jr. Transient oligohydramnios in a severely hypovolemic gravid woman at 35 weeks' gestation, with fluid reaccumulating immediately after intravenous maternal hydration. *The American Journal of Obstetrics and Gynecology.* 1990;162(3):770–771. doi: 10.1016/0002-9378(90)91006-X.
28. Cohen S., Kamarck T., Mermelstein R. A global measure of perceived stress. *Journal of Health and Social Behavior.* 1983;24(4):385–396. doi: 10.2307/2136404.
29. Broadbent E., Petrie K. J., Alley P. G., Booth R. J. Psychological stress impairs early wound repair following surgery. *Psychosomatic Medicine.* 2003;65(5):865–869. doi: 10.1097/01.PSY.0000088589.92699.30.
30. Morfeld M., Kirchberger I., Bullinger M. *SF-36 Fragebogen zum Gesundheitszustand.* Göttingen, Germany: Hogrefe; 2011.

31. Price D. D., McGrath P. A., Rafii A., Buckingham B. The validation of visual analogue scales as ratio scale measures for chronic and experimental pain. *Pain.* 1983;17(1):45–56. doi: 10.1016/0304-3959(83)90126-4.
32. Cork R. C., Isaac I., Elsharydah A., Saleemi S., Zavisca F., Alexander L. A comparison of the verbal rating scale and the visual analog scale for pain assessment. *Internet Journal of Anesthesiology.* 2004;8(1):p. 1.
33. Steyer R., Schwenkmezger P., Notz P., Eid M. *MDMQ Multideimensional Mood Questionnaire.* Göttingen, Germany: Hogrefe; 1997.
34. Nabhan A. F., Abdelmoula Y. A. Amniotic fluid index versus single deepest vertical pocket: a meta-analysis of randomized controlled trials. *International Journal of Gynecology and Obstetrics.* 2009;104(3):184–188. doi: 10.1016/j.ijgo.2008.10.018.
35. Chauhan S. P., Doherty D. D., Magann E. F., Cahanding F., Moreno F., Klausen J. H. Amniotic fluid index vs single deepest pocket technique during modified biophysical profile: a randomized clinical trial. *The American Journal of Obstetrics and Gynecology.* 2004;191(2):661–668. doi: 10.1016/j.ajog.2004.06.078.
36. Cohen J. *Statistical Power Analysis for the Behavior Sciences.* Hillsdale, NJ, USA: Lawrence Earlbaum Associates; 1988.
37. Mayring P. *Einführung in die Gualitative Sozialforschung.* Weinheim, Germany: Beltz; 2002.
38. Sandlin A. T., Chauhan S. P., Magann E. F. Clinical relevance of sonographically estimated amniotic fluid volume: polyhydramnios. *Journal of Ultrasound in Medicine.* 2013;32(5):851–863. doi: 10.7863/ultra.32.5.851.
39. Magann E. F., Sandlin A. T., Ounpraseuth S. T. Amniotic fluid and the clinical relevance of the sonographically estimated amniotic fluid volume: oligohydramnios. *Journal of Ultrasound in Medicine.* 2011;30(11):1573–1585.
40. Bijari B. B., Iranmanesh S., Eshghi F., Baneshi M. R. Gentle human touch and yakson: the effect on preterm's behavioral reactions. *ISRN Nursing.* 2012;2012:6. doi: 10.5402/2012/750363.750363
41. Ditzen B., Neumann I. D., Bodenmann G., et al. Effects of different kinds of couple interaction on cortisol and heart rate responses to stress

in women. *Psychoneuroendocrinology.* 2007;32(5):565–574. doi: 10.1016/j.psyneuen.2007.03.011.

42. Kemper K. J., Shaltout H. A. Non-verbal communication of compassion: measuring psychophysiologic effects. *BMC Complementary and Alternative Medicine.* 2011;11(1, article 132) doi: 10.1186/1472-6882-11-132.
43. Turner J. W., Jr., Fine T. H. Effects of relaxation associated with brief Restricted Environmental Stimulation Therapy (REST) on plasma cortisol, ACTH, and LH. *Biofeedback and Self-Regulation.* 1983;8(1):115–126. doi: 10.1007/BF01000542.
44. Kjellgren A., Sundequist U., Norlander T., Archer T. Effects of flotation-REST on muscle tension pain. *Pain Research and Management.* 2001;6(4):181–189.
45. Björnsdotter M., Morrison I., Olausson H. Feeling good: on the role of C fiber mediated touch in interoception. *Experimental Brain Research.* 2010;207(3-4):149–155. doi: 10.1007/s00221-010-2408-y.
46. Morrison I., Löken L. S., Olausson H. The skin as a social organ. *Experimental Brain Research.* 2010;204(3):305–314. doi: 10.1007/s00221-009-2007-y.
47. Vleeming A., Albert H. B., Östgaard H. C., Sturesson B., Stuge B. European guidelines for the diagnosis and treatment of pelvic girdle pain. *European Spine Journal.* 2008;17(6):794–819. doi: 10.1007/s00586-008-0602-4.
48. Morgan P. M., Salacinski A. J., Stults-Kolehmainen M. A. The acute effects of flotation restricted environmental stimulation technique on recovery from maximal eccentric exercise. *The Journal of Strength & Conditioning Research.* 2013;27(12):3467–3474. doi: 10.1519/JSC.0b013e31828f277e.
49. Schoedinger P. Watsu in aquatic rehabiltiation. In: Cole A. J., Becker B. E., editors. *Comprehensive Aquatic Therapy.* 3rd. chapter 5. Pullman, DC, USA: Washington State University Publishing; 2011.
50. Hannah M. E., Hannah W. J., Hewson S. A., Hodnett E. D., Saigal S., Willan A. R. Planned caesarean section versus planned vaginal birth for breech presentation at term: a randomised multicentre trial. *The Lancet.* 2000;356(9239):1375–1383. doi: 10.1016/S0140-6736(00)02840-3.

51. Hutton E. K., Hofmeyr G. J. External cephalic version for breech presentation before term. *Cochrane Database of Systematic Reviews.* 2006;25(1)CD000084
52. Hutton E. K., Hannah M. E., Ross S. J., et al. The early external cephalic version (ECV) 2 trial: an international multicentre randomised controlled trial of timing of ECV for breech pregnancies. *BJOG.* 2011;118(5):564–577. doi: 10.1111/j.1471-0528.2010.02837.x.
53. Kok M., Cnossen J., Gravendeel L., van der Post J., Opmeer B., Mol B. W. Clinical factors to predict the outcome of external cephalic version: a metaanalysis. *The American Journal of Obstetrics and Gynecology.* 2008;199(6):630.e1–630.e7. doi: 10.1016/j.ajog.2008.03.008.
54. Kok M., Cnossen J., Gravendeel L., van der Post J. A., Mol B. W. Ultrasound factors to predict the outcome of external cephalic version: a meta-analysis. *Ultrasound in Obstetrics and Gynecology.* 2009;33(1):76–84. doi: 10.1002/uog.6277.
55. Brace R. A. Physiology of amniotic fluid volume regulation. *Clinical Obstetrics and Gynecology.* 1997;40(2):280–289. doi: 10.1097/00003081-199706000-00005.
56. Ross M. G., Idah R. Correlation of maternal plasma volume and composition with amniotic fluid index in normal human pregnancy. *Journal of Maternal-Fetal and Neonatal Medicine.* 2004;15(2):104–108. doi: 10.1080/14767050310001650770.
57. Dertkigil M. S. J., Cecatti J. G., Sarno M. A. C., Cavalcante S. R., Marussi E. F. Variation in the amniotic fluid index following moderate physical activity in water during pregnancy. *Acta Obstetricia et Gynecologica Scandinavica.* 2007;86(5):547–552. doi: 10.1080/00016340601181649.
58. Thisted D. L. A., Noergaard L. N., Aabakke A. J. M., Meyer H., Secher N. J. Water immersion and changes in the fetoplacental circulation. *International Journal of Gynecology & Obstetrics.* 2012;119(supplement 2):S500–S501.
59. Magann E. F., Bass J. D., Chauhan S. P., Young R. A., Whitworth N. S., Morrison J. C. Amniotic fluid volume in normal singleton pregnancies. *Obstetrics and Gynecology.* 1997;90(4):524–528. doi: 10.1016/S0029-7844(97)00351-7.

Chapter 10

The Safety and Feasibility of a Halliwick Style of Aquatic Physiotherapy for Falls and Balance Dysfunction in People with Parkinson's Disease: A Single Blind Pilot Trial

Aan Fleur Terrens[1,2], Sze-Ee Soh[2,3], Prue Morgan[2]

[1] Movement Disorder Program, Peninsula Health, VIC, Australia
[2] Department of Physiotherapy, Monash University, VIC, Australia
[3] Department of Epidemiology and Preventative Medicine, Monash University, VIC, Australia

ABSTRACT

Background

There is growing evidence that aquatic physiotherapy may be effective for people with Parkinson's Disease (PD) but most studies have investigated

Citation: The safety and feasibility of a Halliwick style of aquatic physiotherapy for falls and balance dysfunction in people with Parkinson's Disease: A single blind pilot trial. Aan Fleur TerrensID1, Sze-Ee Soh2, Prue Morgan. PLoS ONE 15(7: e0236391. https://doi.org/10.1371/journal.pone.0236391.
0.

land based type exercises in the aquatic environment. Few studies have examined customised aquatic therapies such as the Halliwick concept which focuses on trunk rotation and core stabilisation.

Objective

The primary aim was to determine the feasibility of a Halliwick style aquatic physiotherapy intervention for people with PD. The secondary aim was to compare the Halliwick intervention with traditional aquatic and land based physiotherapy in terms of disease severity, balance and fear of falling.

Methods

Halliwick style aquatic, traditional aquatic and land based physiotherapy were trialled in a single blind pilot study. All interventions ran for 60 minutes per week over 12 weeks. Feasibility outcomes were safety, adherence and attrition. Secondary outcomes included the Unified Parkinson's Disease Rating Scale motor subsection (UPDRS-III), Berg Balance Scale (BBS), Mini BESTest and modified Falls Efficacy Scale (mFES).

Results

30 participants with moderate PD were recruited. Participant mean age was 72 years (SD 8.4; range 51–86) with moderate disease severity (median Hoehn & Yahr score 3; IQR 1).No falls occurred during intervention sessions, however ten participants reported falls during the study period. No other adverse consequences were reported. All groups had adherence over 85%. No within group significant differences were found in UPDRS-III, BBS or mFES scores post-intervention for all groups, but the Halliwick aquatic group improved significantly in the Mini BESTest post-intervention ($p = 0.011$, 95% CI -7.36,-1.31, $t\,(10) = -2.98$).

Conclusions

Despite people with PD being a vulnerable population, aquatic physiotherapy, including the Halliwick style is a safe treatment option. Promising results for balance in the Halliwick aquatic group were observed, but further studies with larger sample sizes is required to increase confidence in the results.

INTRODUCTION

Parkinson's disease (PD) is a progressive neurological disorder with common motor manifestations such as stooped posture, tremor, bradykinesia, rigidity, festination, freezing of gait and postural instability [1–4]. Postural instability results in a high proportion of falls in people with PD, with studies estimating around 60–70% of people with PD fall at least once a year, and 39–50% have recurrent falls [5–7]. Recent treatment approaches to reduce falls in people with PD include movement strategy training by using attention cues [6], progressive resistance strength training [8], falls education [9], Tai Chi and dancing [6] but the impact of these treatment techniques on postural stability has not been conclusive across the disease spectrum. Recent systematic reviews have indicated that aquatic physiotherapy may be an effective treatment option in this population, and may improve balance and disability outcomes [10–12], however it is not clear what elements of an aquatic physiotherapy program lead to best outcomes. Previous studies have generally focused on traditional aquatic physiotherapy programs that have a mix of aerobic, strengthening and balance exercises [12]. Trunk rigidity in PD may lead to impairment in the vestibular system and therefore postural control [13, 14], and prior research has shown that land based trunk rotational exercises lead to improved balance in other neurological cohorts such as stroke [15]. Little is known about whether specific aquatic therapies such as the Halliwick concept which focuses on complex trunk rotations and core stabilisation may be beneficial for this population.

The principals of aquatic physiotherapy utilise the unique hydrostatic and hydrodynamic properties of water to influence human movement [16]. Aquatic physiotherapy, or hydrotherapy, has been shown to be an effective treatment for individuals post hip and knee joint replacements, those with lower back pain and stroke [16, 17]. Given its successful use in older adults with other conditions, aquatic physiotherapy has the potential to be a useful treatment for adults with PD. Previous research has shown that decreasing axial rigidity, a common symptom in PD, has a positive impact on balance performance [18]. This suggests that the Halliwick style of aquatic therapy may be of value in this population. The Halliwick concept is a 10 step, three stage system that teaches individuals to become completely independent with their movement in the water through different positioning and progressive exercises [19, 20]. The first phase focuses on adjusting to the water environment whilst the second phase teaches individuals to balance and control various types of rotations. Phase two of the Halliwick concept, which has a focus on balance and rotations, may improve axial rigidity, core strength and therefore postural stability in those with PD. Lastly, the third stage uses propulsive movements that are tailored to the individual that allow them to achieve independence in the water environment.

One study has examined the effect of the Halliwick program and found this treatment to significantly improve balance in a small group of people following stroke [21]. A study by Zhu, Yin, Cui, et al. [22] compared parts of the Halliwick concept in a general aquatic program to aquatic obstacle training, but the focus of this study was to reduce freezing of gait. Another study has utilised the Halliwick method with PD patients, but did not use any standardised outcome measures to monitor change in function nor report any safety outcomes [23]. We currently do not know if using the complete 10 step Halliwick concept in people with PD is safe. As there is no previous research using the complete Halliwick concept in people with PD, and there have been no studies that have singularly examined whether it has an impact on balance in this population, further research is warranted to determine whether this concept is a feasible treatment option.

A potential risk associated with aquatic physiotherapy is the challenge to both the cardiovascular and respiratory systems associated with partial or complete immersion in water, potentially amplified when applied to older adults with cardiovascular or respiratory co-morbidities, such as those with PD [24]. Previous research has also shown that safety data is under reported in literature [11]. Therefore, the primary aim of this study was therefore to determine whether aquatic physiotherapy, in particular the complete 10 step Halliwick concept, is a feasible and safe treatment modality for people with PD. The secondary aim was to compare the Halliwick style intervention with traditional aquatic and land based physiotherapy in terms of disease severity, balance and fear of falling.

METHODS

Design and Setting

This study was a single blind pilot study completed between June 2016 and July 2018 in an outpatient public health facility in Melbourne, Australia. Ethics approval was gained from the Peninsula Health Human Research Ethics Committee (HREC) (Project 15/PH/32) and Monash University HREC (Project CF16/1341–2016000731). This study was registered with the Australian New Zealand Clinical Trials Registry (ACTRN12616000834459).

Participants

Participants were required to have a diagnosis of idiopathic PD confirmed by a neurologist, transfer and walk without assistance with or without gait aid (as participants are required to independently transfer in and out of the pool via steps), and have a Mini Mental State Exam (MMSE) score of 24 or above so that they can follow instructions. Those with unstable medical conditions or a self-reported history of any musculoskeletal, cardiothoracic, other neurological or psychological condition that might potentially affect participation were excluded. If there were any doubt as to medical stability, the participant's local doctor provided a signed medical assessment form deeming suitability for aquatic and land based physiotherapy. Participants were recruited from local Movement Disorders clinics, private neurologists and from local support groups to ensure the sample would capture the broad characteristics of people with PD. There were no restrictions on disease severity, disease duration or how sedentary participants were. Participants were also sent information regarding the study if they had been involved previously with the Movement Disorders Program at Peninsula Health or if they had telephoned to enquire about the study. A sample size calculation based on a 60% recruitment rate of all participants approached [25], allowing for an attrition rate of 20%, a precision estimate of 20% and 95% confidence interval (CI), resulted in a minimum sample size of 28 participants required for this study. A target sample of 36 was established to aim for equal numbers of participants in the intervention groups.

Procedure

Participants who expressed interest were screened by telephone, and attended the initial assessment where the blinded assessor confirmed study eligibility and obtained informed consent. Initial testing consisted of cognitive and aquatic safety screening to ensure participants were cognitively and medically suitable for study participation. Block randomisation was used for group allocation, with randomisation order developed by hand, via a third party unrelated to the study. Researchers opened an opaque, sealed and numbered envelope which determined the next intervention group. Following the initial assessment, the participant was given a sealed envelope, prepared by a research assistant, containing their allocated intervention group. There was a maximum of six participants per group, and two cohorts per intervention type. Randomisation determined the order of intervention groups, as only one group ran at a time. Details regarding demographic and clinical characteristics of participants were obtained at baseline testing.

Interventions

All interventions were delivered by a physiotherapist and allied health assistant experienced in treating people with PD. All interventions were of 60 minutes duration per week for 12 weeks. Pre-intervention testing by the blinded assessor was undertaken one week prior to commencement of the intervention, and follow up post-intervention testing by the blinded assessor occurred one-week after conclusion of the intervention. When participants were on oral medication for their PD, testing and intervention occurred during the"on" stage of their medication cycle.

There were three intervention groups: (1) Halliwick aquatic exercises (core specific exercises and exercises from the Halliwick concept); (2) traditional or current practise aquatic physiotherapy exercises; and (3) land based exercises (control group). Both aquatic interventions took place in a hydrotherapy pool (6m x 10m), with a depth ranging from 1.1m to 1.5m. The water temperature was approximately 34.7 degrees Celsius for all aquatic interventions, with relative humidity ranging from 63% to 76% and pool deck temperature ranging from 25 to 31 degrees Celsius. The characteristics of the exercises delivered in the three intervention groups are described in Table 1.

Table 1: Intervention characteristics

	INTERVENTION GROUP		
	Halliwick Aquatic	***Traditional Aquatic***	***Land Based***
Walking	✓	✓	✓
Lower Limb	×	✓	✓
Upper Limb	×	✓	✓
Strength	×	✓	✓
Aerobic	×	✓	✓
Balance	✓	✓	✓
Trunk Mobility	✓	✓	✓

Complex Rotations (sagittal, transverse, longitudinal)	✓	×	×
Core Stabilisation	✓	×	×
Stretching	✓	✓	✓

All intervention groups completed walking exercises as a warm up and both upper and lower limb stretching as a cool down. The Halliwick aquatic intervention group completed trunk mobility, core stabilisation and rotational exercises as the primary exercises whereas the other intervention groups completed a range of strength, balance and aerobic exercises. The Halliwick aquatic group followed the Halliwick concept, with participants progressing through each of the 10 steps as able combined with core specific exercises. Most notably, the Halliwick concept contains complex rotational movements where the participant is fully supported by the water, and subsequently require significant core control. The trunk mobility subsection has two standing exercises, where participants' feet are fixed on the ground. Land based exercises were matched with the traditional aquatic intervention exercises as much as possible in terms of the number of balance and cardiorespiratory exercises, types of stretches and muscle groups targeted. Exercise intensity was measured using the Borg rating of perceived exertion scale [26] during each exercise. Participants were advised to exercise to a level 13–14 on the Borg rating scale, which indicates that they were working somewhat hard in intensity.

Outcome Measures

Study feasibility was evaluated by examining safety, adherence and attrition. Safety was measured by the absence or presence of adverse events, either participant reported or observed during the intervention sessions. This included falls during the intervention sessions and the intervention period, chlorine allergies, muscle stiffness, pain and fatigue. Adherence was recorded as the number of sessions that each participant attended in the intervention period expressed as a percentage of total sessions offered. If the participant cancelled or did not attend their scheduled session the treating physiotherapist followed up with a phone call, with each reason recorded for every missed session. An adherence rate of more than 70% is deemed acceptable, as this figure has been used in other feasibility studies for people with PD [27]. Attrition was recorded as the number of participants that dropped out of the study during the intervention period. All participants who dropped out were asked to provide a reason if not already given. To be considered satisfactory, we expected an attrition rate of 20%, which is conservative compared to other studies [28, 29] to allow for the fragility of this population.

Secondary outcome measures for disease severity, balance and fear of falling were also recorded. The motor subsection of the Unified Parkinson's Disease Rating Scale (UPDRS-III) [30] was used to measure the severity and disability associated with the motor manifestations of PD, with higher scores indicating greater level of disability. This section of the UPDRS was developed to measure and rate the different motor manifestations of the disease and has been shown to have good scale reliability and construct validity [31, 32]. Balance was measured using the Berg Balance Scale (BBS) [33] and the Mini BESTest [34]. The BBS was developed to assess balance in the older population, in particular people with PD. The BBS is a reliable and valid measure of balance in this population [35, 36]. The BBS has also been found to have strong correlations with the UPDRS-III, which indicates as disease severity increases there is a correlating decrease in balance [35]. Lower BBS scores indicate reduced balance ability. The Mini BESTest is a valid and reliable predictor of falls in older individuals, including those with PD [36], with lower scores on the scale indicating reduced balance. The modified Falls Efficacy Scale (mFES) [37] assessed how concerned participants were about falling when completing a range of functional tasks. Lower scores on the mFES indicate lower self-confidence and it has been shown to have good construct validity and test-retest reliability within the PD population [38].

Statistical Analysis

All quantitative analyses were completed using SPSS statistical software version 25.0 (SPSS Inc, Chicago, Illinois). Descriptive statistics were used to summarise demographic and clinical characteristics of each group as well as feasibility outcomes of safety, adherence and attrition. Data distribution were visually examined and tested with the Shapiro-Wilk test for normality.

To determine whether there was a change in the secondary outcome measures (i.e. disease severity, balance, fear of falling) over the intervention period for each group, paired sample t-tests were used if data were normally distributed. Where data were not normally distributed, Wilcoxon signed-rank tests were used. Differences between groups for secondary outcome measures were explored using one-way analysis of covariance (ANCOVA) to account for potential differences at baseline between intervention groups. A square root transformation of BBS was undertaken to improve the normality and linearity of the residuals. A Bonferroni adjustment was made for post-hoc comparisons between groups to control for the increased risk of Type I errors. There were two main comparisons made in this study; between the Halliwick aquatic and traditional aquatic intervention groups, and between the Halliwick aquatic and land based intervention groups. For an alpha level of 0.05, the *p*-value had to fall below 0.025 to achieve significance. Any missing data from participants who were randomised to the intervention were analysed using intention to treat analysis by using the last data point captured [39].

RESULTS

Participants

A total of 208 individuals with PD were initially assessed for eligibility with 59 individuals randomised: 24 to the Halliwick aquatic intervention, 16 to the traditional aquatic intervention and 19 to the land based intervention. 30 participants (24 males, 6 females) received the allocated intervention (11 Halliwick, 10 traditional, 9 land based) and post intervention measures were obtained from 25 participants (Fig 1). Participants ranged in age from 51 to 86 years (mean 72; SD 8.4) and had moderate disease severity (median Hoehn & Yahr score of 3; IQR 1). 16 participants had fallen once or more in the past 12 months and 21 had a musculoskeletal comorbidity. Participant characteristics are described in Table 2. More males than females were recruited into this study, and this is reflective of the general population of PD [40].

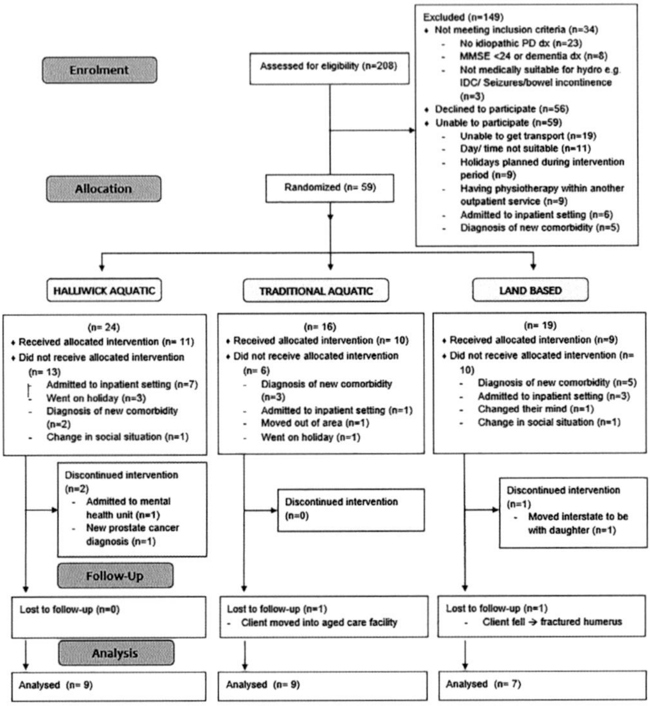

Figure 1: Consort diagram.

Table 2: Characteristics of participants at baseline who received intervention

	Halliwick Aquatic Intervention Group (n = 11)	Traditional Aquatic Intervention Group (n = 10)	Land Based Intervention Group (n = 9)	**All participants (n = 30)**
Age, years (mean, SD)	74.1 ± 6.6	65.6 ± 7.7	76.4 ± 7.4	72.0 ± 8.4
Males, n (%)	10 (90)	7 (70)	7 (78)	24 (80)
MMSE (mean, SD)	27.5 ± 2.1	27.6 ± 1.6	27.2 ± 2.1	27.5 ± 1.9
Hoehn & Yahr	3 (1–3)	2 (1–3)	3 (2–3)	3 (1–3)
Disease duration, years (mean, SD)	6.7 ± 6.3	5.2 ± 7.1	4.2 ± 3.1	5.5 ± 5.6
Levodopa use, n (%)				
Yes	11 (100)	9 (90)	7 (78)	27 (90)
No	0	1 (10)	2 (22)	3 (10)

Living situation, n (%)				
Alone	3 (27)	2 (20)	2 (22)	7 (23)
Not alone	8 (73)	8 (80)	7 (78)	23 (77)
Marital status, n (%)				
Single	3 (27)	3 (30)	2 (22)	8 (27)
Married	8 (73)	7 (70)	7 (78)	22 (73)
Co-morbidities[1], n (%)				
None	1 (9)	1 (10)	0 (0)	2 (7)
Genitourinary	1 (9)	1 (10)	3 (33)	5 (17)
Respiratory	3 (27)	0 (0)	0 (0)	3 (10)
Circulatory	6 (55)	3 (30)	6 (67)	15 (50)
Musculoskeletal	9 (82)	4 (40)	8 (89)	21 (70)
Neoplasms	2 (18)	1 (10)	3 (33)	6 (20)
Mental	2 (18)	2 (20)	3 (33)	7 (23)
Other	3 (27)	6 (60)	4 (44)	13 (43)
Number of falls in past 12 months, n (%)				
None	5 (46)	5 (50)	4 (44)	14 (47)
1 fall:	0	3 (30)	1 (11)	4 (13)
≥2 falls:	6 (55)	2 (20)	4 (44)	12 (40)
UPDRS-III	55 (42.5–60)	46 (41.5–54.8)	37 (32–58)	50.5 (35.5–58.8)
Mini BESTest	22 (14–27)	22 (20.5–28)	18 (18–23)	22 (17.3–27)
Berg Balance Scale	47 (45–51.5)	53 (51.3–54.8)	50 (45–52)	50.5 (45.5–53)
Modified Falls Efficacy Scale	7.6 (4.9–8.5)	6.9 (6.2–8.3)	7.9 (6.8–8.9)	7.1 (5.9–8.6)

All numbers are medians and inter quartile ranges unless otherwise stated.

[1]Classified according to the International Statistical Classification of Diseases and Related Health Problems (ICD-10), MMSE = Mini Mental State Exam, UPDRS-III = Unified Parkinson's Disease Rating Scale-Part 3

Primary Outcome Measures

Safety

There were no falls during the intervention sessions for all intervention groups, however, a third of participants (n = 10) reported falls at home during the 12-week intervention period (Table 3). Fallers were equally represented in the Halliwick aquatic and land based intervention groups (n = 4 each) with only two fallers in the traditional aquatic intervention group. There were 32 falls reported across the intervention period. Main reasons for falling were tripping (n = 11), followed by unknown reasons (n = 10), freezing of gait (n = 5), dizziness (n = 4) and legs giving way (n = 2).

Table 3: Feasibility data on safety, adherence and attrition on those who received intervention

	HALLIWICK AQUATIC (n = 11)	TRADITIONAL AQUATIC (n = 10)	LAND BASED (n = 9)	**TOTAL (n = 30)**
Falls during 12 week intervention period, n (%) *				
0 falls:	7 (64)	8 (80)	5 (67)	20 (67)
1 fall:	1 (9)	1 (10)	3 (33)	5 (17)
≥2 falls	3 (27)	1 (10)	1 (11)	5 (17)
Other adverse events, n (%)				
Muscle Pain	0 (0)	1 (10)	2 (22)	3 (10)
Stiffness	1 (9)	2 (20)	2 (22)	5 (17)
Chlorine allergy	0 (0)	0 (0)	n/a	0 (0)
Fatigue	2 (18)	2 (20)	8 (89)	12 (40)
Adherence, %	90%	85%	91%	89%
Attrition, n (%)				
Discontinued	2 (18)	0 (0)	1 (11)	3 (10)
Lost to follow up	0 (0)	1 (10)	1 (11)	2 (7)
BORG, mean	12.2	13.1	16.7	14

Across all intervention groups fatigue was a common side effect post-intervention. This appeared to be more common in participants in the land based intervention group (n = 8). Stiffness (n = 5) and muscle pain (n = 3) were reported by a few participants across all groups. Importantly, no participants from the aquatic groups reported a chlorine allergy (Table 3).

Adherence and attrition

Interventions consisted of 12 one-hour weekly classes with an overall adherence rate of 89% (Table 3). Reasons for absences included participant illness, transport issues and planned social events. Of all participants randomised, 54% (n = 13) of individuals in the Halliwick aquatic intervention 38% (n = 6) in the traditional aquatic intervention and 53% (n = 10) in the land based intervention exited the study. Attrition after intervention sessions commenced was due to a variety of reasons including admission to hospital (n = 2) or care facility (n = 1), new medical diagnosis (n = 1) or personal reasons (n = 1).

Secondary Outcome Measures

Disease severity

Disease severity, as measured by UPDRS-III scores, improved in the aquatic Halliwick and land based groups post-interventions, however, these were not statistically significant (Fig 2A) (post intervention Halliwick aquatic median 55; IQR 42.5–60; traditional aquatic median 46; IQR 41.5–54.8; land median 37; IQR 32–58). No significant differences were also observed between any intervention groups after adjusting for baseline scores.

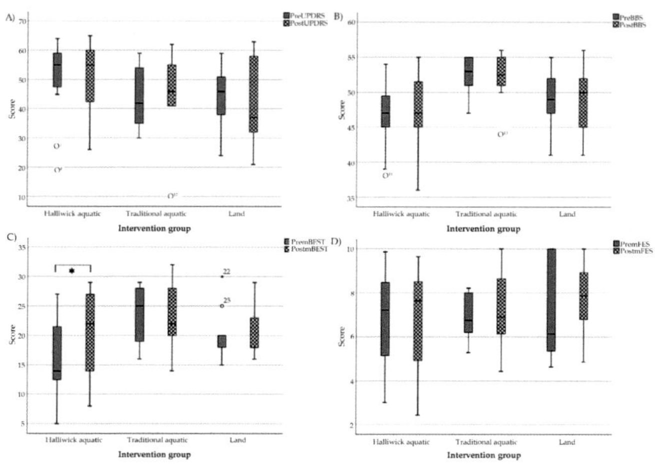

Figure 2: Box plots of outcome measures pre and post intervention for (A) disease severity as measured by the UPDRS-III; (B) balance as measured by

the Berg Balance Scale; (C) balance as measured with the Mini BESTest; and (D) fear of falling (mFES). Significant differences between groups are indicated with a star. A box is drawn from the first quartile to the third quartile, with a line passing through the box at the median.

Balance

There was minimal change in balance when measured using the BBS post intervention in groups after adjusting for baseline scores (Fig 2B) (post intervention Halliwick aquatic median 47; IQR 45–51.5, traditional aquatic median 53; IQR 51.3–54.8, land median 50; IQR 45–52), and no differences were seen between groups post intervention. When balance was measured using the Mini BESTest, a significant improvement was observed in the Halliwick aquatic intervention group (pre-intervention median 14; IQR 12.5–21.5, post-intervention median 22; IQR 14–27) following treatment($p = 0.011$, 95% CI -7.36,-1.31, t (10) = -2.98) (Fig 2C) (post intervention traditional aquatic median 22; IQR 20.5–28, land median 18; IQR 18–23). No improvements in balance as measured by the Mini BESTest were observed in the other groups. There were also no differences in balance improvement following intervention between the groups when measured using the Mini BESTest after adjusting for baseline scores. It is important to note however, that the Halliwick aquatic intervention group had significantly lower Mini BESTest scores compared to the other intervention groups at baseline (pre intervention Halliwick aquatic median 14; IQR 12.5–21.5, traditional aquatic median 25; IQR 19.8–27.5, land median 20; IQR 18–20). Lower scores on the Mini BESTest indicate worse balance.

Fear of falling

There was no significant improvements in fear of falling (as measured by the mFES) post intervention in the Halliwick aquatic intervention group, and there were no differences observed between groups (Fig 2D).

DISCUSSION

The primary aim of this study was to determine whether a Halliwick aquatic physiotherapy intervention that focuses on trunk rotation and core stabilisation (including the Halliwick concept) is a feasible and safe treatment modality in people with PD. Our results indicate that aquatic physiotherapy, including the Halliwick style, is a safe treatment option in this population. Minimal side effects for all participants were observed, and participants in the land based group appeared to report more fatigue compared to those in the aquatic intervention groups. This suggests that the hydrostatic and hydrodynamic properties of the aquatic environment may provide support for joints and aid in muscle relaxation for people with PD [24]. Additionally, none of the participants in the two aquatic intervention groups reported side effects such as chlorine allergies or skin rashes, which is consistent with previous studies [11]. Despite this population being more susceptible to orthostatic hypotension [41], no cardiovascular issues were observed during the trial. Thus, from a safety perspective, aquatic physiotherapy including the Halliwick concept, is just as safe as land based therapy for people with PD.

Although there were no falls observed during sessions, participants in all groups reported a high number of falls (n = 32) throughout the intervention period. Of the 30 participants in this study, 10 reported falls during the 12 week intervention period, with a higher proportion of individuals falling more than twice in the Halliwick aquatic intervention group (n = 3). Amongst the five participants who were multiple fallers, there were 27 falls reported, confirming that people with PD are at a high risk of falls and are a very vulnerable population [7]. This high falls risk needs to be taken into consideration when treating people with PD whether in a land based or aquatic environment.

Whilst adherence to therapy was good, with over 85% adherence for all intervention groups, there was a low uptake of participants into this study. We contacted 208 individuals and although one quarter were not eligible, the majority declined to participate or were unable to attend. The main reason for declining to participate was that participants were not interested in water exercises (n = 36), with the main reasons for being unable to participate being transport (n = 19) and availability (n = 11) difficulties. The proportion of males and females that declined to participate or were unable to participate were reflective of the PD population, with a larger proportion of males compared to females [40, 42]. In addition, after being accepted into the study, another 29 participants were unable to participate. Our recruitment experience indicate that there may be several barriers towards participating in exercise therapy amongst people with PD. This is consistent with existing literature which reports low participation rates in older people for falls prevention programs or home exercise programs [43, 44]. The large number of drop outs, combined with the number of participants unable to commence therapy due to changes in health status highlights the medical fragility of this population, and therefore strategies are required to optimise engagement with this vulnerable population.

Although this study was not powered to show differences in disease severity and balance outcomes, we observed a significant improvement in Mini BESTest scores for the Halliwick aquatic intervention group following intervention. It should be noted that the Halliwick aquatic intervention group had significantly different MiniBESTest scores at baseline although this was adjusted for in the analyses. Recent systematic reviews have shown that aquatic physiotherapy in PD results in improved balance (as measured by the BBS) when compared to land based physiotherapy [10, 45], but it is not clear what elements of aquatic intervention lead to these improvements. Future studies may wish to focus on determining if aquatic physiotherapy broadly results in improvements in balance and fear of falling, or whether specific components of an aquatic program are more effective than others.

Limitations

A number of limitations need to be noted. Firstly, this study was not designed to determine the efficacy of aquatic physiotherapy including the Halliwick concept for the secondary outcome measures and an adequately powered study is warranted to explore potential effects on outcomes including balance, fear of falling or disease severity. For instance, in order to confirm whether the Halliwick aquatic intervention program was more effective compared to traditional aquatic programs, a sample size of 56 will be required in each group to detect a clinically significant difference of 5.52 points [36] in balance as measured by the MiniBESTest with 80% power and alpha set at 5%. Another limitation to note is that exercise intensity was only measured using the BORG scale, which is a self-reported measure of perceived exertion, and therefore may not be an accurate measure as participants may under or overestimate their level of exertion. We also acknowledge that care needs to be taken when generalising study findings given the small sample size and that we only included participants with mild to moderate PD. Nevertheless, this is the first study to examine whether a Halliwick aquatic therapy approach can be used safely in this population. We have ensured that all safety criteria relating to aquatic physiotherapy in PD were reported and used a minimum dataset that includes the UPDRS-III allow for accurate study evaluation and replication [11]. As the focus of this study was on feasibility, and in particular safety, we did not follow the European Physiotherapy Guidelines for Parkinson's Disease [46] in terms of exercise prescription. This may explain why no significant effects were observed in the secondary outcome measures. It is recommended that future studies meet these training guidelines to determine treatment efficacy.

CONCLUSIONS

People with PD are a vulnerable population and are at high risk of falls. Aquatic physiotherapy, including a novel approach such as the Halliwick concept, is a feasible and safe treatment option for people with PD and may have a positive impact on balance and fear of falling. However, a larger scale study is required to explore this further.

ACKNOWLEDGMENTS

The authors would like to thank Lee Silverman Voice Treatment (LSVT) for their support with a small student grant [2017].The authors would also like to thank all participants of this trial, and Wendy Toogood who tirelessly assisted in the recruitment and organisation of this trial.

REFERENCES

1. Heisters D. Parkinson's: Symptoms, treatments and research. British journal of nursing (Mark Allen Publishing). 2011;20(9):548–54.
2. Ossig C, Reichmann H. Treatment strategies in early and advanced Parkinson disease. Neurol Clin. 2015;33(1):19–37. pmid:25432721
3. Beaulne-Seguin Z, Nantel J. Conflicting and non-conflicting visual cues lead to error in gait initiation and gait inhibition in individuals with freezing of gait. Gait Posture. 2016;49:443–7. pmid:27525821
4. Zhang H, Yin X, Ouyang Z, Chen J, Zhou S, Zhang C, et al. A prospective study of freezing of gait with early Parkinson disease in Chinese patients. Medicine. 2016;95(26):e4056. pmid:27368041
5. Kalilani L, Asgharnejad M, Palokangas T, Durgin T. Comparing the Incidence of Falls/Fractures in Parkinson's Disease Patients in the US Population. PLoS ONE. 2016;11(9):e0161689. pmid:27583564
6. Morris ME, Iansek R, Kirkwood B. A randomized controlled trial of movement strategies compared with exercise for people with Parkinson's disease. Mov Disord. 2009;24(1):64–71. pmid:18942100
7. Bloem BR, Marinus J, Almeida Q, Dibble L, Nieuwboer A, Post B, et al. Measurement instruments to assess posture, gait, and balance in Parkinson's disease: Critique and recommendations. Mov Disord. 2016;31(9):1342–55. pmid:26945525
8. Ramazzina I, Bernazzoli B, Costantino C. Systematic review on strength training in Parkinson's disease: an unsolved question. Clin Interv Aging. 2017;12:619–28. pmid:28408811
9. Morris TR, Cho C, Dilda V, Shine JM, Naismith SL, Lewis SJ, et al. A comparison of clinical and objective measures of freezing of gait in Parkinson's disease. Parkinsonism Relat Disord. 2012;18(5):572–7. pmid:22445248
10. Cugusi L, Manca A, Bergamin M, Di Blasio A, Monticone M, Deriu F, et al. Aquatic exercise improves motor impairments in people with Parkinson's disease, with similar or greater benefits than land-based exercise: a systematic review. Journal of physiotherapy. 2019;65(2):65–74. pmid:30904467
11. Terrens AF, Soh SE, Morgan PE. The efficacy and feasibility of aquatic physiotherapy for people with Parkinson's disease: a systematic review. Disability and rehabilitation. 2018;40(24):2847–56. pmid:28793808

12. Pinto C, Salazar AP, Marchese RR, Stein C, Pagnussat AS. The Effects of Hydrotherapy on Balance, Functional Mobility, Motor Status, and Quality of Life in Patients with Parkinson Disease: A Systematic Review and Meta-analysis. PM R 2019;11(3):278–91. pmid:30884205
13. Hubble RP, Naughton GA, Silburn PA, Cole MH. Trunk muscle exercises as a means of improving postural stability in people with Parkinson's disease: a protocol for a randomised controlled trial. BMJ Open. 2014;4(12):e006095–e. pmid:25552609
14. Artigas NR, Franco C, Leão P, Rieder CRM. Postural instability and falls are more frequent in Parkinson s disease patients with worse trunk mobility. Arquivos de Neuro-Psiquiatria. 2016;74:519–23. pmid:27487370
15. Karthikbabu S, Solomon JM, Manikandan N, Rao BK, Chakrapani M, Nayak A. Role of Trunk Rehabilitation on Trunk Control, Balance and Gait in Patients with Chronic Stroke: A Pre-Post Design. Neuroscience and Medicine. 2011;Vol. 02No. 02:7.
16. Geytenbeek J. Evidence for effective hydrotherapy. Physiotherapy. 2002;88(9):514–29.
17. Bartels EM, Lund H, Hagen KB, Dagfinrud H, Christensen R, Danneskiold-Samsoe B. Aquatic exercise for the treatment of knee and hip osteoarthritis. The Cochrane database of systematic reviews. 2007(4):Cd005523. pmid:17943863
18. Bartolić A, Pirtošek Z, Rozman J, Ribarič S. Postural stability of Parkinson's disease patients is improved by decreasing rigidity. European Journal of Neurology. 2005;12(2):156–9. pmid:15679705
19. McMillan J. The role of water in rehabilitation. Fysioterapeuten. 1978;45:87–90.
20. Garcia MK, Joares EC, Silva MA, Bissolotti RR, Oliveira S, Battistella LR. The Halliwick Concept, inclusion and participation through aquatic functional activities. Acta Fisiátrica. 2012;19(3):142–50.
21. Tripp F, Krakow K. Effects of an aquatic therapy approach (Halliwick-Therapy) on functional mobility in subacute stroke patients: a randomized controlled trial. Clin Rehabil. 2014;28(5):432–9. pmid:24177711
22. Zhu Z, Yin M, Cui L, Zhang Y, Hou W, Li Y, Zhao H. Aquatic obstacle training improves freezing of gait in Parkinson's disease patients: a randomized controlled trial. Clinical Rehabilitation. 2017;32(1):29–36. pmid:28629269

23. Zotz TGG, Souza EA, Israel VL, Loureiro APC. Aquatic physical therapy for Parkinson's disease. Advances in Parkinson's Disease. 2013;02(04):102–7.
24. APA. Australian guidelines for aquatic physiotherapists working in and/or managing hydrotherapy pools. : Australian Physiotherapy Association; 2015.
25. Thabane L, Ma J, Chu R, Cheng J, Ismaila A, Rios LP, et al. A tutorial on pilot studies: the what, why and how. BMC Medical Research Methodology. 2010;10(1):1.
26. Borg GA. Psychophysical bases of perceived exertion. Medicine and science in sports and exercise. 1982;14(5):377–81. pmid:7154893
27. Blandy LM, Beevers WA, Fitzmaurice K, Morris ME. Therapeutic Argentine Tango Dancing for People with Mild Parkinson's Disease: A Feasibility Study. Frontiers in Neurology. 2015;6(122).
28. Goodwin VA, Richards SH, Henley W, Ewings P, Taylor AH, Campbell JL. An exercise intervention to prevent falls in people with Parkinson's disease: a pragmatic randomised controlled trial. Journal of Neurology, Neurosurgery & Psychiatry. 2011;82(11):1232.
29. Carroll LM, Volpe D, Morris ME, Saunders J, Clifford AM. Aquatic Exercise Therapy for People With Parkinson Disease: A Randomized Controlled Trial. Archives of Physical Medicine and Rehabilitation. 2017;98(4):631–8. pmid:28088380
30. Goetz CG, Tilley BC, Shaftman SR, Stebbins GT, Fahn S, Martinez-Martin P, et al. Movement Disorder Society-sponsored revision of the Unified Parkinson's Disease Rating Scale (MDS-UPDRS): scale presentation and clinimetric testing results. Mov Disord. 2008;23(15):2129–70. pmid:19025984
31. Lim S-Y, Tan AH. Historical perspective: The pros and cons of conventional outcome measures in Parkinson's disease. Parkinsonism Relat Disord. 2018;46:S47–S52. pmid:28793970
32. Martinez-Martin P, Jeukens-Visser M, Lyons KE, Rodriguez-Blazquez C, Selai C, Siderowf A, et al. Health-related quality-of-life scales in Parkinson's disease: critique and recommendations. Mov Disord. 2011;26(13):2371–80. pmid:21735480

33. Berg K, Wood-Dauphine S, Williams JI, Gayton D. Measuring balance in the elderly: preliminary development of an instrument. Physiotherapy Canada. 1989;41(6):304–11.
34. Franchignoni F, Horak F, Godi M, Nardone A, Giordano A. Using psychometric techniques to improve the Balance Evaluation Systems Test: the mini-BESTest. Journal of rehabilitation medicine. 2010;42(4):323–31. pmid:20461334
35. Qutubuddin AA, Pegg PO, Cifu DX, Brown R, McNamee S, Carne W. Validating the Berg Balance Scale for patients with Parkinson's disease: A key to rehabilitation evaluation. Arch Phys Med Rehabil. 2005;86(4):789–92. pmid:15827933
36. Leddy AL, Crowner BE, Earhart GM. Functional Gait Assessment and Balance Evaluation System Test: Reliability, Validity, Sensitivity, and Specificity for Identifying Individuals With Parkinson Disease Who Fall. Physical Therapy. 2011;91(1):102–13. pmid:21071506
37. Edwards N, Lockett D. Development and validation of a modified falls-efficacy scale. Disability and rehabilitation Assistive technology. 2008;3(4):193–200. pmid:19238720
38. Nilsson MH, Drake A-M, Hagell P. Assessment of fall-related self-efficacy and activity avoidance in people with Parkinson's disease. BMC Geriatrics. 2010;10(1):78.
39. Streiner D, Geddes J. Intention to treat analysis in clinical trials when there are missing data. Evidence-based mental health. 2001;4(3):70–1. pmid:12004740
40. Picillo M, Nicoletti A, Fetoni V, Garavaglia B, Barone P, Pellecchia MT. The relevance of gender in Parkinson's disease: a review. J Neurol. 2017;264(8):1583–607. pmid:28054129
41. Senard JM, Raï S, Lapeyre-Mestre M, Brefel C, Rascol O, Rascol A, et al. Prevalence of orthostatic hypotension in Parkinson's disease. J Neurol Neurosurg Psychiatry. 1997;63(5):584–9. pmid:9408097
42. Hirsch L, Jette N, Frolkis A, Steeves T, Pringsheim T. The Incidence of Parkinson's Disease: A Systematic Review and Meta-Analysis. Neuroepidemiology. 2016;46(4):292–300. pmid:27105081
43. Simek EM, McPhate L, Haines TP. Adherence to and efficacy of home exercise programs to prevent falls: a systematic review and meta-analysis of the impact of exercise program characteristics. Preventive medicine. 2012;55(4):262–75. pmid:22813920

44. McPhate L, Simek EM, Haines TP. Program-related factors are associated with adherence to group exercise interventions for the prevention of falls: a systematic review. Journal of physiotherapy. 2013;59(2):81–92. pmid:23663793
45. Pinto C, Salazar AP, Marchese RR, Stein C, Pagnussat AS. Is hydrotherapy effective to improve balance, functional mobility, motor status, and quality of life in subjects with Parkinson's disease? A systematic review and meta-analysis. PM R 2018.
46. Keus SH, Munneke M, Graziano M, Paltamaa J, Pelosin E, Domingos J, et al. European Physiotherapy Guideline for Parkinson's Disease. KNGF/ ParkinsonNet: the Netherlands; 2014.

Index

C

D

E

F

G

H

I

K

L

U

V

W

FSC
www.fsc.org
MIX
Papier | Fördert gute Waldnutzung
FSC® C083411

Kolibri 360 GmbH
Ferdinand-Jühlke-Straße 7
99095 Erfurt, Deutschland
produktsicherheit@kolibri360.de